BTEC N

BTEC National Engineering

Mike Tooley and Lloyd Dingle

Newnes

OXFORD AMSTERDAM BOSTON LONDON NEW YORK PARIS
SAN DIEGO SAN FRANCISCO SINGAPORE SYDNEY TOKYO

Newnes
An imprint of Elsevier
Linacre House, Jordan Hill, Oxford OX2 8DP
30 Corporate Drive, Burlington, MA 01803

First published 2002
Reprinted 2003, 2004, 2006

British Library Cataloguing in Publication Data
A catalogue record for this book is available from the British Library

Library of Congress Cataloguing in Publication Data
A catalogue record for this book is available from the Library of Congress

ISBN 0 7506 5166 0

For information on all Newnes publications
visit our website at www.newnespress.com

Typeset by Replika Press Pvt Ltd., Delhi 110 040, India
Printed and bound in Great Britain by Biddles Ltd., Kings Lynn, Norfolk

Contents

Introduction

Welcome to the challenging and exciting world of engineering! This book has been written to help get you through the core units of the BTEC National Certificate and Diploma awards in Engineering. It provides the essential underpinning knowledge required of a student who wishes to pursue a career in any branch of engineering.

The book has been written by two highly experienced further and higher education lecturers. With over 40 years of practical teaching experience they have each contributed their own specialist knowledge which has been designed to cover the core units of the revised and updated BTEC Engineering programme. Throughout the book we have adopted a common format and approach with numerous student activities, examples, problems and key points.

About the BTEC National Certificate and Diploma

The BTEC National Certificate and National Diploma qualifications have long been accepted by industry as appropriate qualifications for those who are about to enter industry or who are receiving training at the early stages of employment in industry. At the same time, these qualifications have become increasingly acceptable as a means of gaining entry into higher education.

BTEC National programmes in Engineering attract a very large number of registrations per annum such that there is in excess of 35 000 students currently studying these qualifications in the UK by both part-time and full-time modes of study.

The BTEC National syllabus was recently reviewed and extensively updated and new programmes have been launched with effect from September 2002. The new scheme is likely to be adopted by *all* institutions that currently offer the programme as well as a number of others who will now be able to mix and match parts of the BTEC qualification with vocational GCSE and AVCE awards.

Many organizations have contributed to the design of the new BTEC National Engineering programme including the Qualifications and Curriculum Authority (QCA), the Engineering Council, and several National Training Organizations (NTO).

The Engineering Council continues to view the BTEC National Certificate/Diploma as a key qualification for the sector. They also recognize that BTEC National qualifications are frequently used as a means of entry to higher education courses, such as HNC/HND programmes and Foundation Degree courses.

In revising and updating the Engineering BTEC National programme, Edexcel has taken into consideration a number of issues, including:

- Occupational standards and NTO requirements.
- Professional requirements (particularly with regard to the engineering technician).
- Progression into employment.
- Progression to Higher National qualifications (with a particularly close match in the core units of engineering science and mathematics).
- A flexible course structure that is commensurate with the broad aims of Curriculum 2000.
- Relevant QCA criteria.
- External assessments (as required).
- Key skills signposting.
- Several other issues, including the ways in which the study of the area can contribute to an understanding of spiritual, moral, ethical, social and cultural issues.

To assist tutors and lecturers, the book is supported by an A4 format curriculum resource pack containing photocopiable resource material and information on assessment, grading and practical activities.

How to use this book

This book covers four of the core units that are common to BTEC National Engineering programmes. The remaining core unit (Project) is covered in the Curriculum Support Pack. Each unit contains Text, 'Key points', 'Test your knowledge' questions, Examples, Activities and Problems.

The 'Test your knowledge' questions are interspersed with the text throughout the book. These questions allow you to check your understanding of the preceding text. They also provide you with an opportunity to reflect on what you have learned and consolidate this in manageable chunks.

Most 'Test your knowledge' questions can be answered in only a few minutes and the necessary information, formulae, etc., can be gleaned from the surrounding text. Activities, on the other hand, make excellent vehicles for gathering the necessary evidence to demonstrate that you are competent in Key Skills. Consequently they normally require a significantly greater amount of time to complete. They may also require additional library or resource area research time coupled with access to computing and other information technology resources.

Many tutors will use 'Test your knowledge' questions as a means of reinforcing work done in class while Activities are more likely to be 'set work' for students to do outside the classroom. Whether or not this approach is taken, it's important to be aware that this student-centred work is designed to complement a programme of lectures and tutorials based on the BTEC syllabus. Independent learners (i.e. those not taking a formal course) will find complete syllabus coverage in the text.

In the units on Mathematics and Science, the worked examples not only show you how to solve simple problems but also help put the subject matter into context with typical illustrative examples. In order to successfully tackle this work you will need to have a good scientific calculator (and get to know how to use it). In the units on Business Systems and Communications you will require access to a computer with word processing, spreadsheet, drawing and CAD software.

Finally, here are a some general points to help you with your studies:

- Allow regular time for reading – get into the habit of setting aside an hour, or two, at the weekend. Use this time to take a second look at the topics that you have covered during the week or that you may have not completely understood.
- Make notes and file these away neatly for future reference – lists of facts, definitions and formulae are particularly useful for revision!
- Look out for the inter-relationship between subjects and units – you will find many ideas and a number of themes that crop up in different places and in different units. These can often help to reinforce your understanding.
- Don't expect to find all subjects and topics within the course equally interesting. There may be parts that, for a whole variety of reasons, don't immediately fire your enthusiasm. There is nothing unusual in this; however, do remember that something that may not appear particularly useful now may become crucial at some point in the future!
- However difficult things seem to get – don't be tempted to give up! Engineering is not, in itself, a difficult subject, rather it is a subject that *demands* logical thinking and an approach in which each new concept builds upon those that have gone before.
- Finally, don't be afraid to put your new ideas into practice. Engineering is about *doing* – get out there and *do* it!

Good luck with your BTEC Engineering studies!

Mike Tooley and Lloyd Dingle

Unit 1 Business systems in engineering

This unit is designed to provide you with an introduction to the business and commercial aspects of engineering. It aims to broaden and deepen your understanding of business, industry and the effects of engineering on the environment. It also aims to provide you with a firm foundation for employment in the engineering industry together with an understanding of the financial, legal, social and environmental constraints within which an engineering company operates.

To achieve this unit you must show that you can:

- describe external factors that affect business and the economic environment in which engineering companies operate;
- describe the impact of relevant legislation, environmental and social constraints on typical engineering companies;
- explain and use basic project planning and control methods used in typical engineering companies;
- select and use costing techniques used in typical engineering companies.

This unit is assessed by portfolio assessment and you must include in your portfolio a full range of appropriate evidence. This can include assignment and case study work as well as reports, course notes and solutions to class-set problems. You should begin work on your portfolio as you start the unit and add to it systematically as you progress through the unit. You will also find it valuable to cross-reference the evidence in the portfolio to the individual learning outcomes and the list of evidence requirements. Your tutor will help you to do this.

This unit has strong links with the core units Communications for Technicians and Project. Wherever possible, you should apply the techniques that you have developed in the communications unit to work that you undertake in this unit. There are also links to several of the optional units including Quality Assurance and Control and Production Planning and Scheduling.

As you study this unit you will need to gain access to appropriate information and communications technology resources (including the Internet and the World Wide Web). You will also benefit from access to a well-stocked library of reference material. Your tutor will provide you with case study material featuring data on real or invented engineering companies involved with engineering production and manufacture, engineering services, materials handling and transportation. He or she will also ensure that you have access to the relevant UK and EU Acts as well as relevant health and safety literature and other materials.

Case studies (based on real or invented engineering companies) are an important part of this unit. When you carry out a case study you will be presented with sample data to analyse. You might find it useful to relate your experience of employment or work experience periods in industry to the case study as well as to work covered elsewhere in the unit.

Introduction

All engineering companies must operate as commercial enterprises in order to survive. In this unit you will look at how external factors affect engineering companies and how engineering activities as a whole contribute to the economy. You will also learn about the importance of effective planning and control and about how and why companies make financial decisions. In addition, you will learn about environmental legislation and how it affects the engineering industry, and about how engineering companies design their operations to minimize adverse effects on the environment.

Some of the engineering sectors, engineered products and engineering companies with which you are probably familiar include:

Chemical engineering	Fertilizers, pharmaceuticals, plastics, petrol, etc. Companies in this field include Fisons, Glaxo, ICI and British Petroleum.
Mechanical engineering	Bearings, agricultural machinery, gas turbines, machine tools and the like from companies such as RHP, GKN and Rolls-Royce.
Electrical and electronic engineering	Electric generators and motors, consumer electronic equipment (radio, TV, audio and video), power cables, computers, etc., produced by companies such as GEC, BICC and ICL.
Civil engineering	Concrete bridges and flyovers, docks, factories, power stations, dams, etc., from companies such as Bovis, Wimpey and Balfour-Beatty.
Aerospace engineering	Passenger and military aircraft, satellites, space vehicles, missiles, etc., from companies such as British Aerospace, Westland and Rolls-Royce.
Telecommunications	Telephone and radio communication, data communications equipment, etc., from companies such as Nokia, GEC, Plessey and British Telecom.
Motor vehicle engineering	Cars, commercial vehicles (lorries and vans), motorcycles, tractors and specialized vehicles from companies such as Rover, Vauxhall UK and McLaren.

As you work through this unit it will help you to put things into context by relating the topics to those engineering companies with which you are familiar. This will give you an appreciation of the factors that affect their operation as well as the constraints under which they operate.

The production workflow and some of the functions in an engineering firm are shown in Figure 1.1. This production workflow starts with suppliers that provide an input to the various engineering processes. The output of the engineering processes is delivered to the customers. You may find this easier to recall by remembering the acronym *SIPOC*.

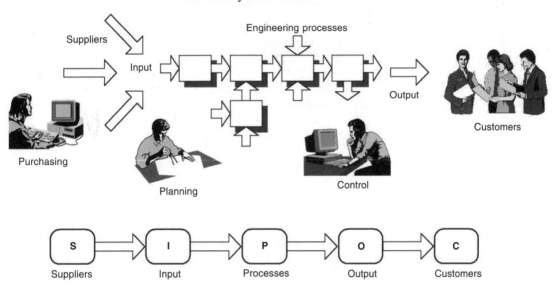

Figure 1.1 *Simplified view of the flow of work in an engineering company from its suppliers to its customers*

The three functions that we have included in the diagram (there are many more in a real engineering firm) operate as follows:

Purchasing	The purchasing function ensures that supplies are available as and when required by the engineering processes.
Planning	The planning function ensures that the correct engineering processes are in place and also that the workflow is logical and timely.
Control	The control function ensures the quality of the output and the cost effectiveness of the processes.

We will revisit these functions in some detail later in this unit but first we will look at the environment in which an engineering firm operates.

External factors and the economic environment

External factors affect the operation of all engineering companies. Some factors may be foreseen in which case they are predictable (and can be planned for) but others are seemingly random occurrences. Some external factors can be regarded as *threats* whilst others can be regarded as *opportunities*. For example, the availability of a new manufacturing process that significantly reduces costs can be considered to be an opportunity. Falling consumer demand resulting from high interest rates can be considered a threat. Perhaps somewhat confusingly, some external factors can be considered to be either a threat or an opportunity depending upon whether or not a company is ready and able to respond!

External factors that you need to be aware of include:

- markets and the general state of the economy
- consumer demand
- demographic and social trends
- competitive products and services
- consumer confidence and customer/client relationships
- innovation and technological change.

It's worth noting that, to some extent, all of these factors are present to a varying degree all of the time.

Figure 1.2 shows factors that affect a typical engineering firm. Some factors are considered to be internal (and part of the task environment) whilst others are external (and considered to be part of the wider environment). Internal factors are relatively easy to influence. External factors, on the other hand, may be difficult if not impossible to influence. Note that Figure 1.1 fits inside this diagram with the SIPOC stages fitting between 'Suppliers' and 'Customers' in the region that we have marked 'Core engineering processes'. Because it has far-reaching implications, it is important that you understand this diagram!

Test your knowledge 1.1

Classify each of the following as a threat or an opportunity for a small engineering company:

- availability of low-cost business development loans
- interest rates climb to an all-time high
- high salaries offered by a major competitor
- a local college starting new courses in advanced manufacturing technology.

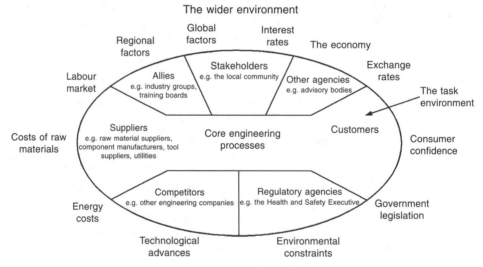

Figure 1.2 *Factors that affect a typical engineering firm. Note that the entire 'SIPOC' diagram (Figure 1.1) fits into the core of this diagram*

Activity 1.1

A small high-tech engineering company based in Kent manufactures a range of colour photocopiers. The company has a customer base in the south-east of England as well as northern France and Belgium. In common with other suppliers, the company offers a

Test your knowledge 1.2

A change of government may have consequences for the economy and in turn this may affect interest rates. Explain the likely consequences of (a) low and (b) high interest rates on the operation of a small engineering company.

Key point

External factors affect the operation of all engineering companies. Some of these factors can be predicted and planned for whilst others cannot. Some factors can be considered to be opportunities and some can be considered to be threats.

standard warranty and an extended service contract of a six-hour 'fix or replace' response time when one of its copiers breaks down.

Assess each of the following scenarios on a scale of 1 to 5 (1 = negligible and 5 = major) on its likely impact on the operation of the company. In each case briefly explain your answer and suggest which of these scenarios represent a potential threat to the company and which can be regarded as an opportunity.

Present your work in the form of a brief word-processed report to the company's chairman.

(a) An earthquake in the Far East results in the destruction of a major semiconductor manufacturing plant.
(b) A major fire causing long-term suspension of all services through the Channel Tunnel.
(c) Funded by a substantial European Community grant, a French manufacturer sets up a colour photocopier production plant in Calais.
(d) A substantial fall in price of desktop colour laser printers.
(e) A Japanese company patents a new low-cost high-quality colour printing process.
(f) A series of strikes by workers in cross-channel ferry companies brings chaos to the channel ports.
(g) A progressive fall in interest rates results in an exceptionally large number of business start-ups in the south-east of England.
(h) A major high-street chain of newsagents decides to provide low-cost in-store colour photocopying.
(i) Well known for its high salaries and excellent benefits packages, a North American electronic equipment manufacturer sets up its European headquarters on the same industrial estate.
(j) New European environmental legislation makes manufacturers of electronic equipment responsible for its disposal and/or recycling.

The economic environment

The economic environment is crucial in determining the financial context in which an engineering company operates. The profitability of an engineering company often goes hand-in-hand with the state of the economy as a whole but this is not always the case.

In order to understand the more important economic issues it is necessary to have a grasp of some basic concepts and terminology including gross national product (GNP), gross domestic product (GDP) and balance of payments.

Figure 1.3 shows how the production of goods and services and their consumption can be regarded as forming a closed circle. The firms make the goods and provide the services consumed by the people living in the households. The people living in the households are the same ones that own and work in the production and service firms. The households create the demand for the goods and services. This, in turn, is satisfied by the production firms that pay the wages of the householders. Hence, the flow of goods and services and production resources is in one direction and this flow is balanced by the flow of money in the other direction.

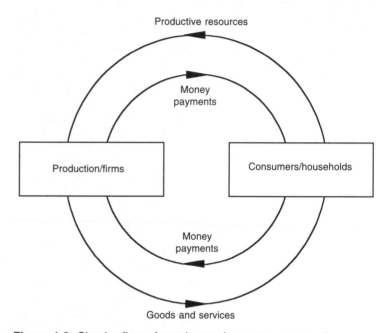

Figure 1.3 *Circular flow of goods, services, resources and money*

It's worth noting that many goods and materials used in engineering processes are not consumed for their own sake but are actually used in the production of other goods. They are known as *intermediate goods*. For example, sheet steel is used in the production of cars, wooden planks in the construction of buildings, copper wire in the manufacture of transformers, and so on. You may now be wondering how this all works in the knowledge that GDP only accounts for goods that are in their finished state. What about intermediate goods?

To explain this, consider the use of sheet steel in the production of cars (see Figure 1.4). Here, the output of the steel producer is sheet steel and this is regarded as intermediate goods in the car production process. The steel producer purchases iron ore (a raw material) from a mining company and uses it to produce sheet steel. This is then sold to the car manufacturer who produces the finished product. Only the final sale of cars is included in the calculation of GDP.

An alternative but equivalent way of measuring GDP is to add up the value added at each stage of the production process. The value added by the steel producer is the difference between the value of output (sheet steel) and input (iron ore). The value added by the car manufacturer is the difference between the value of output (finished cars) and input (sheet steel and other materials).

It is important to note that there will often be several stages in the process of intermediate goods production but the sum of value added at each stage of production must always be equal to the value of the final output.

> **Key point**
>
> The sum of the value added at each stage of an engineering production process is equal to the value of the final output. This gives us an alternative way of determining the GDP contribution of an engineered product.

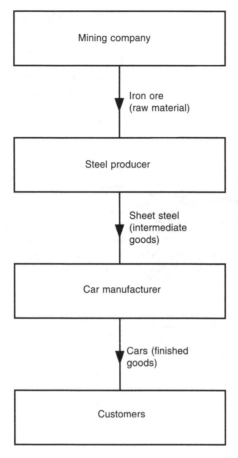

Figure 1.4 *Intermediate and finished goods in a car production process*

Gross national product (GNP)

The total output of the UK can be measured by adding together the value of all goods and services produced by the UK. This figure is called the gross national product. The word 'gross' implies that no deduction has been made for the loss in value of the country's capital equipment, which helps to make the national product, caused by normal wear and tear.

The word 'national' in this context does not mean that the GNP is the total output produced within the borders of the UK. The gross product of the UK includes some output taking place and produced by resources within the UK but owned by people from other countries. Therefore, this particular element of output cannot be regarded as part of our national income. At the same time, some sources of output are located in other countries but owned by UK citizens. Therefore the GNP is defined as the value of the total output of all resources owned by citizens of the UK wherever the resources themselves may be situated.

Gross domestic product (GDP)

If we measure the value of the total output produced within the borders of the UK, it is called the gross domestic product. So, the

GDP is defined as the value of the output of all resources situated within the UK wherever the owners of the resources happen to live. In many ways the GDP is a more important measure than the GNP.

The UK Government issues detailed statistics that indicate the performance of the different sectors of the economy. You can obtain more information from the National Statistics website (StatBase) at http://www.statistics.gov.uk

Gross national income (GNI)

Gross national income (GNI) was traditionally called gross national product (GNP), even though it was a measure of income rather than output (or production). The term gross national product is no longer used within the official definition of economic accounts.

GNI is a measure of the income coming to the UK from production wherever in the world it occurs. For example, if a British-owned company operating in North America sends some of its profits back to the UK this *adds to* the UK's GNI. Similarly, a British subsidiary of a Japanese company sending profits to Japan will *reduce* the UK's GNI.

Strictly, gross national income is GDP *plus* the net income from the rest of the world, see Figure 1.5.

> **Key point**
>
> The difference between GNP and GDP is that the latter includes income earned domestically by foreign nationals and excludes income earned by UK nationals in other countries. In other words, GNP is GDP plus net income earned overseas.

Gross domestic product (GDP)

less

Income payable to overseas divisions

plus

Income receivable from the rest of the world

equals

Gross national income (GNI)

Figure 1.5 *Relationship between gross domestic product (GDP) and gross national income (GNI)*

> **Key point**
>
> GDP per head is a measure of the productivity of a region or a nation. It relates output to the number of people employed producing that output.

> **Key point**
>
> Economists often use the terms 'output' and 'income' interchangeably. If this seems odd just remember that, for the economy as a whole, total production is equal to total income.

Activity 1.2

The UK's GDP for the 11-year period from 1989 to 1999 is shown in Table 1.1. Plot this data as a line graph. Also determine:

(a) the percentage increase in GDP over the ten year period from 1990 to 1999
(b) the average annual percentage increase in GDP over the ten year period from 1990 to 1999
(c) the year in which the actual increase in GDP was least
(d) the year in which the actual increase in GDP was greatest.

Present your results in a brief written report. Hint: You may find a spreadsheet useful for determining annual changes and also for plotting graphs.

Table 1.1 *UK GDP (at current basic prices) for the 11 years from 1989 to 1999*

Year	GDP (£ million)
1989	461 925
1990	501 473
1991	523 137
1992	545 487
1993	573 377
1994	606 720
1995	635 498
1996	674 029
1997	715 127
1998	755 297
1999	787 385

Key point

The circular flow of income and expenditure leads us to conclude that there are two ways of measuring this flow – by adding up all of the income or by adding up all of the expenditure. According to the rules of accounting these two must be equal!

Activity 1.3

Dragon Data Controls (DDC) is the UK subsidiary of a Japanese machine tool manufacturer. In one year, DDC receives a total income of £25 million from the sale of CNC equipment. Of this, £8 million worth of income comes from sales to other European countries. DDC returns £5 million of profit to its Japanese parent company. DDC imports £7 million of components from Japan and its UK costs (employment and overheads) amount to £13 million.

Explain the contribution that DDC makes to the UK's GNP and illustrate your answer using a pie chart. Present your work in the form of a brief word-processed report.

Gross domestic product per head

Gross domestic product per head is a measure of *productivity*. It relates output to the number of people employed producing that output. The formula used is:

output per head (or per capita) = output produced divided by the number of people producing it

Put another way, we can write:

output per head = output/employment

Often we are concerned not with absolute figures for output and employment but more with trends. For this reason, both the output and employment figures are usually quoted as *index* figures (i.e. they are stated relative to a base figure of 100 for a particular year). Table 1.2 shows how the UK output per head and cost of labour have changed over the eleven year period from 1990 to 2000.

Regional variations

There are significant variations in the economic performance of the various UK regions. Table 1.3 shows the GDP by UK region in

Table 1.2 *UK productivity and labour costs per unit for the period 1990 to 2000*

Year	Output (manufacturing jobs)	Output (whole economy)	Unit labour costs (whole economy)
1990	83.7	86.7	90.5
1991	86.2	88.4	96.8
1992	91.7	91.3	99.1
1993	96.2	94.5	99.5
1994	100.5	98.2	98.4
1995	100.0	100.0	100.0
1996	99.4	101.6	101.8
1997	100.3	103.1	104.8
1998	101.2	104.6	108.2
1999	105.2	105.8	112.3
2000	111.3	108.2	114.8

1996 and 1999. Given the different sizes and populations of the regions this table is not particularly meaningful in terms of the contribution of each person who lives and works in that region. To address this problem, Table 1.4 shows how the GDP varies on a 'per capita' (i.e. 'per head') basis.

Table 1.3 *Regional GDP in the UK for 1996 and 1999*

Region	GDP (£ million) 1996	GDP (£ million) 1999
United Kingdom	674 029	787 386
North East	23 755	25 875
North West	68 937	77 562
Yorkshire and the Humber	50 043	57 554
East Midlands	44 184	50 906
West Midlands	54 851	63 495
East	66 484	81 793
London	99 490	122 816
South East	100 614	121 956
South West	50 128	58 151
England	558 483	660 108
Wales	27 017	30 689
Scotland	57 338	64 050
Northern Ireland	14 936	17 003

Activity 1.4

Present the data shown in Table 1.4 in the form of two bar charts in descending order of per capita GDP. Use the data to determine:

(a) the best and worst performing regions in each of the two years
(b) the region that has shown the greatest increase in per capita GDP over the four-year period

(c) the region that has shown the greatest decline in per capita GDP over the four-year period.

Present your results in a brief written report. Hint: You may find a spreadsheet useful for determining the four-year changes and also for plotting graphs.

Table 1.4 *Regional variation of GDP per capita in the UK for 1996 and 1999*

Region	Per capita GDP (UK = 100)	
	1996	1999
North East	81.5	77.3
North West	89.2	86.9
Yorkshire and the Humber	88.8	87.9
East Midlands	95.4	93.6
West Midlands	92.2	91.7
East	112.5	116.4
London	126.1	130.0
South East	114.1	116.4
South West	92.5	90.8
Wales	82.6	80.5
Scotland	99.8	96.5
Northern Ireland	80.1	77.5

Key point

Employment and unemployment statistics are among the most watched of all economic data. One of the main reasons for this is that a well-functioning economy will make use of all of its available resources. High levels of unemployment often imply wasted resources and hence problems in the functioning of the economy.

Key point

The balance of payments is the difference between all money coming into a country and all the money going out. It includes trade, property income flows and other transfers and goes wider than the trade balance that's simply the difference between exports and imports.

Direct employment

This is the term used to describe the mode of employment of people that are actually working to produce a product. For example, the people working in a factory making furniture could all be regarded as direct labour being directly employed on the manufacture of the furniture.

Indirect employment

Following on from the previous example, the people concerned with the transportation of the production materials to the furniture factory and the finished furniture to the shops which sell the furniture are regarded as indirect labour and are in indirect employment.

Note, however, that people may be directly employed by their own trade but indirectly by another. For example, the people felling the trees and producing the raw timber used to make the furniture are directly employed by the timber trade but are indirectly employed by the furniture making trade.

Exports

Exports are goods and services that we sell to other countries. Visible exports are the hard goods which are physically transported abroad and for which we receive payment. Invisible exports are those services that we provide for foreigners and for which they pay. Invisible exports include the payments we receive for insurance and financial services, technical or military training and the like. Tourism is also an important part of our invisible export trade. A German tourist in London will spend money that he has earned in Germany. This provides a useful income to the UK no different from, say, visibly exporting and selling the German a bottle of Scotch whisky in Berlin. Note that when we take our holidays abroad, the goods and services we buy and consume abroad are in fact imports to the UK.

Imports

Imports flow in the opposite direction to that of exports. Imports are the goods and services that we purchase from abroad.

Balance of payments

For a nation to remain financially sound, it must ensure that its expenditure does not exceed its income. The difference between the total income we earn from abroad for our exports is balanced against the total payment we must make abroad for our imports. We compare or *balance* the two figures and hopefully the export figure is at least as large as and preferably larger than the import figure.

Very often the UK balance of visible exports and imports shows an excess of import value over exports. However, we have a relatively healthy sale of our services abroad and our *invisibles* usually give us a favourable trading balance. The government lets us know what our overseas trading position is by issuing monthly balance of payment figures.

Local economy

For the first half of the twentieth century, engineering was generally located within cities. Since then there has been a tendency for any new engineering enterprise to be located in an industrial estate on the periphery of a town rather than in the town itself. Usually this is because:

- the town centre is already too congested to allow for additional new industry
- of the advantages of being located in a ready, purpose-built industrial accommodation on a site having good road links with the national motorway network
- engineering activities which may involve noise and other pollutants are best kept away from the commercial and domestic centres of towns.

In general, the engineering industries that remained in the city centres have slowly become outdated and, in many cases, have closed down. The impact of this migration from the city centres to the suburbs has been to leave derelict buildings, unemployment and social deprivation for the city residents. For the outer suburbs receiving the new engineering industries, the impact has not always been positive. The decentralization of engineering from the city centres has contributed to urban sprawl, and this has led to conflict for land on the cities' boundaries between engineering, farming and recreation. Also, it has tended merely to move the problem of engineering pollution from the city centre to its suburbs.

Regional economy

The regional economy comprises many local economies but the change in the engineering pattern is much the same. While there is still a great deal of engineering activity to be found in and around many large cities and built-up areas, there is a definite migratory move towards the small town and rural areas. This trend is to be found in most economically developed countries and has been a consistent feature of the last 25 years.

National economy

At the national level the uneven spread of engineering is between the different regions. The processes that caused this variation are historic. Very often they are directly connected to the availability of natural resources. For example, in the nineteenth century, regions rich in coal were favoured with engineering expansion because of the local availability of coal to fire boilers to drive the steam engines that powered the factories. The technical skills acquired by the workers in the coal bearing regions were the same skills required for other industries and enterprises and cumulative expansion took place. This expansion, and the highly paid work it created, attracted labour from the less industrialized regions so exacerbating the regional disparities.

However, over the last 50 years there has been a shift of engineering away from the old industrial regions such as the North East and Midlands of England and parts of Scotland to more convenient locations such as the Thames Valley along the M4 motorway and along the M11 motorway north of London. The reasons for the regional shift are many and varied and include such factors as:

- because of its cost and record of causing pollution, coal is no longer a popular fuel
- with natural gas and electrical power being available almost anywhere in the country, new engineering activities can be located in regions having pleasant natural and social environments
- because of the ubiquitous motor car, good roads and frequent air services, commuter and business communication to most regions is no longer a major problem
- the availability of a pool of technologically skilled labour in places where high-technology companies are clustered together.

Within the UK there are significant differences in weekly earnings in different regions. The official statistics for 1999 are shown in Table 1.5.

Table 1.5 *Regional variation in average weekly earnings for 1999*

Region	1999 average weekly earnings (all persons)
East Midlands	£361.7
East	£396.6
England	£405.4
London	£520
North East	£349.6
North West	£372.6
Northern Ireland	£344.9
Scotland	£364.9
South East	£423.2
South West	£364.9
Wales	£353.6
West Midlands	£375.6
Yorkshire and Humberside	£361

Activity 1.5

1. Photocopy the map shown in Figure 1.6 (your tutor may be able to supply you with this) and use it to identify and shade in the following economic regions:

(a) Yorkshire and the Humber
(b) East Midlands
(c) West Midlands
(d) London
(e) South East
(f) South West.

2. Mark the following motorways on the photocopied map:

(a) M1
(b) M3
(c) M4
(d) M6
(e) M5
(f) M40
(g) M45.

3. Which motorway(s) link the following pairs of economic regions:

(a) East Midlands and West Midlands?
(b) South East and South West?
(c) London and Yorkshire and the Humber?
(d) London and East Midlands?
(e) London and West Midlands?

4. A steel producer in Sheffield supplies a transformer manufacturer in Coventry with sheet steel. The finished transformers are transported to Southampton for shipping to Bilbao in Spain.

(a) Explain the effect this has on the GDP of various regions in the UK and name the economic regions concerned.
(b) Explain the effect this has on the UK balance of payments and say whether it is positive or negative.
(c) State which motorway routes are used for the transport of raw materials and which are used for the shipping of finished goods.

Activity 1.6

Plot the data from Table 1.5 in the form of a bar chart in ascending order of average weekly earnings. Use the data to determine:

1. The region in which average weekly earnings are the greatest.
2. The region in which average weekly earnings are the least.
3. The average weekly earnings in the North East expressed as a percentage of those in the South East.
4. The average weekly earnings in London expressed as a percentage of those in Yorkshire and Humberside.
5. A large production plant is to be set up in the north of England. In which one of the three regions is the labour cost likely to be the least?

Present your work in the form of a set of word-processed notes and an accompanying bar chart produced using a spreadsheet.

European union

The performance of 15 European economies is shown in Table 1.6. Note from this the large variation in the number of people in employment and the sectors in which they are employed. Within the European Union (EU) engineering activities have a similar varied pattern. The countries that tend to perform better in terms of industry and the economy tend to be those that were the first to become industrialized in the eighteenth and nineteenth centuries. Britain, Germany, France and Italy are predominant in Europe with the main concentration lying within a rough triangle formed by London, Hamburg and Milan. Ireland, Spain, Southern Italy and Greece lie outside this triangle and tend to be less industrialized.

Global factors

The past 30 years has seen a shift in some of the major engineering activities that used to be concentrated in Europe, North America

Figure 1.6 *Map of the UK (see Activity 1.5)*

and Japan. In particular, much of the electronics and printing industries have migrated to the *Pacific rim* countries such as Hong Kong, Singapore, Taiwan, Thailand and more recently Indonesia. The main reason for this shift is the low labour costs to be found in the Far East.

Another prime example of the shift of engineering activities out of Europe is that of shipbuilding. The UK's contribution in particular

Table 1.6 *Performance of 15 European economies in 1999*

	Persons in employment in 1999 (thousands)	Percentage in 1999 employed in:			Unemployment rate (percentages), 1999	Long-term unemployed as a percentage of the unemployed, 1999	Gross domestic product per head (PPS) Europe=100, 1998	Estimates of the percentage of GDP in 1996 derived from		
		Agriculture	Industry	Services				Agriculture	Industry	Services
Europe	152 494	4.7	29.6	65.5	9.4	49.1	100	–	–	–
Austria	3678	6.2	29.8	64.0	4.0	37.1	112	1.5	31.6	66.9
Belgium	3987	2.4	25.8	71.8	8.8	59.3	111	1.4	27.3	71.3
Denmark	2708	3.3	26.8	69.5	5.6	18.6	119	4.1	27.6	68.3
Finland	2333	6.4	27.6	65.7	11.5	23.6	102	5.5	33.0	61.5
France	22 755	4.3	26.3	69.4	11.4	41.3	99	2.4	27.4	70.1
Germany	36 089	2.9	33.8	63.3	8.9	50.6	108	1.1	34.9	64.0
Greece	3967	17.8	23.0	59.2	11.7	55.3	66	14.9	25.0	60.0
Ireland	1593	8.5	28.3	62.5	5.9	56.0	108	7.2	40.4	52.4
Italy	20 618	5.4	32.4	62.2	11.7	60.8	101	3.5	28.9	67.6
Luxembourg	176	1.9	21.9	75.8	2.4	32.2	176	1.3	23.9	74.8
Netherlands	7605	–	–	–	3.3	41.5	113	3.2	26.8	70.0
Portugal	4830	12.6	35.3	52.1	4.7	39.9	75	4.1	33.9	62.0
Spain	13 773	7.4	30.6	62.0	16.1	45.0	81	4.5	29.0	66.5
Sweden	4054	3.0	25.0	72.0	7.6	29.1	102	2.2	30.0	66.9
United Kingdom	27 107	1.6	26.0	72.3	6.1	30.3	102	1.8	29.8	68.4

Test your knowledge 1.6

Use the data in Table 1.6 to answer the following questions:

1. Which European country has the largest population?
2. Which European country has the smallest population?
3. In which European country was industry the most significant employment sector?
4. In which European country was industry the least significant employment sector?
5. Which European country had the highest unemployment rate?
6. Which European country had the lowest unemployment rate?
7. Which European country had the least per capita GDP?
8. Which European country had the greatest per capita GDP?

has fallen and is now virtually non-existent except for the manufacture of oil platforms and ships for the Royal Navy.

However, the traffic in engineering activities has not been all negative. The Japanese, wanting to sell their cars in Europe, have established several engineering production plants in the UK. The firms of Toyota (Deeside and Burnaston), Nissan (Sunderland) and Honda (Swindon) are three good examples. All occupy rural sites and have access to skilled and well-educated workforces. Road communications are good and, in the case of Nissan, the site is in an assisted area where substantial government grants are available.

Activity 1.7

A company based in the Far East has asked you to carry out some research in order to help them investigate some investment opportunities in Europe. They have asked you to produce a broad comparison of the performance of the various countries that comprise the European Union (EU).

(a) Photocopy the map of Europe shown in Figure 1.7 (your tutor may be able to supply you with this).
(b) Shade in the countries that constitute the EU.
(c) Mark in the name of each EU country.
(d) List the countries of the EU in descending order of GDP. What is the position of the UK?

Produce your findings in the form of a set of overhead projector transparencies and handouts to be used at a board meeting.

Legislation and the environment

Modern engineering processes and systems are increasingly designed and implemented to minimize environmental affects. Engineering companies must ensure that the negative effects of engineering activities on the natural and built environment are minimized. You need to be able to identify how individual engineering companies seek to do this, such as through:

- design of plant and products which optimizes energy use and minimizes pollution
- good practice such as the efficient use of resources and recycling and the use of techniques to improve air and water quality
- management review and corrective action
- relevant legislation and regulations.

Material processing

Many engineering activities involve the processing of materials. Such materials may appear in the product itself or may be used in the manufacturing process. Some of these materials occur naturally and, after extraction from the ground, may require only minimal treatment before being used for some engineering purpose. Examples

Figure 1.7 *Map of Europe (see Activity 1.7)*

are timber, copper, iron, silicon, water and air. Other engineering materials need to be manufactured. Examples are steel, brass, plastic, glass, gallium arsenide and ceramic materials. The use of these materials produces effects; some beneficial, some not.

Economic and social effects stem from the regional wealth that is generated by the extraction of the raw material and its subsequent processing or manufacture into useful engineering materials. For example, the extraction of iron ore in Cleveland and its processing into pure iron and steel has brought great benefit to the Middlesbrough region. The work has attracted people to live in the area and the money they earn tends to be spent locally. This benefits trade at the local shops and entertainment centres, and local builders must provide more homes and schools, and so on. The increased numbers of people produce a growth in local services which includes a wider choice of different amenities, better roads and communications, and arguably, in general, a better quality of life.

On the debit side, the extraction of raw materials can leave the landscape untidy. Heaps of slag around coal mines and steelworks together with holes left by disused quarries are not a pretty sight. In recent years much thought and effort has been expended on improving these eyesores.

Slag heaps have been remodelled to become part of golf courses and disused quarries filled with water to become centres for water sports or fishing. Disused mines and quarries can also be used for taking engineering waste in what is known as a landfill operation prior to the appropriate landscaping being undertaken.

Other potential problems can arise from having to transport the raw materials used in engineering processes from place to place. This can have an adverse affect on the environment causing noise and pollution.

Environmental effects

Engineering activities are a major source of *pollutants* and many types of *pollution*. Air, soil, rivers, lakes and seas are all, somewhere or other, polluted by waste gases, liquids and solids discarded by the engineering industry. Because engineering enterprises tend to be concentrated in and around towns and other built-up areas, these tend to be common sources of pollutants.

Electricity is a common source of energy and its generation very often involves the burning of the *fossil fuels*: coal, oil and natural gas. In so doing, each year billions of tonnes of carbon dioxide, sulphur dioxide, smoke and toxic metals are released into the air to be distributed by the wind. The release of hot gases and hot liquids also produces another pollutant: heat. Some electricity generating stations use nuclear fuel that produces a highly radioactive solid waste rather than the above gases.

The generation of electricity is by no means the only source of toxic or biologically damaging pollutants. The exhaust gases from motor vehicles, oil refineries, chemical works and industrial furnaces are other problem areas. Also, not all pollutants are graded as *toxic*. For example, plastic and metal scrap dumped on waste tips, slag heaps around mining operations, old quarries, pits and derelict land are all *non-toxic*. Finally, pollutants can be further defined as *degradable* or *non-degradable*. These terms simply indicate whether the pollutant will decompose or disperse itself with time. For example, smoke is degradable but dumped plastic waste is not.

Carbon dioxide

Carbon dioxide in the air absorbs some of the long-wave radiation emitted by the earth's surface and in so doing is heated. The more carbon dioxide there is in the air, the greater the heating or greenhouse effect. This is suspected as being a major cause of global warming causing average seasonal temperatures to increase. In addition to causing undesirable heating effects, the increased quantity of carbon dioxide in the air, especially around large cities, may lead to people developing respiratory problems.

Oxides of nitrogen

Oxides of nitrogen are produced in most exhaust gases and nitric oxide is prevalent near industrial furnaces. Fortunately, most oxides of nitrogen are soon washed out of the air by rain. But if there is no rain, the air becomes increasingly polluted and unpleasant.

Sulphur dioxide

Sulphur dioxide is produced by the burning of fuels that contain sulphur. Coal is perhaps the major culprit in this respect. High concentrations of this gas cause the air tubes in people's lungs to constrict and breathing becomes increasingly difficult. Sulphur dioxide also combines with rain droplets eventually to form dilute sulphuric acid or *acid rain*. This is carried by the winds and can fall many hundreds of miles from the sulphur dioxide source. Acid rain deposits increase the normal weathering effect on buildings and soil, corrode metals and textiles and damage trees and other vegetation.

Smoke

Smoke is caused by the incomplete burning of fossil fuels, such as coal. It is a health hazard on its own but even more dangerous if combined with fog. This poisonous combination, called *smog*, was prevalent in the early 1950s. It formed in its highest concentrations around the large cities where many domestic coal fires were then in use. Many deaths were recorded, especially among the elderly and those with respiratory diseases. This led to the first Clean Air Act which prohibited the burning fuels that caused smoke in areas of high population. So-called 'smokeless zones' were established.

Dust and grit

Dust and grit (or *ash*) are very fine particles of solid material that are formed by combustion and other industrial processes. These are released into the atmosphere where they are dispersed by the wind before falling to the ground. The lighter particles may remain in the air for many hours. They form a mist, which produces a weak, hazy sunshine and less light.

Toxic metals

Toxic metals such as lead and mercury are released into the air by some engineering processes and especially by motor vehicle exhaust gases. Once again the lead and mercury can be carried over hundreds of miles before falling in rainwater to contaminate the soil and the vegetation. Motor vehicles are now encouraged to use lead-free petrol in an attempt to reduce the level of lead pollution.

Ozone

Ozone is a gas that exists naturally in the upper layers of the earth's atmosphere. At that altitude it is one of the earth's great protectors but should it occur at ground level it is linked to pollution. *Stratospheric ozone* shields us from some of the potentially harmful excessive ultraviolet radiation from the sun. In the 1980s it was discovered that emissions of gases from engineering activities were causing a 'hole' in the ozone layer. There is concern that this will increase the risk of skin cancer, eye cataracts and damage to crops and marine life.

At ground level, sunlight reacts with motor vehicle exhaust gases to produce ozone. Human lungs cannot easily extract oxygen (O_2) from ozone (O_3) so causing breathing difficulties and irritation to the respiratory channels. It can also damage plants.

This ground level or *tropospheric ozone* is a key constituent of what is called photochemical smog or summer smog. In the UK it has increased by about 60 per cent in the last 40 years.

Heat

Heat is a waste product of many engineering activities. A typical example being the dumping of hot coolant water from electricity generating stations into rivers or the sea. This is not so prevalent today as increasingly stringent energy saving measures are applied. However, where it does happen, river and sea temperatures can be raised sufficiently in the region of the heat outlet to destroy natural aquatic life.

Chemical waste

Chemical waste dumped directly into rivers and the sea, or on to land near water, can cause serious pollution that can wipe out aquatic life in affected areas. There is also the long-term danger that chemicals dumped on soil will soak through the soil into the ground water which we use for drinking purposes and which will therefore require additional purification.

Radioactive waste

Radioactive waste from nuclear power stations or other engineering activities that use radioactive materials poses particular problems.

Test your knowledge 1.7

Name a pollutant that fits each of the following categories:

(a) toxic and degradable
(b) toxic and non-degradable
(c) non-toxic and degradable
(d) non-toxic and non-degradable.

Not only is it extremely dangerous to people – a powerful cause of cancer – its effects do not degrade rapidly with time and remain dangerous for scores of years. Present methods of disposing of radioactive waste, often very contentious however, include their encasement in lead and burial underground or at sea.

Derelict land

Derelict land is an unfortunate effect of some engineering activities. The term derelict land may be taken to mean land so badly damaged that it cannot be used for other purposes without further treatment. This includes disused or abandoned land requiring restoration works to bring it into use or to improve its appearance. Land may be made derelict by mining and quarrying operations, the dumping of waste or by disused factories from bygone engineering activities.

Environmental legislation

Engineering activities can have harmful effects on the physical environment and therefore on people. In order to minimize these effects, there is a range of legislation (rules and regulations) that all engineering companies must observe.

The appropriate United Kingdom Acts of Parliament include Deposit of Poisonous Wastes Act, Pollution of Rivers Act, Clean Air Act, Environmental Protection Act, Health and Safety at Work etc. Act and the like. Additionally, not only are there local by-laws to be observed there are also European Union (EU) directives that are activated and implemented either through existing UK legislation in the form of Acts of Parliament or mandatory instructions called Statutory Instruments (SI).

New Acts and directives are introduced from time to time and industry needs to be alert to and keep abreast of these changes. Typical of these new initiatives is the European Electromagnetic Compatibility (EMC) legislation. This states that, with effect from 1 January 1996, it is a requirement that all products marketed must conform with the new legislation. This new EMC legislation, at last, officially recognizes the well-known problem of unwanted electromagnetic wave radiation that emanates from most pieces of electrical equipment. The unwanted radiation can interfere with nearby electronic equipment causing it to malfunction.

In the case of UK Acts of Parliament, the above legislation is implemented by judgement in UK Courts of Justice in the normal manner but based on EU legislation, if more appropriate, or by judgement of the European Court of Justice.

The purpose of this legislation is to provide the following functions:

- *prevent* the environment being damaged in the first place;
- *regulate* the amount of damage by stating limits, for example the maximum permitted amount of liquid pollutant that a factory may discharge into the sea;

- *compensate* people for damage caused, for example, from a chemical store catching fire and spreading wind borne poisonous fumes across the neighbourhood;
- *impose sanctions* on those countries or other lesser parties that choose to ignore the legislation;
- *define who is responsible* for compliance with legislation to persons who can be named and their precise area of responsibility documented.

For the purposes of showing an understanding of the above, you are *not* expected to have a detailed understanding of the various Acts; however, you *should be aware of the general provisions* of the legislation and what it is trying to achieve. Your school, college or local library will be able to provide you with more details.

The effects of the above legislation on engineering activities have, in general, made them more difficult and more expensive to implement. A few simple examples of this follow.

- *Chemical factories* can no longer discharge their dangerous waste effluent straight into the river or sea without first passing it through some form of purification.
- *Coal fuelled power stations* must ensure that their chimney stacks do not pollute the neighbourhood with smoke containing illegal limits of grit, dust, toxic gases and other pollutants. A system of smoke filtration and purification must be (expensively) incorporated.
- *Motor car* exhaust gases must be sufficiently free of oxides of nitrogen, carbon monoxide and other toxic gases. This can only be achieved by, among other things, replacing the crude petrol carburettor with a more sophisticated petrol injection system and fitting a catalytic converter in the exhaust system. All this has added to the price of the motor car and has made it more difficult for the DIY motorist to service his or her vehicle.
- All *electrical equipment* including TVs, PCs, power hand tools, electro-medical machines, lighting and the like must be tested and certified that they comply with the EMC legislation. So, in addition to the cost of reducing any excessive radiation from the product itself, the purchase or hire of expensive EMC test equipment and the training of people in its use must also be taken into account. Further, because of delays in obtaining an official EMC examination and a supporting EC certificate, the introduction of new product designs can also be delayed and this may have adverse marketing effects.

Test your knowledge 1.8

Explain what is meant by EMC and why the legislation that relates to it is important.

Test your knowledge 1.9

Explain the effect of environmental legislation on the cost of motoring in the UK.

Finally, it is important to be clear about the difference between *waste products* and *by-products*. The by-products from one process can be sold as the raw materials for other processes. For example, natural gas is a by-product of oil extraction and a useful fuel used in the generation of electricity. Waste products are those that cannot be sold and may attract costs in their disposal. Nuclear power station waste is a typical example.

Activity 1.8

Find out what happens to the domestic waste produced in your locality.

- What items of domestic waste are recycled?
- Is any of the waste burnt to produce useful heat?
- If the waste is transported to another site for processing, where does it go and what processes are used?
- What arrangements are there for disposing of hazardous waste?

Present your findings in the form of a brief word-processed report. Illustrate your report with the aid of a simple flowchart.

Activity 1.9

Investigate the effects of each of the following engineering activities on the environment:

- manufacture of motor vehicles
- service and repair of motor vehicles
- road transport of bulk materials.

In your investigation consider the environmental effects of materials and consumables (such as fuel and lubricants) as well as the short-term and long-term environmental effects of any waste products.
Present your work in the form of a brief written report.

Activity 1.10

Visit your local museum and obtain information concerning the engineering activities and companies that were active in your area (city, town or county) in:

(a) 1900
(b) 1950.

Compare these activities with those of the present day. Identify any new industries and explain why they have developed in your area. In the case of any industries or companies that have been active in your area for the full period, identify ways in which they have changed their operation in order to:

(a) remain competitive
(b) safeguard the environment.

Present your findings in the form of a brief class presentation with appropriate visual aids.

Planning

The essential business activities performed in an engineering company can be grouped together under the general headings of planning, controlling and organizing (see Figure 1.8). The first of these activities, planning, is absolutely fundamental to the correct functioning of an engineering company. If no planning is done then activities are almost certainly going to be very ineffective. What is planning? It is the sum of the following activities:

- setting the *goals* for an engineering company
- forecasting the *environment* in which the engineering company will operate
- determining the *means* to achieve goals.

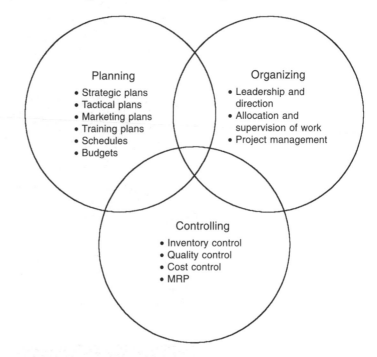

Figure 1.8 *Relationship between planning, controlling and organizing in an engineering business*

Setting goals or objectives is the first step in planning. It determines the direction an engineering company is going. It encourages all engineering company members to work toward the same ends, otherwise members are likely to set their own objectives which will conflict with each other. Good objectives make for rational engineering companies that are more coordinated and effective. The objectives must therefore be set by the most senior management group in an engineering company so that all of its staff can be given clear direction. If the goals are clearly stated, logical and appropriate for the business then they act both as motivators and yardsticks for measuring success.

Once the engineering company has clear direction the next step is to analyse the environment and forecast its effect on the business. For example, an engineering company that makes lawn mowers may set objectives so that, within five years, it will:

- achieve a 30 per cent share of the market
- be acknowledged as the market leader

> **Key point**
>
> Planning is about setting goals, forecasting the environment and determining the means to achieve the goals.

- be the accepted technological leader
- be highly competitive on price
- operate internationally.

When it forecasts its environment it may conclude that:

- new designs will be marketed by competitors
- new battery technology will become available to support cordless mowers
- there will be a sharp decline in demand for manual lawn mowers.

Its designs are technically very sound but are being threatened by new rotary 'hover' models that are proving attractive to customers. It has to decide how to deal with the threat, either to improve its existing design concept so that customers continue to find them attractive or to follow the new trend and produce products based on new design concepts.

Forecasting the environment allows the company to set new objectives and to prepare plans to meet its revised goals. Companies that fail to go through this process will go into decline in the long run because they are ignoring the changing world around them.

Once the goals have been refined and changed in the light of environmental forecasting then plans can be made to achieve the goals. Some plans will not change that much, others will be dramatically affected by the changing environment. For this reason plans can be classified as follows:

- standing plans
- single use plans
- strategic plans
- tactical plans.

Standing plans are those that are used many times and remain relatively unaffected by environmental change. Examples are employment, financial, operating and marketing policies and procedures. For example, hiring new employees involves standard procedures for recruitment and selection. Another example would be the annual routines for establishing budgets.

Single use plans are those that are used once only, such as those for the control of a unique project or specific budgets within an annual budget. (Budgets themselves are single use plans, even though the procedures used for producing them are standing plans.)

Strategic plans are the broad plans related to the whole engineering company and include forecasting future trends and overall plans for the development of the engineering company. They are often in outline only and very often highly subjective, involving judgements made by top managers. For example, a plan may be made to build an entirely new factory based on forecasts of demand. This plan is strategic, and if it is wrong and the sales forecasts on which it is based do not materialize, the results for a company could be devastating.

Tactical plans operate within the strategic plan. The new factory has to be brought into commission and production has to be scheduled and controlled. Plans for the latter are tactical, since they focus on how to implement the strategic plan.

Test your knowledge 1.10

Audio Power Systems (APS) manufactures a range of loudspeakers for audio hi-fi, music and sound-reinforcing applications. The board of directors at APS has set the following strategic goals for the company to be achieved over a three-year period:

1. Introduce two new loudspeaker designs each year.
2. Increase market share by 5 per cent each year.
3. Double the number of European distributors by the end of the three-year period.

Which THREE of the following objectives are most likely to be instrumental in delivering these strategic goals?

1. Employing more staff in research and development.
2. Appointing an assistant to the marketing manager with a good command of European languages.
3. Introducing measures that will reduce pay-costs by around 10 per cent.
4. Increasing expenditure in targeted marketing campaigns.
5. Building a new anechoic chamber test facility.

Control

In an engineering business, control is required for a variety of reasons including ensuring that a project or product:

- remains within budget
- is delivered on time
- complies with relevant legislation
- meets appropriate quality standards
- operates according to the agreed specification
- fully satisfies the expressed and implied needs of the customer or client.

The prerequisite of control is planning. Controlling involves comparing events with plans, correcting deviations and ensuring that the planned events happen. Sometimes deviations are so fundamental that they require a revision to the plan so that later events are controlled against a new plan. For example, the original sales forecast may turn out to be too optimistic, and production plans may have to be reduced to bring output into line with what sales are possible.

There are various ways in which control can be exercised. It can be predictive as in the case of a cash-flow forecast. This forecast may indicate a shortfall of cash in August but a surplus in September. The finance manager may need to arrange additional finance with the bank in August and then in September he might deposit surplus funds onto the money market. The point here is that variances are predicted in advance, thereby promoting cash control.

In the case of monthly comparisons between budgeted expenditures and actual expenditures an overspend might be revealed. This triggers action that holds back expenditure until spending comes back into line with budget. This is historical control since action is based on a report of events in the recent past.

Concurrent control is *real time* such as that which might occur in controlling a continuous process or a production line. In this case the system has built-in *feedback* which enables it to remain in balance by regulating inputs and outputs. An example of concurrent control would be where a production process requires temperature regulation. The control system is designed to switch off heating when temperature reaches a threshold or switch on heating when it drops to a minimum level. The 'feedback' is information on temperature. The same principle applies in some stock control systems, where stocks are maintained at predetermined minimum and maximum levels, with supplies being switched on and off to maintain equilibrium.

Key point

In an engineering firm, control is required to ensure that:

- costs remain within budget
- a project is delivered on time
- stages in the project comply with relevant legislation
- appropriate company and sector quality standards are met
- customers' and clients' needs are satisfied and there is effective liaison at all stages of a project
- modifications (and any other subsequent changes) meet the customer's specification and remain within budget.

Leadership and direction

Planning and control activities are the tasks of engineering management. However, they are only achieved through people. People will work effectively if they are led and directed properly. This implies that top managers must be in touch with the engineering processes as well as business issues and they should be 'visible'. They must have a clear vision for the future reinforced by specific objectives that are effectively communicated to their employees.

This approach to leadership is apparent in some of our best companies as exemplified by British Airways and Marks & Spencer. Such companies have a clear mission and objectives, and have a visibly committed top management. This philosophy permeates the whole engineering company stimulating better performance from all employees.

Motivating good performance from all employees is the responsibility of all managers. What motivates individuals and groups within commercial engineering companies is a complex and important subject, the detail of which is well beyond the scope of this book. However, it is still worth saying that managers must discover what it is that will stimulate employees to work productively.

In general people respond best to 'considerate' styles of management, whereby their personal contributions are fully recognized. It is also true that there has to be an atmosphere of discipline coupled with a work-oriented culture. The task has to be accomplished, and being considerate does not extend to the toleration of slack or sloppy practices and behaviour.

> **Key point**
>
> Clear direction, sound and explicit guidelines, and proven work processes and techniques (all of which must be effectively communicated to employees) jointly contribute to the successful operation of an engineering business.

Allocation and supervision of work

This is the practical implementation of all that we have discussed in this section. An engineering company exists to fulfil the goals of its owners. It has to function in a coordinated and rational way. The people who are its members have to work together and understand their specific roles and functions. They need to receive directions and work has to be allocated. There has to be supervision of these activities. An engineering company is analogous to a machine or a living organism. In order to function properly everything has to work together smoothly. The company's managers have the task of ensuring that this work takes place according to plan and within the engineering company's stated objectives.

> **Test your knowledge 1.11**
>
> Explain the following types of engineering production:
>
> (a) Continuous flow.
> (b) Intermittent or batch.
> (c) Project-based.

Process and facilities planning

Decisions have to be made in relation to location of the factory and the design and layout of production facilities. The design of production processes is interactive with product design, requiring close cooperation with research and product development functions.

Selecting the process of production is important and is strategic in nature. This means that it has a wide impact on the operation of the entire business. Decisions in this area bind the company to particular kinds of equipment and labour force because of the large capital investment that has to be made in order to make changes. For example, a motor manufacturer has to commit very large expenditures to lay down plant for production lines to mass produce cars. Once in production the company is committed to the technology and the capacity created for a long time into the future. There are three basic methods for production processes:

> **Test your knowledge 1.12**
>
> Which form of production is most appropriate to each of the following engineering processes:
>
> (a) Construction of a bridge.
> (b) Manufacture of copper wire.
> (c) Production of military aircraft.
> (d) Manufacture of music CDs.
> (e) Production of blank floppy disks.

- line flow

- intermittent flow
- project.

Line flow is the type of system used in the motor industry for assembly lines for cars. It also includes continuous type production of the kind that exists in the chemicals and food industries. Both kinds of line flow are characterized by linear sequences of operations and continuous flows, and tend to be highly automated and highly standardized.

Intermittent flow is the typical batch production or job shop, which uses general-purpose equipment and highly skilled labour. This system is more flexible than line flow, but is much less efficient than line flow. It is most appropriate when a company is producing small numbers of non-standard products, perhaps to a customer's specification.

Finally *project-based production* is used for unique products that may be produced one at a time. Strictly speaking there is not a flow of products, but instead there is sequence of operations on the product that has to be planned and controlled. This system of production is used for prototype production and in some engineering companies that produce major machine tool equipment for other companies to use in their factories.

Capacity planning

Once facilities for production have been put in place the next step is to decide how to flex the capacity to meet predicted demand. Production managers will use a variety of ways to achieve this from maintaining excess capacity to making customers queue or wait for goods to having stocks to deal with excess demand. The process is complex and may require the use of forecasting techniques, together with careful planning.

Scheduling activities are different for each process method and require the use of a variety of techniques. The objectives of good scheduling are:

- meeting customer delivery dates
- correct loading of facilities
- planning starting times
- ensuring jobs are completed on time.

Later we will look at scheduling in relation to the management of engineering projects.

Test your knowledge 1.13

List four outcomes of good scheduling.

Inventory control

With any manufacturing facility good inventory control is an absolute essential. It is estimated that it costs up to 25 per cent of the cost value of stock items per year to maintain an item in stock. Proper control systems have to be used to ensure that there is sufficient stock for production while at the same time ensuring that too much stock is not held. If stock levels are high there are costs associated with damage, breakage, pilferage and storage that can be avoided.

Workforce management

Workforce management is required in order to ensure that there is a suitably trained and experienced labour force that can apply the engineering processes, tools and facilities. The important aspects of workforce management are:

- work and method study
- work measurement
- job design
- health and safety.

The production manager has to establish standards of performance for work so that the capacity of the factory can be determined and so that the labour costs of products can be calculated. Work study, method study and work measurement activities enable this to be done, as well as helping to promote efficient and safe methods of working. The design of jobs is important in respect of worker health as well as effective work. Good job design can also make the work more interesting and improves employee job satisfaction, which in turn can improve productivity.

Quality control

Quality is a key objective for most engineering companies. It is especially important to the production function that is manufacturing the product for the customer.

Quality is generally defined as 'fitness for purpose'. In effect this means meeting the identified needs of customers. Thus it is really the customer that determines whether or not a company has produced a quality product, since it is the customer who makes an assessment of the perceived value of a product and is then either satisfied or dissatisfied.

This does bring problems for manufacturers since customer perceptions of quality vary. As a consequence some customers will be satisfied with a product more than other customers (the 'acid test' is probably whether a customer will make a repeat purchase or decide to purchase from another supplier). Because of the subjective nature of quality assessment, manufacturers often attempt to use more objective criteria for assessing fitness for purpose. This often includes:

- design quality
- conformance quality
- reliability
- service/maintenance.

Design quality is usually the joint responsibility of a company's marketing or customer liaison function and its research and development function. Design quality relates to the development of a specification for the product that meets a customer's identified needs.

Conformance quality means producing a product that conforms to the design specification. A product that conforms is a quality product, even if the design itself is for a cheap product. That seems contradictory, but consider the following example. A design is

Test your knowledge 1.14

Explain what is meant by each of the following terms:

(a)   Design quality.
(b)   Conformance quality.
(c)   Reliability.

Test your knowledge 1.15

List FOUR activities that are performed within the context of quality control.

drawn up for a budget camera, which is made from inexpensive materials and has limited capability. If the manufacture conforms to the specification then the product is of high quality, even though the design is of low quality compared with other more up-market cameras.

Reliability includes things like continuity of use measured by things like *mean time between failure* (MTBF). Thus a product will operate for a specified time, on average, before it fails. It should also be maintainable when it does fail, either because it can easily and cheaply be replaced or because repair is fast and easy.

When an engineered product fails or becomes unreliable it will require *service*. Service relates to aftersales service, the realization of guarantees and warranties, as well as the need for ongoing maintenance in order to ensure that performance remains within specification.

Quality control is concerned with administering *all* of these aspects. In the UK there are general standards for *quality systems,* the most relevant one here is BS 5750; the international counterpart is ISO 9000. The activities that make up a quality control system include the following:

- inspection, testing and checking of incoming materials and components
- inspection, testing and checking of the company's own products
- administering any supplier quality assurance systems
- dealing with complaints and warranty failures
- building quality into the manufacturing process.

While many of these activities are performed in order to monitor quality *after* the event, others may be carried out to prevent problems *before* they occur and some may be carried out to determine causes of failure that relate to design rather than manufacturing faults.

Activity 1.11

Skylane Aircraft Manufacturing (SAM) builds a range of light aircraft. SAM is about to manufacture a new twin-engine light aircraft with customers being given the choice of seating configuration and type of engine to be used. A successful prototype has been produced and full production is expected to start with two aircraft a month increasing to one aircraft per week after six months. SAM's managing director has asked you to advise the company on the following issues:

1.   The most appropriate production process.
2.   The advantages of production scheduling.
3.   Quality control of incoming materials and components.
4.   The reasons for introducing inventory control.

Present your answer in the form of a set of overhead projector transparencies accompanied by brief word-processed notes to be used at a presentation to SAM's board of directors.

Project planning and scheduling

A project can be defined as a series of activities with a definite beginning and ending and with a series of actions that will lead to the achievement of a clearly defined goal. You need to be able to apply basic project planning and scheduling methods, including establishing timescales and resource requirements and the relationships that exist between the various activities that make up a project.

Project planning is different from other forms of planning and scheduling simply because the set of activities that constitute a project are unique and occur only once. Production planning and scheduling relate to a set of activities that may be performed a large number of times.

Project planning involves considering the full set of activities necessary to achieve the project goal. Typical of these activities are:

- appointing consultants
- appointing suppliers
- forming a team to be responsible for carrying out the project
- preparing a budget for the project
- preparing a detailed costing
- producing drawings
- producing itemized parts lists
- producing specifications
- obtaining management approval
- obtaining planning permission
- scheduling the phases of the project.

Project goals must be clearly defined at the outset. If these are not clear, not properly understood or agreed by all members of the project team then the chances of a successful outcome (or any outcome at all, for that matter) can be significantly reduced. In addition, you need to be able to:

- establish the project resources and requirements
- produce a plan with an appropriate timescale for completing the project
- identify human resource needs
- identify approximate costs associated with each stage of the project.

Programme evaluation and review technique

The programme evaluation and review technique (PERT) was developed for the US Navy in 1958 for planning and control of the Polaris nuclear submarine project. This project involved around 3000 contractors and the use of PERT was instrumental in reducing the project completion time by two years. PERT is widely used today as a fundamental project management tool both by governments and in industry.

PERT requires that project activities should be discrete and have definite start and end points. The technique provides most benefit when projects have a very large number of interrelated activities where it can be very effective in helping to identify the most effective sequence of activities from a variety of possibilities.

Key point

The critical path is the path through the network diagram for which the activity time is the greatest.
There are two reasons for this:

- problems or delays with the activities on a project's critical path can potentially jeopardize the entire project
- the overall time spent on the project can be reduced only by reviewing (and hopefully reducing) the time spent on the activities that appear on the critical path.

Key point

There are two reasons for being concerned with the critical path:

- problems or delays with the activities on a project's critical path can potentially jeopardize the entire project
- the overall time spent on the project can be reduced only by reviewing (and reducing) the time spent on the activities that appear on the critical path.

Test your knowledge 1.17

Explain what is meant by:

(a) Programme evaluation and review technique (PERT).
(b) Critical path method (CPM).

One important aspect of PERT is that it allows us to identify the path through the network for which the total activity times are the greatest. This is the *critical path*.

Critical path method

The critical path method (CPM) is also widely used by government departments and industry. The CPM and PERT are very similar and the critical path is important for several reasons:

1. Since it represents the most time-critical set of activities, the total time to reach the project goal can be reduced *only* by reducing time spent in one or more of the activities along the critical path. In other words, the critical path highlights those activities that should be critically reviewed to see whether they can be shortened in any way. Putting extra resources into one or more of the critical path activities can usually be instrumental in reducing the overall project time.
2. The critical path is unique for a particular set of activities and timings. If any of these are changed (e.g. by directing extra resources into them) a new critical path will be revealed. We can then apply PERT evaluation to this new critical path, critically reviewing the activities that it points us to. This process is iterative – in a large project we can continue to reduce overall project time making changes as the project develops.
3. The critical path shows us where the most risky and potentially time threatening activities occur. Since any problems or delays with activities on the critical path may jeopardize the entire project it is in our interests to focus particular attention on these tasks.

Applying PERT

PERT is straightforward to apply. It comprises:

1. Identifying all of the activities that make up the project.
2. Identifying the sequence of the activities in step 1 (and, in particular, the order of precedence of these activities).
3. Estimating the timing of the activities.
4. Constructing a diagram that illustrates steps 1, 2 and 3.
5. Evaluating the network and, in particular, identifying (and clearly marking) the critical path.
6. Monitoring actual performance as the project is carried out against the schedule produced in 5, revising and re-evaluating the network as appropriate.

Network diagrams

The network diagram used in PERT comprises a series of events which forms the nodes in the network. Events are linked together by arrows that denote the activities. Figure 1.9 shows two events, 0 and 1, linked by a single activity, A, which would normally be

Test your knowledge 1.18

Explain the term 'critical path' in relation to PERT and CPM. Give TWO reasons why the critical path is important.

described in words. The two events (0 and 1) could be stated on the diagram or referred to elsewhere in a table. The expected time for the activity (here shown as 3 units) would normally be specified in hours, days, weeks, months, etc.

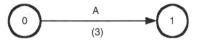

Figure 1.9 *A network diagram showing two events linked by a single activity*

Figure 1.10 shows a slightly more complicated network diagram where there are three events (0, 1 and 2) linked by two activities (A and B). Note that activity A must be completed before activity B (another way of saying this is that 'activity A precedes activity B'. The expected times for activities A and B are respectively 3 and 2 units of time and so it takes 5 units of time to reach event 2 from event 0.

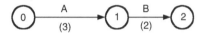

Figure 1.10 *A network diagram with three events and two activities*

The network diagram shown in Figure 1.11 shows four events, 0, 1, 2 and 3, linked by four activities, A, B, C and D. In this network, activity A precedes activity B, while activity D precedes activity A. Note that event 3 is not reached until activities C and D have *both* been completed. The expected times for activities A, B, C and D are respectively 2, 3, 3 and 2 units of time. There are several other things to note about this network:

- the events occur in the following order: 0, 1, 2, 3
- activity B is performed at the same time as activity A
- activity D is performed at the same time as activity C
- the total time to reach event 3, whichever route is chosen, amounts to 5 units of time.

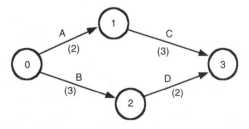

Figure 1.11 *A network diagram with four activities*

Another network diagram is shown in Figure 1.12. In this network, activities A and B both precede activity C (in other words, activity C cannot start until *both* activities A and B have been completed). The total time to reach event 3, whichever route is chosen, amounts to 5 units of time.

Within a network diagram, the activities that link two events must be unique. Consider Figure 1.13. This shows that event 2 can

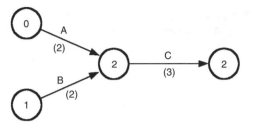

Figure 1.12 *A network diagram in which activity C cannot start until both activities A and B have been completed*

be reached via activities B and C (where the expected time for activity B is 2 units while the expected time for activity C is 3 units).

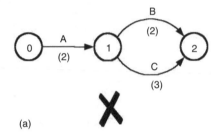

(a)

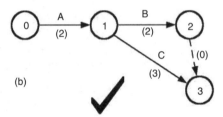

(b)

Figure 1.13 *A network diagram with a dummy activity*

To avoid potential confusion, we have introduced a *dummy activity* between event 2 and event 3. This activity requires no time for completion and thus its expected time is 0 (note that we have adopted the convention that dummy activities are shown as a dashed line).

Finally, you should see that event 2 is reached before event 3 and that the total expected time through the network amounts to 5 units and that there is *slack time* associated with activity B amounting to 1 unit of time (in other words, activity B can be performed up to 1 unit of time late without affecting the expected time through the network).

The critical path

Within a network diagram, the *critical path* is the path that links the activities that have the greatest expected time. In other words, it is the longest route through the network in terms of time.

Consider the network diagram shown in Figure 1.14. This diagram shows five events linked by five real activities plus one dummy

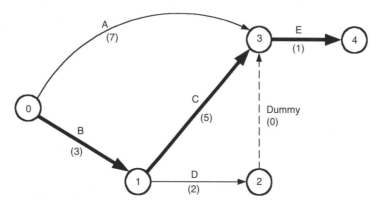

Figure 1.14 *A network diagram showing the critical path*

activity. The relationship between the activities and their expected times can be illustrated in the form of a table:

Activity	Preceding activity	Expected time
A	none	7
B	none	3
C	B	5
D	B	2
E	A, C, D	1

The critical path (shown as a thick line) constitutes activities B, C and E that produce a total expected time, between event 0 (the start of the project) and event 4 (the completion of the project) of nine units.

The critical path allows us to identify those activities that are critical. Remember that, by reducing the time spent on activities along the critical path, we can reduce the expected time for the complete project. If, for example, we can direct extra resources into the project we would be able to reduce the overall time only by concentrating our efforts on activity B, C or E. Reducing the time spent on activity A or D would have no effect on the overall time spent on the project!

Now let's put this into context by taking a practical example of using network diagrams and applying critical path analysis. Let's assume that APS (the loudspeaker manufacturer that we first met in Test your knowledge 1.10) are about to produce a prototype of one of their new loudspeaker designs. This particular design is to use two drivers: a bass unit (or 'woofer') for low-frequency signals and a treble unit (or 'tweeter') for high-frequency signals. In order to feed the correct frequency range of sounds to the appropriate driver, the signal applied to the loudspeaker is to be split into two bands by means of a cross-over unit.

The loudspeaker enclosure will house the drivers together with the cross-over unit. The design of the enclosure and cross-over unit (both are designed using CAD) will depend upon the specification and the drivers used.

Before they proceed with the project, APS has produced a network diagram so that everyone involved with the project knows the timescale and where any potential problems may arise. The network diagram is based on a total of nine activities and is shown in Figure 1.15. The events associated with the project are listed below:

Event number	Event
0	Commencement of project
1	Specifications and budget agreed
2	Materials and parts specified
3	Design of cross-over completed
4	Design of enclosure completed
5	Components (i.e. enclosure, cross-over and driver) ready for assembly
6	Loudspeaker assembled
7	Testing complete

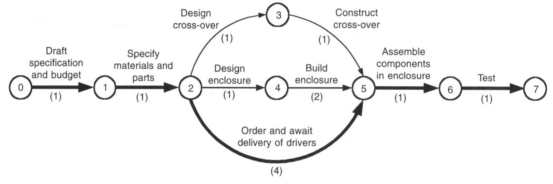

Figure 1.15 *Network diagram for the APS prototype loudspeaker*

Test your knowledge 1.20

The following data refers to the activities that make up a project. Use this information to construct a network diagram and identify the critical path. Also determine the expected time to complete the project.

Activity	Preceding activity	Expected time
A	none	1
B	A	2
C	none	2
D	C	3
E	D	2
F	F	1

From Figure 1.15 it is easy to see that the design and construction of the cross-over network takes place at the same time as designing and building the enclosure. These activities are quite separate and can be performed by different people using different resources. Note, however, that it is not possible to start assembling the loudspeaker until *both* the cross-over and the enclosure have been designed and built *and* the drivers have been delivered.

The critical path shows that APS should pay particular attention to the ordering and delivery of the driver units. It also shows that there is slack time amounting to two days associated with the design and construction of the cross-over unit and one day associated with the design and building of the enclosure.

As with most projects, there is a trade-off between time and cost. By deploying additional resources (assuming, of course, that APS has the money available to do this!) along the critical path it can make significant reductions in the overall time taken. We will look at this next.

Activity 1.12

The Head of Engineering in your school or college has asked you to produce a portable display stand to advertise

Test your knowledge 1.21

The following data refers to the activities that make up a project. Use this information to construct a network diagram and identify the critical path. Also determine the expected time to complete the project.

Activity	Preceding activity	Expected time
A	none	1
B	none	1
C	B	1
D	A, C	2
E	D	2
F	A, C	5
G	B	2
H	G	3
I	E, F, H	1

engineering courses and the BTEC National Diploma in Engineering in particular. You have been asked to include photographs of students' work, details of school/college resources and some example of students' projects. The display stand is to fold up for transport into a container that will fit in the boot of a car and it should be equipped with leaflet storage and display lighting.

(a) List the activities that will make up this project and estimate the expected time for each activity.
(b) Construct a network diagram showing how the activities are related.
(c) Identify the critical path on the network diagram.
(d) Determine the expected time to complete the project.

Present your answer in the form of a word-processed report and a printed chart to be submitted to the Head of Department for approval. Use a computer aided drawing package to produce the network diagram.

Project costs

Projects involve two types of cost: *indirect project costs* and *direct activity costs*. Indirect costs include items such as administrative overheads and facilities costs (heating, lighting, etc.). Direct costs are concerned with additional labour costs, equipment leasing, etc. We can spend extra money to reduce the time taken on the project; however, this only makes sense up to the point where further direct cost expenditure (such as the cost of employing additional contract staff) becomes equal to the savings in indirect project costs (heating, lighting and other overheads).

To examine the trade-off between project time and costs, we need to have the following information:

1. A network diagram for the project showing expected times and indicating the initial critical path. We also need to know the minimum time for each activity when there are no resource constraints (this is known as the *crash time*).
2. Cost estimates for each project activity expressed in terms of indirect expenditure per unit time.
3. The costs of providing additional resources for each project activity and the consequent time saving expressed in terms of expenditure per unit time reduction.

With the above information we can reduce the critical path activity times, beginning with the activity that offers the least expenditure per unit time reduction. We can then continue with the second least costly, and continue until we are left with the most costly until either we reach the target minimum time for the project or the additional direct cost expenditure becomes equal to the savings in indirect costs.

Next we shall look at time analysis in a little more detail.

Project time analysis using PERT

PERT defines a number of important times in the project life cycle. These are as follows:

Expected time t_e — The expected time for an activity is simply the average time for the activity.

Optimistic time t_o — This is fastest time for the completion of the activity. This time will rarely be achieved and will only be bettered under exceptionally favourable circumstances.

Pessimistic time t_p — This is slowest time for the completion of the activity. This time will nearly always be bettered and will only be exceeded under exceptionally unfavourable circumstances.

Most likely time t_m — This time represents the 'best guess' time for the completion of the activity. This time is the statistical mode of the distribution of the times for the activity.

Estimates of project times are often based on previous experience of performing similar tasks and activities. The expected time, t_e, is usually calculated from the formula:

$$t_e = \frac{(t_o + 4t_m + t_p)}{6}$$

The expected time to reach a particular event, T_E, can be found by adding up all of the values of expected time, t_e, that lead up to the event in question. To explain how this works take a look at Figure 1.16. This shows a network diagram with five activities: A, B, C, D and E. The critical path is through A, B, C and E.

Activity	Preceding activity	Time (weeks)			
		t_o	t_m	t_p	t_e
A	none	1	2	4	2.2
B	A	1	2	4	2.2
C	B	2	3	5	3.2
D	A	2	4	8	4.3
E	C, D	1	2	4	2.2

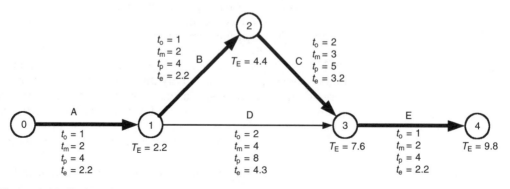

Figure 1.16 *Project time analysis using a network diagram*

Adding together the values of t_e for the critical path activities (A, B, C and E) gives a total time for the project of 2.2 + 2.2 + 3.2 + 2.2 = 9.8 weeks.

The values of T_E can be found by adding together the values of t_e for the preceding activities. For example, the time taken to reach event 2 will be 2.2 + 2.2 = 4.4 weeks while that to reach event 3 will be 2.2 + 2.2 + 3.2 = 7.6 weeks. You should also see that the slack time for activity D is 7.6 − 4.3 = 3.3 weeks.

Activity 1.13

SAM (whom we first met in Activity 1.11) has engaged a consultant to advise on the modification of its new twin-engined aircraft for traffic spotting and aerial surveillance. The consultant has suggested that the aircraft should be fitted with a gyro-stabilized camera platform and has identified the following sequence of activities prior to supplying a prototype to a major client:

Ref.	Activity	Preceding activity	Time estimate (days)		
			t_o	t_m	t_p
A	Design and costing	none	3	5	10
B	Client review	A	2	7	14
C	Specify camera and optical equipment	B	1	2	4
D	Order camera and await delivery	C	7	14	21
E	Detailed design of camera platform	D	1	3	6
F	Construct camera platform	E	3	7	11
G	Assemble platform and camera	D, F	1	2	4
H	Design camera controller	B	3	4	7
I	Construct camera controller	H	3	7	11
J	Fit platform and controller to aircraft	G, I	1	2	4
K	Client acceptance testing	J	3	5	10

(i) Draw the network diagram for the project.
(ii) Determine the expected time for each activity, t_e, and mark this on the network diagram.
(iii) Determine the critical path and mark this in the network diagram.
(iv) Determine the expected time to each event, T_E, and the time for the complete project. Mark these on the network diagram.

Present your work in the form of a network diagram produced using a computer aided drawing package.

Gantt charts

A Gantt chart is simply a bar chart that shows the relationship of activities over a period of time. When constructing a Gantt chart, activities are listed down the page while time runs along the horizontal axis. The standard symbols used to denote the start and end of activities, and the progress towards their completion, are shown in Figure 1.17.

Symbol	Meaning
[	Start of an activity
]	End of an activity
[———]	Actual progress of an activity
[▢▢]	(alternative representation)
V	Time now

Figure 1.17 *Symbols used in Gantt charts*

A simple Gantt chart is shown in Figure 1.18. This chart depicts the relationship between four activities A to D that make up a project. The horizontal scale is marked off in intervals of one day, with the whole project completed by day 14. At the start of the sixth day (see *Time now*) the following situation is evident:

- activity A has been completed
- activity B has been partly completed and is on schedule
- activity C has not yet started and is behind schedule
- activity D is yet to start.

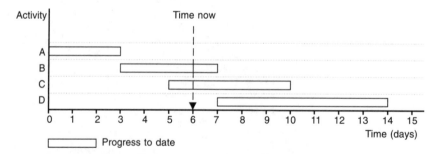

Figure 1.18 *A simple Gantt chart*

Another example is shown in Figure 1.19. This chart depicts the relationship between six activities A to F that make up a project. The horizontal scale is marked off in intervals of one day, with the whole project completed by day 18. At the start of the eighth day (again marked *Time now*) the following situation is evident:

- activity A has been completed
- activity B has been partly completed but is running behind schedule by two days
- activity C has been partly completed and is running ahead of schedule by one day
- activity D is yet to start
- activity E has started and is on schedule
- activity F is yet to start.

Activity
Time now
A
B
C
D
E
F
0 1 2 3 4 5 6 7 8 9 10 11 12 13 14 15 16 17 18
Time (days)
Progress to date

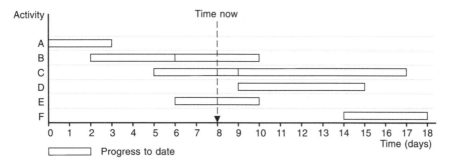

Figure 1.19 *Another simple Gantt chart*

As an example of how a Gantt chart is used in practical situations take a look at Figure 1.20. This shows how APS (the loudspeaker manufacturer we met earlier) could use this technique to track the progress of its project to produce a new loudspeaker design. The chart shows the situation at the beginning of day 4 with all activities running to schedule. You might like to compare this with the network diagram for the same project.

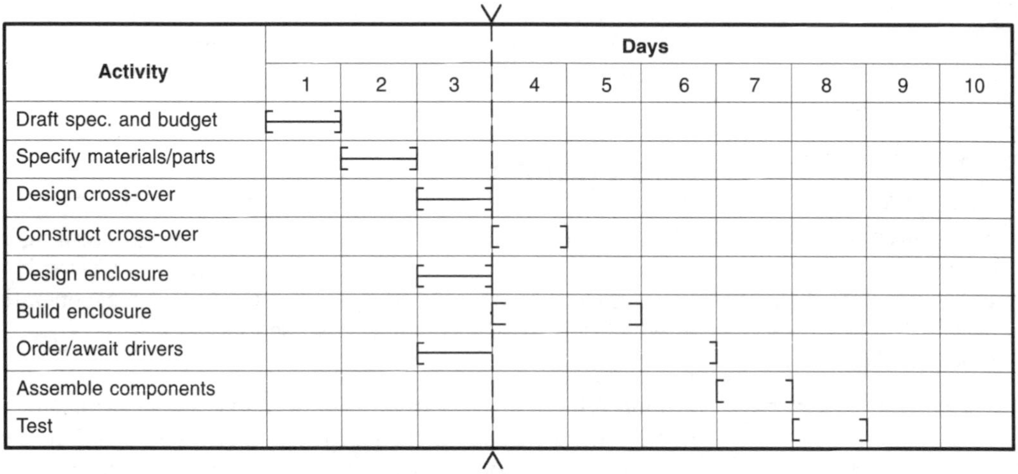

Activity
Days
1 2 3 4 5 6 7 8 9 10
Draft spec. and budget
Specify materials/parts
Design cross-over
Construct cross-over
Design enclosure
Build enclosure
Order/await drivers
Assemble components
Test

Figure 1.20 *Gantt chart for the new APS loudspeaker (see also Figure 1.15)*

Activity 1.14

Use a computer aided drawing package to create a Gantt chart for the construction of the Engineering Department display stand in Activity 1.12. Present your work in the form of a printed chart to be submitted to the Head of Department for approval. The chart should show each of the activities plotted against a common timescale.

Costing

To ensure profitability, engineering companies need to have in place an effective costing system that take into account the real cost of manufacturing the product or delivering the service that it provides. Without such a system in place it is impossible to control costs and determine the overall profitability of the business operation.

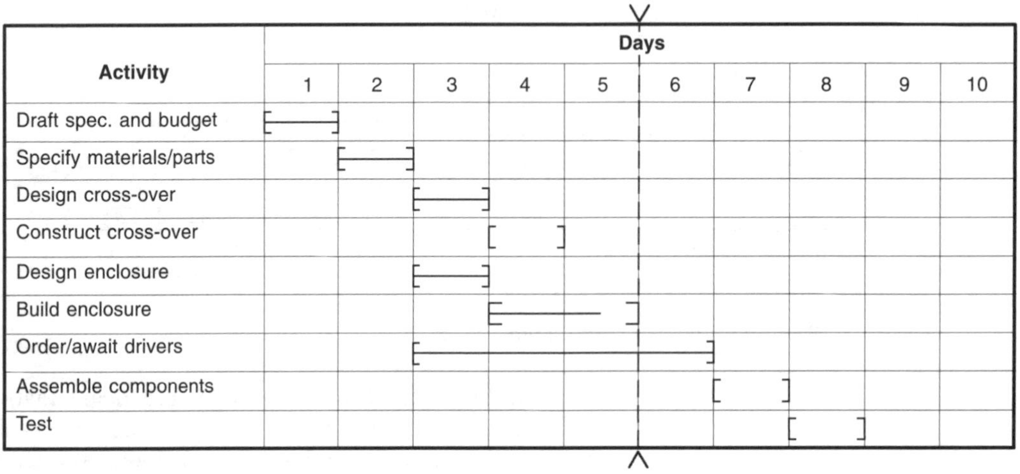

Figure 1.21 *See Test your knowledge 1.22*

Test your knowledge 1.22

Figure 1.21 shows the Gantt chart for the APS loudspeaker design project with the situation at the beginning of day 6.
Which activities are:

(a) On schedule.
(b) Behind schedule by 1 day, or less.
(c) Behind schedule by more than 1 day.
(d) Ahead of schedule.

Cost accounting is necessary for a company to be able to exercise control over the actual costs incurred compared with planned expenditure. From the point of view of cost control, a costing system should not only be able to identify any costs that are running out of control but should also provide a tool that can assist in determining the action that is required to put things right.

Job costing

The most simple costing technique is known as 'job costing'. As its name implies, it relates to a unique operation, such as replacing a part or carrying out a modification to a product. Typical operations in which job costing is commonly used include:

- supplying a unique or 'one-off' item
- modifying a product in order to improve its performance
- adapting a product to meet a particular customer's requirements
- providing a service that is only required on an intermittent or irregular basis.

Job costing involves estimating the cost of materials used (sometimes referred to as a *bill of materials*) and then adding this to the cost of labour in order to determine the total amount that will be charged to the customer. Let's take an example.

A small company supplies and fits entertainment systems for use in coaches and buses. The company has been asked to supply a one-off entertainment system for use in a luxury coach. The quotation for parts and materials required for this job is as follows:

Item	Quantity	Price per unit	Cost
Display screens	2	£475	£950
DVD player	1	£275	£275
Rack unit	1	£125	£125
Cable	25 m	£1 per metre	£25
Control box	1	£75	£75
Sundry items			£30
Total for parts and materials:			£1480

To this should be added the cost of labour. Let's assume that this amounts to 10 hours at £35 per hour (this figure includes the overheads associated with employment, such as National Insurance contributions). Hence the cost of labour is:

Item	Quantity	Price per unit	Cost
Labour	10 hours	£35/hour	£350
Total for labour:			£350

We can add the cost of labour to the total *bill of materials* to arrive at the final cost for the job which amounts to £1830. Note that parts and materials *may* be supplied 'at cost' or 'marked up' by a percentage which can often range from 10 to 50 per cent (and sometimes more).

Total cost: £1830

Finally we would add VAT at the current rate to the total cost. This is the amount payable by the customer.

VAT: (at 17.5%) £320.25

Amount payable: £2150.25

Activity 1.15

Use a spreadsheet to prepare a job costing for the Engineering Department display stand in Activity 1.12. Calculate your labour costs based on £12.50 per hour. Present your work in the form of an itemized costing to be submitted to the Head of Department for approval.

Contract costing

Large companies also use job costing when they produce a variety of different, and often unique, products at the same time. These products are often referred to as *custom built* and each is separately costed as a 'job' in its own right. This type of production is described as *intermittent* (and traditionally referred to as *job shop* production) to distinguish it from the *continuous* or *assembly-line* production associated with the manufacture of a large number of identical units.

In jobbing production, individual manufactured units are normally produced to meet an individual customer's requirements and production is not normally speculative. Costs are agreed before manufacturing starts and they form the basis of a contract between the manufacturer and the customer.

Parts costing

Parts costing is fairly straightforward and is simply a question of determining the cost of all of the physical parts and components

Test your knowledge 1.23

Explain the following terms:

(a) job costing
(b) contract costing
(c) parts costing.

used in a manufactured or engineered product. Parts costing works from the 'bottom up' – in other words, the cost of each individual component (i.e. the *per unit cost*) is determined on the basis of the given *standard supply multiple*. For example, if plastic case parts are purchased in quantities of 1000 then the per unit cost is simply the total cost of purchasing 1000 items from the supplier divided by the supply multiple (i.e. 1000). You should note that the cost of an individual item purchased from a supplier is usually very much greater than the per unit cost when the item is purchased in quantity. As an example of parts costing, consider the following example.

It is often useful to group together individual component parts under groupings of similar items. The reason for this is that such groupings tend to be subject to the same fluctuation in cost. We can thus quickly determine the effect of market fluctuations by examining the effect of changes on particular groups of parts.

Activity 1.16

The bill of materials for the prototype of the new APS loudspeaker (see Figure 1.22) is as follows:

Item	Quantity	Cost
18 mm plywood sheet	4 m^2	£2.50/m^2
20 × 20 mm timber batten	4 m	50p/m
30 × 40 mm pine	1.8 m	£2/m
Grille and mounting hardware	1	£4.40
10 cm plastic port tubes	2	£1.00
Acoustic wadding	3 m	£1.50
Chrome corners	8	£0.60
Heavy duty handles	2	£2.10
Screws and sundry items	1	£2.50
Jack socket and recess plate	1	£4.50
Low-frequency driver unit	1	£26.00
High-frequency driver unit	1	£16.00

Use a spreadsheet to determine the total cost of the parts used in the prototype loudspeaker. Also determine the total cost of the loudspeaker if the cost of timber increases by 20 per cent and the supplier of the driver units offers a discount of 15 per cent. Present your work in the form of two spreadsheet printouts.

Process costing

Process costing takes into account the cost of each manufacturing process and apportions part of the cost of each process to an individual product. Typical processes might be:

- forming, bending or machining of metal and plastic parts
- flow soldering of printed circuit boards
- heat treatment of metal parts
- paint spraying and finishing.

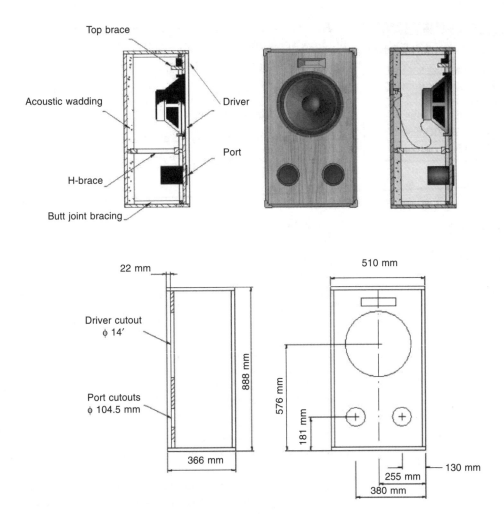

Figure 1.22 *The new APS speaker system (see Activity 1.16)*

The following example illustrates the use of process costing. Note that, when determining the total cost of manufacturing a product, it is essential to take into account the notional cost of all of the processes involved. Alternatively, where there may be a number of products sharing the same process, it may be more convenient to amalgamate these costs into a production overhead cost. The danger with this, as we shall see later, is that we usually need to apportion costs more precisely so that we can determine the relative profitability of *each* product that we manufacture. If this is beginning to sound a little complex, an example will show you how process costing works.

Let's assume that our loudspeaker manufacturer, APS, has invested in a flow soldering plant to automate the manufacture of their cross-over units. The flow soldering plant operates at a rate of 60 units per hour and its operating cost (including capital cost recovery calculated over a nominal eight-year asset life) amounts to £5000/week plus £10 material and energy costs per hour. Let's determine the cost of producing each cross-over unit using the flow soldering process (this is known as the *unit cost* of the process) assuming that it operates for a total of 40 hours each week.

Test your knowledge 1.24

Determine the cost per unit of the APS cross-over if the flow soldering plant operates for 75 hours per week and the hourly operating costs increase to £15 per hour.

Based on 40 hours' operation per week, the total cost of operating the flow soldering process for a week will be given by:

Total cost = £5000 + (40 × £10) = £5400

At 60 units per hour, the total weekly production will be given by:

Total production = 60 × 40 = 2400

The cost, per unit, will thus be given by

Cost per unit = £5400/2400 = £2.25

Costing techniques

Any engineering company will incur a variety of costs. These will typically include:

- rent for factory and office premises
- rates
- energy costs (including heating and lighting)
- material costs
- costs associated with production equipment (purchase and maintenance)
- salaries and National Insurance
- transport costs
- postage and telephone charges
- insurance premiums.

Given the wide range of costs above, it is often useful to classify costs under various headings, including fixed and variable costs, overhead and direct costs, average and marginal costs, and so on.

In order to be able to control costs, it is, of course, vital to ensure that all of the costs incurred are known. Indeed, the consequences of not being fully aware of the costs of a business operation can be dire!

This section examines a number of different methods used by businesses to determine the total cost of the product or service that they deliver. The prime objective of these techniques is that of informing commercial decisions such as:

- how many units have to be produced in order to make a profit?
- is it cheaper to make or buy an item?
- what happens to our profits if the cost of production changes?
- what happens to our profits if the cost of parts changes?

Absorption costing

One method of determining the total cost of a given product or service is that of adding the costs of overheads to the direct costs by a process of *allocation*, *apportionment* and *absorption*. Since *overheads* (or *indirect costs*) can be allocated as whole items to production departments, it is possible to arrive at a notional amount that must be added to the cost of each product in order to cover the production overheads.

$$\text{make-up} = \frac{\text{total of fixed and variable costs attributable to the product}}{\text{total number of units produced}}$$

In absorption costing, each product manufactured is made (at least in theory) to cover all of its costs. This is achieved by adding a notional amount to the total unit cost of each product.

Marginal costing

Marginal costing provides us with an alternative way of looking at costs that provides an insight into the way costs behave by allowing us to observe the interaction between costs, volumes and profits. The marginal cost of a product is equal to the cost of producing one more unit of output.

Accountants define marginal cost as:

The amount by which aggregate costs change if the volume of output is increased or decreased by one unit.

Figure 1.23 shows the behaviour of costs based on the marginal cost concept. The vertical axis of Figure 1.23 shows aggregate costs and revenues in £. The horizontal axis shows the units of output. The fixed costs are shown as a straight line across all levels of output. Fixed costs are costs which are unaffected by activity, at least in the short term. Figure 1.23 assumes that such costs do not vary for the whole range of output possible.

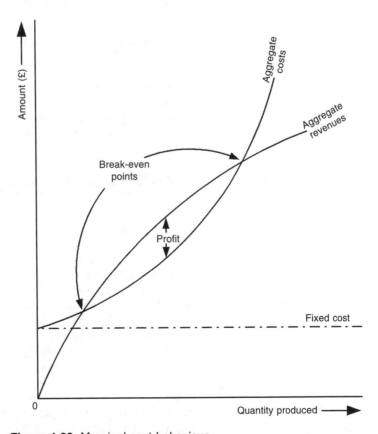

Figure 1.23 *Marginal cost behaviour*

Fixed costs include things such as:

- insurance premiums
- business rates
- subscriptions
- audit fees
- rental charges
- fixed elements of power and telephone charges.

There are a number of advantages of using marginal costing, notably:

- Marginal costing systems are simpler to operate than absorption costing systems because they do not involve the problems associated with overhead apportionment and recovery.
- It is easier to make decisions on the basis of marginal cost presentations. Where several products are being produced, marginal costing can show which products are making a contribution and which are failing to cover their variable costs.

The disadvantages of marginal costing include:

- The effect of time (and its effect on true cost) tends to be overlooked in marginal costing. For example, two jobs may have the same marginal cost but one may use significantly more plant time than another (thus increasing its true cost).
- There is a temptation to spread fixed costs and real dangers in neglecting these in favour of more easily quantified variable costs.

The choice of whether to use absorption costing or marginal costing is usually determined by factors such as:

- The system of financial control used within a company (e.g. *responsibility accounting* is consistent with absorption costing)
- the production methods used (e.g. marginal costing is easier to operate in simple processing applications whereas absorption costing is usually preferred when several different products require different plant and processing techniques)
- the significance of the prevailing level of overhead costs.

Activity-based costing

Activity-based costing is an attempt to assess the 'true' cost of providing a product or service. Knowledge of the 'true' cost is not only important in helping us to identify opportunities for cost improvement but it also helps us to make strategic decisions that are better informed.

Activity-based costing focuses on indirect costs (overheads). It does this by making costs that would traditionally be considered indirect into direct costs. In effect, it traces costs and overhead expenses to an individual *cost object*. The basic principles of activity-based costing are shown in Figure 1.24.

Activity-based costing is particularly useful when the overhead costs associated with a particular product are significant and where a number of products are manufactured in different volumes. Activity-based costing is particularly applicable where competition is severe and the margin of selling price over manufacturing cost has to be precisely determined.

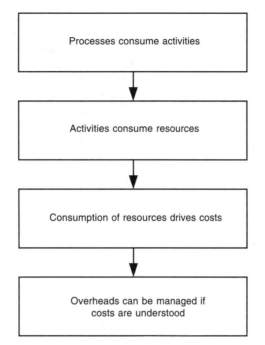

Figure 1.24 *Principles of activity-based costing*

The steps required to carry out activity-based costing are:

1. Identify the activities.
2. Determine the cost of each activity.
3. Determine the factors that drive costs.
4. Collect the activity data.
5. Calculate the product cost.

The use of activity-based costing is best illustrated by taking an example. Let's assume that APS has decided to go ahead with the production of its new loudspeaker but is also manufacturing a high-quality demountable loudspeaker stand. The company is interested in knowing how these two products compare using activity-based costing.

Activity	Loudspeakers	Cost	Stands	Cost	Total
Set-up	1 @ £45 000	£45 000	1 @ £25 000	£25 000	£70 000
Manufacture	3000 @ £20/unit	£60 000	1000 @ £12/unit	£12 000	£72 000
Assembly	3000 @ £10/unit	£30 000	1000 @ £6/unit	£6 000	£36 000
Inspection	3000 @ £2/unit	£6 000	1000 @ £2/unit	£2 000	£8 000
Packaging	3000 @ £2/unit	£6 000	1000 @ £2/unit	£2 000	£8 000
Total		£147 000		£47 000	£194 000

The activity-based product cost for each loudspeaker amounts to £147 000/3000 = £49 while the activity-based product cost for each stand amounts to £47 000/1000 = £47. To these two costs we must add the direct (material) costs of each product. Assuming that this amounts to £60 for the loudspeaker and £20 for the demountable stand, we would arrive at a cost of £129 for the loudspeaker and £67 for the stand.

It's worth pointing out that traditional cost accounting would have arrived at two rather different figures for the costs of these two products. Let's assume that 7000 hours of direct labour are used in the manufacture of the loudspeakers and stands. Dividing the total overhead cost of £194 000 by this figure will give us the hourly direct labour cost of £27.71 per hour. If loudspeakers require 2 hours of direct labour and stands require 1.5 hours of direct labour the allocation of costs would be £55.42 per loudspeaker and £41.56 per stand. Adding the same direct (material) costs to this yields a cost of £115.42 for the loudspeaker and £61.56 for the stand.

Financial control methods

An engineering company will normally make use of a number of different control methods to ensure that its operation is profitable. These control methods include making forecasts of overall profitability, determining the contribution made by each individual activity towards overheads and fixed costs, and performing a 'what-if' analysis to determine the effects of variations in cost and selling price. We shall start by describing the most simple method, break-even analysis.

Break-even charts

Break-even charts provide a simple (and relatively unsophisticated) method for determining the minimum level of sales that a company must achieve in order for the business to be profitable. Consider the simple relationship illustrated in Figure 1.25. Here total income

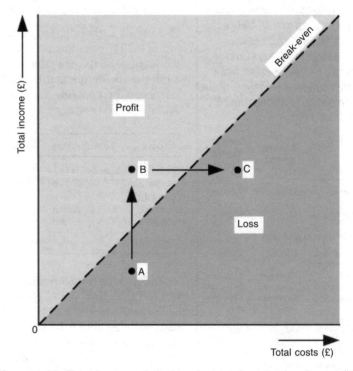

Figure 1.25 *Total income plotted against total costs showing profit and loss regions*

has been plotted against total costs using the same scale for each axis. At point A, total costs exceed total income and the operation is not profitable, i.e. it makes a *loss*. If we charge more for the product, whilst keeping the costs fixed, we would move from point A to B. At a certain point, total income exceeds total costs and we move into *profit*. Finally, let's assume that our total costs increase whilst the total income from sales remains unchanged. We would then move from profit (point B) to loss (point C).

The *break-even point* is the volume of sales at which the operation becomes profitable and it marks the transition from loss into profit. A *break-even chart* takes the form of a graph of costs plotted against volume of product sold. At this point, it is important to recall that the total costs of the business operation are the sum of the constant fixed and overhead costs with the variable costs of production. Thus:

total cost = fixed cost + overhead cost + variable cost

The income derived from the sale of the product (assuming a constant pricing structure) will simply be the product of the quantity sold (i.e. the volume of product sold) and the price at which it is sold (i.e. the per unit selling price). This relationship (a straight line) can be superimposed on the break-even chart and the point of intersection with the total cost line can be identified. This is the break-even point and the corresponding production volume can be determined from the horizontal axis, see Figure 1.26.

> **Key point**
>
> The break-even point is the volume of output at which an engineering operation becomes profitable. In other words, it marks the transition from loss into profit.

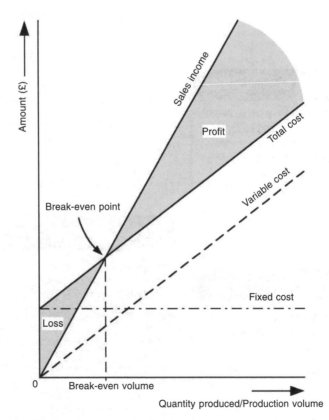

Figure 1.26 *Fixed and variable costs plotted against production volume showing break-even*

The *break-even quantity* can be determined from:

$$\text{break-even quantity} = \frac{\text{fixed cost}}{\text{selling price} - \text{variable cost}}$$

(Note that, in the above formula, *selling price* and *variable cost* are per unit.)

It is also possible to use the break-even chart to determine the profit that would result from a particular production quantity.

Profit can be determined from:

profit = (selling price × quantity sold)

− (fixed cost + (variable cost × quantity sold))

Note that, in the above formula, *selling price* and *variable cost* are again per unit.

Let's assume that our loudspeaker manufacturer, APS, has established the following costs of manufacturing its demountable loudspeaker stand:

Fixed costs = £25 000

Variable costs = £20 per unit

If the stand is to be sold for £100, determine the break-even quantity and the profit that would be returned from sales of 5000 units.

Now

$$\text{break-even quantity} = \frac{\text{fixed cost}}{\text{selling price} - \text{variable cost}}$$

Thus

break-even quantity = £25 000/(£100 − £20) = 3125

The profit based on sales of 10 000 units will be given by:

profit = (selling price × quantity sold)

− (fixed cost + (variable cost × quantity sold))

profit = (£100 × 10 000) − (£25 000 + (£20 × 10 000))

= £100 000 − £45 000

Thus the profit on 10 000 units will be £55 000.

Test your knowledge 1.27

Explain, briefly, what is meant by break-even and illustrate your answer using a graph showing costs plotted against production volume.

Test your knowledge 1.28

The following figures relate to the production of a small component:

Fixed cost: £30 000
Variable cost per unit: £15
Selling price: £25

Determine the break-even point in sales turnover.

Activity 1.17

VoltMax (see Test your knowledge 1.26) has analysed its fixed and overhead costs relating to its standard charger which together amount to £100 000 while the variable costs of its standard charger amount to £12 per unit manufactured.

Construct a break-even chart and use this to determine:

(a) the break-even production volume when the charger is sold at:
   (i) £20 per unit
   (ii) £30 per unit
(b) the profit for a production quantity of 10 000 units if the selling price is £30 per unit.

Present your work in the form of a brief word-processed report to VoltMax's board of directors.

It is important to realise that simple break-even analysis has a number of serious shortcomings. These may be summarized as follows:

- The sales income line (i.e. the product of the volume produced and its selling price) takes no account of the effect of price on the volume of sales. This is important as it is likely that the demand for the product will fall progressively as the selling price increases and the product becomes less competitive in the open market.
- The assumption that fixed costs remain fixed and variable costs increase linearly with production are somewhat dangerous. The reality is that both of these will change!

For the foregoing reasons it is important to regard break-even analysis as a 'rule-of-thumb' method for evaluating product pricing. Before making any business decisions relating to pricing and targets for production volume it is important to undertake further research into effect of pricing on potential sales as well as the pricing of competitive products.

Average cost

The fixed costs associated with production has to be shared between the entire volume produced. Hence, a proportion of the final cost of a product will be attributable to the fixed costs of manufacture. The larger the quantity produced, the smaller this proportion will be. In other words, the *average cost* of the product will fall as the volume increases. We can illustrate this in the form of a graph (see Figure 1.27). Note that:

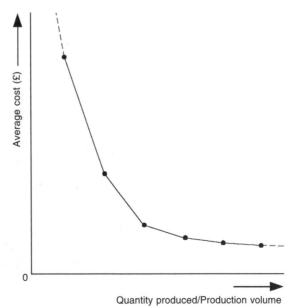

Figure 1.27 *Average cost plotted against production volume*

$$\text{average cost} = \frac{\text{total cost}}{\text{quantity produced}}$$

As an example of how this works, let's calculate the average costs of VoltMax's standard charger for production levels of 5000, 7500, 10 000 and 15 000 units. Based on the information given in Activity 1.23, VoltMax's fixed costs are £100 000 and the company's variable costs amount to £12 per unit

Now $\quad \text{average cost} = \dfrac{\text{total cost}}{\text{quantity produced}}$

and $\quad$ total cost = fixed cost + variable cost

Thus:

Quantity produced:	5000	7500	10 000	15 000
Fixed cost:	£100 000	£100 000	£100 000	£100 000
Variable cost:	£60 000	£90 000	£120 000	£180 000
Total cost	£160 000	£190 000	£220 000	£280 000
Average cost:	**£32**	**£25.33**	**£22**	**£18.67**

This information is shown in graphical form in Figure 1.28.

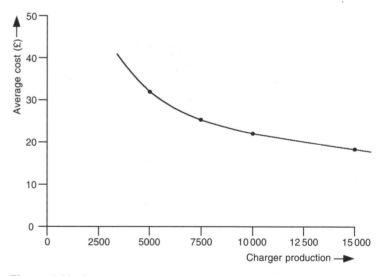

Figure 1.28 *Average cost plotted against production volume for VoltMax's standard charger*

Marginal cost

Once we have established a particular volume of production, the cost of producing one more unit is referred to as the marginal cost. The marginal cost of a product is that cost of the unit that results only from changes in those costs that do not vary with the amount produced. Marginal cost is *not* the same as average cost – the reason for knowing the marginal cost of a product is that it can help us decide whether or not to increase production from an existing level. This is best illustrated with another example.

VoltMax has established a production level of 12 000 units for its standard charger. As before, the fixed cost of the company's standard charger manufacturing operation is £100 000 and the variable costs of producing the standard charger amount to £12 per unit. The charger is normally sold for £30 but a large high-street chain store has offered to take an additional 2000 units at a non-negotiable price of £20 per unit. We can use marginal costing to determine whether this proposition is financially sound.

The total cost associated with a production volume of 12 000 units is found from:

total cost = fixed cost + variable cost

total cost = £100 000 + (12 000 × £12) = £100 000 + £144 000

$\qquad$ = £244 000

Now the average cost (based on 12 000 units) will be given by:

$$\text{average cost} = \frac{\text{total cost}}{\text{quantity produce}} = £244\,000/12\,000 = £20.33$$

Based on an average cost of £20.33, a selling price of £20 per unit does not appear to be sound business sense. However, if we consider the marginal cost of the standard charger based on an *existing* production level of 12 000 units, we arrive at a rather different view. The rationale is as follows.

Let's assume a scenario in which we sell 12 000 standard chargers at £30 and 2000 standard chargers at £20. The total income produced will be given by:

total income = (12 000 × £30) + (2000 × £20)

$\qquad$ = £360 000 + £40 000 = £400 000

The total cost associated with producing 14 000 standard chargers will be:

total cost = £100 000 + (14 000 × £12) = £100 000 + £168 000

$\qquad$ = £268 000

Thus the resulting profit will be given by:

profit = total income − total cost = £400 000 − £268 000

$\qquad$ = £132 000

Had we decided not to accept the order for the extra 2000 units, we would have generated a profit given by:

profit = total income − total cost = £360 000 − £244 000

$\qquad$ = £116 000

Thus, meeting the order for an additional 2000 units at £30 has helped to increase our profits by £16 000. The important thing to note here is that although the selling price of £20 per unit is less than the average cost per unit of £20.33, it is greater than the marginal cost of £12!

Another way of looking at this. In marginal costing, we consider the cost of a product when all of the fixed costs are removed. We arrive at this figure by calculating the cost of producing just one more unit – the difference in the cost of this unit and the previously manufactured one is the variable cost attributable to just one unit (i.e. the variable cost per unit). This assumes that the variable

---

Test your knowledge 1.29

An electric screwdriver manufacturer has a total production volume of 6000 units. The fixed cost of the operation amounts to £150 000 while the variable cost amounts to £20 per unit. Determine the average cost per unit and the total profit if the electric screwdriver sells for £65.

cost per unit is the same for all volumes of production output. This will usually be true for significant production volumes (note how the average cost tends towards a fixed value as the quantity increases in Figure 1.27).

Profitability

Profit, or *return on capital employed* (ROCE), is the aim of every business. Being able to make a realistic forecast of profits is an essential prerequisite to making a financial case for investment. It is also an essential ingredient in any business plan.

The need to maximize profits should be an important factor in decision making. Traditional theory assumes that a company will invest in the most profitable projects first, and then choose projects of descending profitability, until the return on the last project just covers the funding of that project (this occurs when the marginal revenue is equal to the marginal cost).

The process of choosing projects is, however, much more complex. It may, for example, involve strategic issues (such as the need to maintain a presence in a particular market or the need to developing expertise in a particular technology with the aim of improving profits at some later date). Furthermore, many companies do not have sufficient funds available to reach the marginal position. Instead, they will rely on one or two 'hurdle' rates of return for projects. Projects that do not reach these rates of return will be abandoned in favour of those that are considered 'profitable'.

Financial planning

By now you should have begun to understand how costing systems and techniques are applied in a typical engineering company. This next topic introduces you to some important aspects of financial planning and control.

Adequate financial planning is essential if a business is to achieve its objectives and profit targets. The basic procedure required to formulate a financial plan is as follows:

(a) Formulate company policy, profit targets and long-term plans.
(b) Prepare forecasts for sales, production, stocks, costs, capital expenditure and cash.
(c) Compile these separate forecasts into a master forecast.
(d) Consider all the alternatives available and select the plan which gives the best results, for example in terms of profit and long-term financial stability.
(e) Review limiting factors and the principal budget factor. This process takes place concurrently with (d) and enables work to begin on the framing of the budgets in (f).
(f) Prepare individual budgets and finally the master budget which includes a forecasted profit and loss account and balance sheet.

This process is illustrated in Figure 1.29.

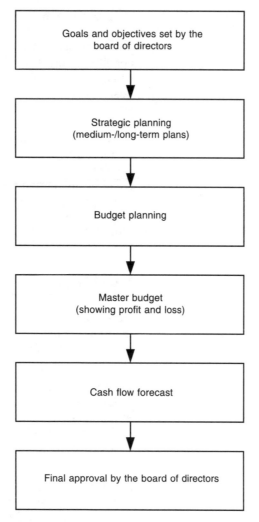

Figure 1.29 *Process of financial planning and control*

Budget plans

The starting point of budget planning is with the board of directors (or equivalent body) who determines the scale and nature of the activities of the company. This policy and objective setting is done within the constraints that exist at the time. For example, plans may have to be made within the current capacities and capabilities of the company, since making changes to the location of operations, the size and composition of the workforce and the product range are usually long-term matters. The budget is essentially for the short term, usually for one year, and created within the framework of long-term corporate planning. Successive budgets will be influenced by the preparation of long-term plans, but will always relate to the current period.

Some organizations prepare outline budgets over much longer periods, perhaps for a five- to ten-year horizon, but such budgets are really part of the long-term corporate planning activity and are subject to major revision before being used as a basis for current period budgetary planning.

External factors will exercise considerable effects on the company in preparing its forecasts and budgets. Government policy, the proximity of a general election, taxation, inflation, world economic conditions and technological development will all combine to constrain or influence the budget planning process. Once the board of directors has settled on a policy within the prevailing situation, then the process of turning the policy into detailed quantitative statements can begin.

We normally assume a budget period of one year, which is usual for most industries. It is therefore recognized that the budget period is fixed in relation to the needs of the organization concerned and could be any period ranging from three months to five years. The shorter the period the more accurate the forecasts will be and that is why most companies find that an annual budgeting procedure is a satisfactory compromise.

Business plans

On occasions, it is necessary to provide a detailed business plan in order to make a case for a particular business venture or project. Before a business plan is written, it is necessary to:

- clearly define the target audience for the business plan
- determine the plan's requirements in relation to the contents and levels of detail
- map out the plan's structure (contents)
- decide on the likely length of the plan
- identify all the main issues to be addressed (including the financial aspects).

Shortcomings in the concept and gaps in supporting evidence and proposals need to be identified. This will facilitate an assessment of research to be undertaken before any drafting commences. It is also important to bear in mind that a business plan should be the end result of a careful and extensive research and development project that must be completed before any serious writing should be started.

A typical business plan comprises the following main elements:

- An *introduction* which sets out the background and structure of the plan.
- A *summary* consisting of a few pages that highlight the main issues and proposals.
- A *main body* containing sections or chapters divided into numbered sections and subsections. The main body should include financial information (including a *profitability forecast*).
- *Market and sales projections* should be supported by valid market research. It is particularly important to ensure that there is a direct relationship between market analysis, sales forecasts and financial projections. It may also be important to make an assessment of competitors' positions and their possible response to the appearance of a rival product.
- *Appendices* should be used for additional information, tabulated data and other background material. These (and their sources) should be clearly referenced in the text.

The financial section of the plan is of crucial importance and, since it is likely to be read in some detail, it needs to be realistic about sales expectations, profit margins and funding requirements ensuring that financial ratios are in line with industry norms. It is also essential to make realistic estimates of the cost and time required for product development, market entry and the need to secure external sources of funding.

When preparing a plan it is often useful to include a number of 'what-if' scenarios. These can help you to plan for the effects of escalating costs, reduction in sales, or essential resources becoming scarce. During a *what-if analysis*, you may also wish to consider the *halve–double* scenario in which you examine the financial viability of the project in the event that sales projections are halved and costs and time are doubled. The results can be sobering!

When writing a business plan it is necessary to:

- avoid unnecessary jargon
- economize on words
- use short crisp sentences and bullet points
- check spelling, punctuation and grammar
- concentrate on relevant and significant issues
- break the text into numbered paragraphs, sections, etc.
- relegate detail to appendices
- provide a contents page and number pages
- write the summary last.

Finally, it can be useful to ask a consultant or other qualified outsider to review your plan in draft form and be prepared to adjust the plan in the light of comments secured and experiences gained.

Budget plans

Budgets are used as a means of achieving planning and control objectives in most businesses and in many non-commercial organizations. A budget has been defined as:

> *A financial or quantitative statement prepared and approved, prior to a defined period of time, of the policy to be pursued during that period for the purpose of attaining given objectives.*

The benefits that derive from budgetary control arise from the ability to coordinate policy, plans and action and to be able to monitor the financial consequences of carrying out the plans.

An engineering company will prepare a number of budgets, each corresponding to a particular functional area. A named manager will normally control each budget although some managers may control several budgets according to the particular management organization employed within the company. In a typical engineering business you will find the following budgets:

- marketing budget
- manufacturing budget
- research and development budget
- administration budget

Test your knowledge 1.30

Describe the main sections found in a business plan.

Key point

A business plan is produced in order to make a case for a particular business venture or project. The financial section of a business plan needs to provide realistic estimates of sales, profit margins and funding requirements.

- capital expenditure budget
- cash budget.

Each of these budgets may be subdivided into further budgets. For example, the manufacturing budget may be subdivided into a budget for direct materials, a budget for direct labour and a budget for factory overheads (heating, lighting and other energy costs).

Each functional manager will forecast his/her own budget; however, there is a need for managers and departments to coordinate their budget activities. For example, the capital expenditure budget may reflect the purchase of major items of capital equipment (such as a fork-lift truck or an overhead crane) that will be shared by several departments.

> **Key point**
>
> A budget is a financial statement that relates to a particular period of time. A budget needs to be prepared and approved and is a means of meeting a company's objectives and financial goals.

Stock control

Stock control systems used within a manufacturing company will affect the way purchasing is done. Economic order quantities may be established which the buyer has to take into account when arranging supplies. Deliveries may have to be phased according to minimum and maximum and reorder stock levels.

The firm's purchasing function (i.e. its buyer of supplies and services) will need a clear understanding of the importance of deliveries that enables the company to control its inventory costs, while at the same time ensuring a reliable supply of materials and components for the various production processes.

The company may operate a *just-in-time* (JIT) system. JIT originated in Japan and is a way of delivering supplies at the point in time they are required by production. JIT avoids the costs of holding buffer stocks of raw materials and components.

JIT works well when suppliers are dependable and when transport systems are good. The buyer will liaise with the factory on the establishment and operation of JIT for given products.

There will also be the routine matters of passing invoices for payment of goods or dealing with returns for credit so that the accounts department can pay for goods received. Materials purchasing will be subject to budgetary constraints like most other company activities. The purchasing department will be involved, either directly or indirectly, in budgets for inventory levels, and in setting up minimum, maximum and reorder levels for stocks.

Monthly monitoring of inventory levels will be done by the accounting function and purchasing activities may be responsible for ensuring that stocks of components and raw materials stay within agreed levels.

> **Key point**
>
> A just-in-time (JIT) system of stock control ensures that supplies of components and materials are made available only when they are required by production. JIT reduces the cost of holding stocks of materials and components.

Depreciation

Depreciation is the estimate of the cost of a fixed asset consumed during its useful life. If a company buys a car, to be used by a sales representative, for £15 000 it has to charge the cost in some reasonable way to the profit being earned. This process is essential, otherwise the whole cost of running the business cannot be obtained, and profit figures would be overstated. If it is estimated that the car will be worth £6000 in three years' time when it is to be sold, then it could be charged to profit at (£15 000 – £6000)/3 = £3000 for each year of use.

The way companies accumulate funds with which to replace fixed assets is to charge depreciation as an overhead cost. In the case of our car this recovers £9000 which together with the sale price of the used car generates a fund of £14 000 towards the purchase of a replacement. It is also possible that the sums so deducted from profit can be invested to offset inflation until the time comes to replace the asset.

Because depreciation is an estimate and is deducted from profit it has the effect of keeping the money available in the business. Many companies use the aggregate depreciation charged as the basis for the fund against which investment appraisal is done.

There are two main methods used to calculate depreciation. We will outline each in turn.

Straight line method

This charges an equal amount as depreciation for each year of the asset's expected life. It is called straight line method because if the annual amounts were plotted on a graph they would form a straight line. The formula is:

$$d = \frac{p - v}{n}$$

where d = annual depreciation, p = purchase price, v = residual value and n = years of asset life. The method used for the car in the above example is straight line. It is a very popular and easy-to-use method.

Reducing balance method

In this method a fixed percentage is applied to the written-down balance of the fixed asset. The formula for establishing this fixed percentage is:

$$r = 1 - \sqrt[n]{\frac{v}{p}}$$

where r = the percentage rate, n = number of years, v = residual value and p = asset purchase price. Note that v must be a significant amount, otherwise the rate will be very large and somewhat meaningless.

Using our car example above the reducing balance rate, r, is:

$$r = 1 - \sqrt[3]{\frac{6000}{15000}} = 1 - 0.7368 = 26.31\%$$

Applying this, the depreciation pattern is as follows:

Purchase price		£15 000
First year	26.31%	£3 946
Reduced balance		£11 054

Test your knowledge 1.31

Use the straight line model of depreciation to determine the asset value after three years of a CNC lathe purchased for £100 000 assuming that it has an asset life of five years.

Second year	26.31%	£2 908
		————
Reduced balance		£8 116
Third year	26.31%	£2 135
		————
Reduced value (approx.)		£5 981

The reducing balance system means that a greater sum is taken in earlier years, but reduces year on year since the percentage rate is applied to the reduced balance. This method is more complex but it is more logical because a new asset gives better service than an old asset and should suffer more depreciation in earlier years. Certainly, with regard to cars the early years' depreciation is very heavy in relation to resale prices.

Investment appraisal

The process of investment appraisal is necessary in order to select the best projects for investment, whether they be replacement machines for the factory, office equipment or new cars for managers and sales representatives.

Major factors to consider are:

- risk and return
- time scale
- time value of money
- evaluating alternative choices.

The first one, that of risk and return, is important. A good general rule is to expect a high return if the investment is risky. That is why interest rates on mortgages are less than interest rates on loans for consumer durables. A bank will regard loans against property as low risk because a house is a better security than a refrigerator if something goes wrong.

For commercial businesses some capital investment projects carry low risk, such as those which merely involve a replacement of a machine that is being used to make a product for which there is a regular demand. However, the risk is much greater for investment in a new factory for a range of new products. These different risk profiles have to be taken into account when comparing different projects for investment.

Timescale is important too. If a project is going to take five years to be completed, then many things could change before completion including inflation, government action, market conditions, competitive pressures, etc. The investment decision must take account of all these risk factors as well and this is not an easy process, because of the need to forecast up to five years into the future.

Money has *time value*. This fact lies at the heart of the investment appraisal process. If you could choose to have £5 today or £5 in a year's time which would you prefer and why?

Hopefully you would take account of the risk of not getting your £5 in a year's time and would prefer to avoid risk and get it now. Also you should have thought about the fact that if you had the £5 now you could invest it for a year, so that it would be worth more than £5 in a year's time.

The time value of money exists because of these two aspects, risk and return. Most people who decide to save money instead of spending it look for a secure and lucrative place to keep it until they wish to spend it.

If your risk preference is low you would probably put your money into a building society account. If you were more adventurous you might buy shares. The risks attached to buying shares are greater than a building society account, but the return may be greater.

The essence of the capital investment decision is no different to this. A company must decide whether or not to keep its money in the bank or invest in the business. If it chooses to invest in the business, the company must expect a better return than merely leaving it in the bank. It is for this reason that certain techniques for evaluating projects have arisen. There are numerous approaches, but the most theoretically sound method is called the *net present value* (NPV) technique.

> **Key point**
>
> Depending on the return on investment, companies must decide whether to keep money in the bank or to reinvest it in the business.

Activity 1.18

Loudspeaker manufacturer APS has asked you to help them assess the financial implications of making two capital purchases: a new flow-soldering plant and a CNC table saw. The details of these two items are as follows:

Flow-soldering plant	Capital cost:	£90 000
	Asset life:	4 years
CNC table saw	Capital cost:	£75 000
	Asset life:	5 years

The company needs to know what the combined residual value of these two items will be after three years using the reducing balance method.

Write a brief word-processed report for the APS board of directors and include all calculations.

Make or buy decisions

Make or buy decisions require the involvement of more than one functional area. They almost certainly require the involvement of the production function, production control and engineers from design and manufacturing.

Before a decision is made certain questions must be addressed:

(a) Could the item be made with existing facilities?
(b) If the answer to (a) is yes, is current production capacity adequate?

If the answer to (a) is no, then it indicates that new plant and equipment may be needed, or indeed that the company has no experience in this area of manufacture, which might therefore mean a move toward vertical integration, which involves a change to its basic business. The latter involves top level policy decisions and will not yet even be considered at the functional level at all.

Theoretically any item, whether already manufactured or purchased outside (i.e. *outsourced*) can be reviewed for alternative sourcing, but in practice only those items will be considered for change that do not involve major changes to the business. Therefore, we will confine our analysis to decisions that can be made within the functional areas we described earlier.

Thus if a purchased item is being considered for manufacture then there must be suitable facilities available and sufficient capacity. If a manufactured item is being considered for purchase there must be alternative uses for the capacity or there should be an intention to reduce that capacity rather than leave it unused.

The other major factors to account for are:

- incremental costs of alternatives
- quality control problems
- multi-sourcing
- costs of tooling
- strategic importance.

Incremental cost is the additional or unavoidable cost incurred for an item purchased outside or being manufactured inside. The point to note is that many overhead costs will be unaffected by the make or buy decision. Thus overhead or indirect costs are almost certainly irrelevant to the decision and only the direct or marginal costs are important. Thus comparison between the outside purchase price and the marginal or incremental manufactured price must be made to give a true picture.

If an item is being outsourced then quality control will need to be considered. Potential suppliers may not have in place suitable quality procedures or may not be part of any supplier quality assurance scheme. If multi-sourcing is necessary then this problem is even greater.

If an item for outside purchase is a special design to the company's own specification then capital investment costs will have to be accounted for. In these circumstances it is common for the customer and supplier to share the investment in tooling and this could be quite significant.

An item or component that is of strategic importance in the company's own manufacturing process presents risks if it is manufactured outside. This may be true if it is a crucial safety item or is something upon which other production processes depend. Thus a cessation of supply or quality problems could have serious consequences. Before arranging outside purchase these risks have to be evaluated and some form of risk reduction undertaken before committing to outside sourcing.

Test your knowledge 1.32

Explain what is meant by a 'make or buy' decision. Also explain why such a decision can be important.

Problems

1.1 Identify and explain the effects of THREE external factors that affect engineering businesses.

1.2 Define and explain the terms local economy, regional economy, national economy, gross national product (GNP), gross domestic product (GDP), balance of payments.

1.3 Explain why Japanese firms sometimes find it advantageous to manufacture their products in the UK as opposed to Japan. Also explain the effect that this has on the UK's GNP.

1.4 (a) Define the following terms:

(i) gross national product (GNP)
(ii) gross domestic product (GDP).

(b) Explain the difference between GNP and GDP.

1.5 The annual percentage change in GDP for the UK and Germany over an eight-year period is shown in the table below:

Year	1991	1992	1993	1994	1995	1996	1997	1998
Germany	5.0	2.2	−1.1	2.3	1.7	0.8	1.5	2.2
UK	−1.5	0.1	2.3	4.4	2.8	2.6	3.5	2.2

(a) Compare the performance of Germany and the UK between 1991 and 1993. Suggest reasons for these trends in relation to the national economies of the two countries.

(b) Determine the average percentage increase in GDP for each country over the full eight-year period. Which country has performed best over this period?

1.6 (a) Distinguish between visible and invisible exports.
(b) Give TWO examples of invisible exports.

1.7 Describe TWO examples of the effect of relevant UK and EU environmental legislation on a typical engineering company.

1.8 (a) Explain the term 'by-product' in relation to an engineering process.
(b) Give TWO examples of by-products produced as a result of typical engineering processes and name the processes.

1.9 Describe, briefly, the effects of the following waste products produced as a result of engineering activities:

(a) carbon dioxide
(b) sulphur dioxide
(c) ozone
(d) smoke.

1.10 Engineering companies must organize their activities in such a way as to comply with relevant environmental legislation. Give TWO examples of environmental legislation that relate to engineering activities. In each case, describe typical measures taken to ensure that the legislation is complied with.

1.11 Explain why quality control is important in an engineering firm. Give examples of FOUR typical activities that are performed as part of a quality control system.

1.12 Explain what is meant by:

(a) programme evaluation and review technique (PERT)
(b) critical path method (CPM).

1.13 The following data refers to the activities that make up a project. Use this information to construct a network diagram and identify the critical path. Also determine the expected time to complete the project.

Activity	Preceding activity	Expected time
A	none	1
B	none	2
C	A	1
D	C	2
E	A	1
F	B	2
G	E, F	2
H	E, F	2
I	D, G	3
J	H	2
K	I, J	1

1.14 Explain, using with the aid of an example, the use of Gantt charts to analyse and monitor the sequence of activities in an engineering project.

1.15 Define the following terms:

(a) fixed cost
(b) overhead cost
(c) variable cost.

1.16 The following figures relate to the production of a small component:

Fixed cost:	£25 000
Variable cost per unit:	£12
Selling price:	£26

Determine the break-even point in sales turnover.

1.17 Explain the concept of 'marginal cost'.

1.18 Sketch a graph showing how costs and revenues vary with units of output. Mark the break-even points on your graph.

1.19 Explain what is meant by incremental cost in relation to a 'make or buy' decision.

1.20 Determine the residual asset value after three years of a CNC milling machine purchased for £120 000 if the machine has an asset life of five years.

1.21 Explain THREE factors that must be considered in investment appraisal.

1.22 State TWO advantages and TWO disadvantages of outsourcing the manufacture of a component used in an engineered product.

Unit 2 Engineering communications

This unit aims to provide you with a foundation for employment in engineering and a basis for further study. It aims to develop your communication skills by diverse means including drawing, sketching, writing and speaking. It also introduces you to a variety of techniques used for obtaining, evaluating, processing and presenting information.

To achieve this unit you must be able to show that you can:

- interpret and produce basic engineering drawings and sketches using appropriate geometry, standards, symbols and conventions and projections;
- demonstrate proficiency in verbal and written communication;
- obtain, evaluate, process and present information from a variety of information sources;
- demonstrate proficiency in the use of standard software packages and hardware devices used in typical engineering communication applications.

This unit is assessed by portfolio assessment and you must ensure that you include a variety of appropriate evidence within your portfolio. This can include assignment and case study work carried out in conjunction with the Business Systems unit as well as sketches and drawings, presentations, data sheets, technical reports and letters that you have produced. You should begin work on your portfolio as soon as you start the unit and then continue to collect evidence systematically as you progress through the unit. You should also cross-reference the evidence in the portfolio to the individual learning outcomes and the list of evidence requirements. Your tutor will be able to help you with this.

This unit has strong links with the Business Systems core unit. Wherever possible, you should apply the techniques that you have developed in this unit to work undertaken in the Business Systems unit. There are also links to the optional unit on Engineering Drawing.

As with the previous unit, you will need to have access to appropriate information and communications technology resources (including the Internet and the World Wide Web) to help you with your studies. You also need to be able to use an appropriate 2-D engineering drawing package (such as AutoCAD or AutoSketch). Your tutor will also provide you with access to a variety of sample written materials (letters, memos, technical reports, data sheets, catalogues), drawings (general arrangement, assembly and detail drawings) and sketches.

Being able to present your work and demonstrate proficiency in your own communications skills is an important aspect of this unit. To do this, you will need to have access to appropriate presentation and graphics software (such as Microsoft PowerPoint, Visio, Corel Draw, etc.) and spreadsheet software (such as Microsoft Excel) as well as hardware including scanners, laser and inkjet printers, optical character recognition and speech recognition software, digital cameras and bar code readers. Your school or college will normally be able to supply you with everything that you need as well as guidance and help with its use.

Introduction

The dictionary definition of 'communication' usually mentions something like 'the means by which we convey information'. There are, of course, many different ways of conveying information. You might consider, for example, the different ways that you could let your tutor know that you won't be attending college next week. You could speak to him in person, telephone him, leave a message on his answering machine, put a note on his desk, send him an e-mail message, or even write a letter and post it to him.

The method that you chose depends on a number of factors (not the least of which might be how friendly he is!). To be certain that he gets the message you might decide to speak to him in person (this might not be so easy if you don't have a good reason for not attending!). Alternatively, your college may require that you explain your absence in writing and they may have a form that you must complete. In deciding how to convey a simple message like this you probably need to think about several factors, including:

- How important is the information?
- How will I know that the information has been conveyed and understood?
- Do I need to keep a record of the information?
- How well do I know the person or persons with whom I am communicating?
- How urgent is the information?
- Is this a 'one-off' message or is it part of something much bigger?

The forms of communication that we use in everyday life can be broken down into four main types, namely:

- written
- graphical
- verbal
- other (non-verbal).

Each of these main types of communication can be further subdivided. For example, graphical communication can take the form of drawings, sketches, block diagrams, exploded views, graphs, charts, etc. Some of these can be further divided. For example, there are many different types of graph and chart. We have shown some of these in Figure 2.1.

> **Key point**
>
> It's important to remember that effective communication is a two-way process. In 'getting the message across' you need to make sure that it has been understood in the way that you intended!

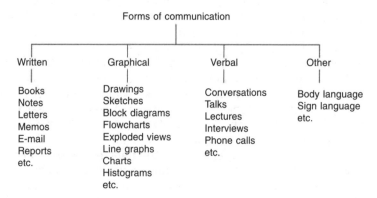

Figure 2.1 *Forms of communication*

In everyday life, we usually convey information by combining different forms of communication. For example, when we speak to

Key point

Methods of communication like those shown in Figure 2.1 are not mutually exclusive. Communication can be made more effective by combining different methods so that they reinforce the message or the information that you wish to convey.

Key point

The use of graphics to communicate ideas is essential for designers, engineers and draftspersons. Technical drawings are graphical representations of an idea or product that is to be processed, manufactured or constructed. Engineers use drawings to specify and transfer technical information.

other people we often combine verbal with non-verbal (body language) forms of communication. Body language can help add emphasis to our words or can be used to convey additional meaning. Presentations to groups of people usually involve verbal communication supported by visual aids such as overhead projector transparencies, handouts or flipcharts. Technical reports invariably combine written text with diagrams and photographs.

Throughout this unit there are numerous opportunities to develop your own skills in obtaining, processing, evaluating and presenting information. In order to do this, you should be prepared to make use of appropriate technology (i.e. hardware and software). For this reason, we will not study hardware and software as a separate topic. Instead, we will assume that you will acquire familiarity with its use as you progress through the topics in this book.

Activity 2.1

From the types of communication shown in Figure 2.1, identify (with reasons) methods of communication that you consider most appropriate in each of the following situations:

(a) Making an appointment to see a doctor.
(b) Apologizing for forgetting your sister's birthday.
(c) Directing a friend to a restaurant in the next town.
(d) Selling your car.

Present your work in the form of a brief set of hand-written notes.

Activity 2.2

Complete the table shown below by placing a tick against the applicability of different forms of communication in relation to fitting a plug to the electric cable on a portable appliance:

Situation	Highly applicable	Possibly applicable	Not applicable
A written instruction sheet			
A verbal commentary supplied on a cassette tape			
A sequence of diagrams with brief text			
A flowchart that lists each of the steps required			
A videotape			

Copy the table into a word-processing package and then print out your work.

Activity 2.3

Describe TWO forms of non-verbal communication and give examples of a situation in which each might be used. Present your work as a single A4 word-processed page.

Information and information handling

Information sources

Engineers use a wide variety of information in their everyday lives. This information is derived from a variety of different sources including:

- books
- application notes
- technical reports
- data sheets and data books
- catalogues
- engineering drawings
- CD-ROM
- databases
- websites.

Books

Books, whether they are paper-based publications or one of the new generation of electronic books (*eBooks*), provide information on an almost infinite number of subjects and a good technical library can be invaluable in any engineering context. All books contain summary information. This typically includes:

- the date of first publication
- the date(s) of any reprint(s) (with or without corrections)
- the date(s) of any subsequent edition(s)
- the date(s) of any reprint(s) of subsequent editions (with or without corrections)
- information concerning copyright
- British Library Cataloguing data
- an ISBN number
- other library (e.g. Library of Congress) cataloguing data
- information relating to the printing and binding of the book.

When using a book as a source of information it is important to ensure that it is up to date. It is also necessary to ensure that the content is reliable and that there are no omissions or errors. Book reviews (often published in the technical press) can be useful here!

Electronic books are presented in electronic rather than paper format. The book actually comprises of a number of files which can either be downloaded into a PC, laptop, or pocket reader device or which can be supplied on a CD-ROM. In order to convert the files that make up the book into something that can be read on the screen, the PC, laptop or pocket reader requires appropriate software such as Adobe's Acrobat Reader or Microsoft's Reader. Popular

Test your knowledge 2.1

Take a look at the information that appears in the first few pages of this book and then answer the following questions:

1. What was the date of first publication?
2. Is this book a reprint?
3. Who owns the copyright?
4. What is the book's ISBN number?
5. Who has published the book?

file formats for eBooks include HTML (HyperText Markup Language – see page 83) and Adobe's PDF (Portable Document File) format (see Figure 2.2).

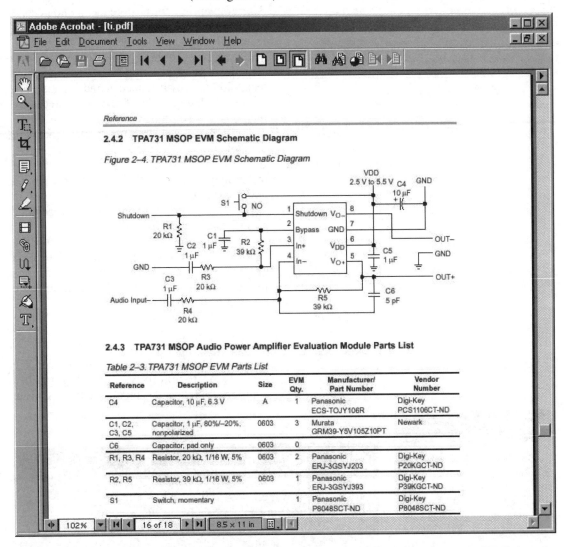

Figure 2.2 *An extract from an electronic document in Adobe's Portable Document File (PDF) format*

Application notes and technical reports

Application notes are usually brief notes (often equivalent in extent to a chapter of a book) supplied by manufacturers in order to assist engineers and designers by providing typical examples of the use of engineering components and devices. An application note can be very useful in providing practical information that can help designers to avoid pitfalls that might occur when using a component or device for the first time.

Technical reports are somewhat similar to application notes but they focus more on the performance specification of engineering components and devices (and the tests that have been carried out on them) than the practical aspects of their use. Technical reports usually include detailed specifications, graphs, charts and tabulated data.

Typical section headings used in application notes and technical reports include:

Summary   A brief overview for busy readers who need to quickly find out what the application note or technical report is about.

Introduction   This sets the context and background and provides a brief description of the process or technology – why it is needed and what is does. It may also include a brief review of alternative methods and solutions.

Main body   A comprehensive description of the process or technology.

Evaluation   A detailed evaluation of the process or technology together with details of tests applied and measured performance specifications. In appropriate cases comparative performance specifications will be provided.

Recommendations   This section provides information on how the process or technology should be implemented or deployed. It may include recommendations for storage or handling together with information relating to health and safety.

Conclusions   This section consists of a few concluding remarks.

References   This section provides readers with a list of sources of further information relating to the process or technology, including (where appropriate) relevant standards and legislation.

Key point

Application notes explain how something is used in a particular application or how it can be used to solve a particular problem. Application notes are intended as a guide for designers and others who may be considering using a particular process or technology for the first time. Technical reports, on the other hand, provide information that is more to do with whether a component or device meets a particular specification or how it compares with other solutions. Technical reports are thus more useful when it comes to analysing how a process or technology performs than how it is applied.

Activity 2.4

Write an application note that explains the use of AA-size NiMh batteries as replacements for the conventional alkaline batteries used in a digital camera. You should carry out some initial research before starting to write your application note but the following are possible headings (these can be combined or expanded if you think it necessary):

- Executive summary
- Introduction
- Basic requirements for batteries used in digital cameras
- Comparison of three battery types (NiCd, NiMH and conventional alkaline types)
- Battery life and charging arrangements
- Analysis of costs
- Suitability
- Recommendations
- Reference data (including a list of manufacturers and suppliers).

Use a word processor to present your work in printed form.

Data sheets and data books

Data sheets usually consist of abridged information on a particular engineering component or device. They usually provide maximum and minimum ratings, typical specifications, as well as information on dimensions, packaging and finish. Data sheets are usually supplied free on request from manufacturers and suppliers. Collections of data sheets for similar types of engineering components and devices are often supplied in book form. Often supplementary information is included relating to a complete family of products. An example of a data sheet is shown in Figure 2.3.

Catalogues

Most manufacturers and suppliers provide catalogues that list their full product range. These often include part numbers, illustrations, brief specifications and prices. While catalogues are often extensive documents with many hundreds or thousands of pages, short-form catalogues are usually also available. These usually just list part numbers, brief descriptions and prices but rarely include any illustrations. A brief extract from a short-form catalogue is shown in Figure 2.4.

Engineering drawings

Engineering drawings can be produced either manually or on a computer using suitable CAD software. Drawings produced manually can range from freehand sketches to formally prepared drawings produced with the aid of a drawing board and conventional drawing instruments. Later in this unit we will introduce you to the most common types of engineering drawing.

CD-ROM

Compact disks can provide storage for around 650 Mbytes of computer data. This is roughly equivalent to 250 000 pages of A4 text. It is therefore hardly surprising that the compact disk has now become firmly established as a storage medium for a variety of different types of information, including text, drawings, sound and video (i.e. multimedia). The latest PCs offer CD-ROM drives that can be used to store data on recordable CDs (CD-R) or rewritable CDs (CD-RW). High-end PCs now also have DVD (Digital Versatile Disk).

Test your knowledge 2.2

Refer to the extract from the Texas Instruments data sheet shown in Figure 2.3 and use it to answer the following questions:

(a) What is the data of issue of the data sheet?

(b) Who owns the copyright of the data sheet?

(c) How many independent logic gates are contained in the device?

(d) What do the letters 'NC' mean?

(e) How many pins are there on an 'FK' package?

(f) Which two devices are suitable for operation over the 'military temperature range'?

(g) What is the absolute maximum supply voltage for these devices?

(h) What is the storage temperature range specified for these devices?

(i) What does the manufacturer specify as typical values of supply voltage and operating free-air temperature for these devices?

Test your knowledge 2.3

An engineering company manufactures a variety of different types of fastener including nuts and bolts, screws and washers. What typical information would appear in a short-form catalogue for this company?

Activity 2.5

Complete the table shown on page 78 by placing a tick in the column against the most appropriate method for communicating the listed information (tick only once in each row):

SN54ALS08, SN54AS08, SN74ALS08, SN74AS08
QUADRUPLE 2-INPUT POSITIVE-AND GATES

SDAS191A – APRIL 1982 – REVISED DECEMBER 1994

- **Package Options Include Plastic Small-Outline (D) Packages, Ceramic Chip Carriers (FK), and Standard Plastic (N) and Ceramic (J) 300-mil DIPs**

description

These devices contain four independent 2-input positive-AND gates. They perform the Boolean functions $Y = A \cdot B$ or $Y = \overline{A} + \overline{B}$ in positive logic.

The SN54ALS08 and SN54AS08 are characterized for operation over the full military temperature range of −55°C to 125°C. The SN74ALS08 and SN74AS08 are characterized for operation from 0°C to 70°C.

FUNCTION TABLE
(each gate)

INPUTS		OUTPUT
A	B	Y
H	H	H
L	X	L
X	L	L

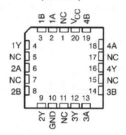

SN54ALS08, SN54AS08 . . . J PACKAGE
SN74ALS08, SN74AS08 . . . D OR N PACKAGE
(TOP VIEW)

```
      1A [ 1      14 ] Vcc
      1B [ 2      13 ] 4B
      1Y [ 3      12 ] 4A
      2A [ 4      11 ] 4Y
      2B [ 5      10 ] 3B
      2Y [ 6       9 ] 3A
     GND [ 7       8 ] 3Y
```

SN54ALS08, SN54AS08 . . . FK PACKAGE
(TOP VIEW)

```
          1B 1A NC Vcc 4B
           3  2  1  20 19
      1Y [ 4              18 ] 4A
      NC [ 5              17 ] NC
      2A [ 6              16 ] 4Y
      NC [ 7              15 ] NC
      2B [ 8              14 ] 3B
           9 10 11 12 13
          2Y GND NC 3Y 3A
```

NC – No internal connection

logic symbol†

```
1A  1
1B  2        &        3   1Y
2A  4
2B  5                 6   2Y
3A  9
3B  10                8   3Y
4A  12
4B  13                11  4Y
```

† This symbol is in accordance with ANSI/IEEE Std 91-1984 and
  IEC Publication 617-12.
Pin numbers shown are for the D, J, and N packages.

logic diagram (positive logic)

```
1A  1
1B  2                3   1Y
2A  4
2B  5                6   2Y
3A  9
3B  10               8   3Y
4A  12
4B  13               11  4Y
```

Copyright © 1994, Texas Instruments Incorporated

TEXAS
INSTRUMENTS

POST OFFICE BOX 655303 ● DALLAS, TEXAS 75265

1

(a)

SN54ALS08, SN54AS08, SN74ALS08, SN74AS08
QUADRUPLE 2-INPUT POSITIVE-AND GATES

SDAS191A – APRIL 1982 – REVISED DECEMBER 1994

absolute maximum ratings over operating free-air temperature range (unless otherwise noted)†

Supply voltage, V_{CC}	7 V
Input voltage, V_I	7 V
Operating free-air temperature range, T_A: SN54ALS08	−55°C to 125°C
SN74ALS08	0°C to 70°C
Storage temperature range	−65°C to 150°C

† Stresses beyond those listed under "absolute maximum ratings" may cause permanent damage to the device. These are stress ratings only, and functional operation of the device at these or any other conditions beyond those indicated under "recommended operating conditions" is not implied. Exposure to absolute-maximum-rated conditions for extended periods may affect device reliability.

recommended operating conditions

		SN54ALS08			SN74ALS08			UNIT
		MIN	NOM	MAX	MIN	NOM	MAX	
V_{CC}	Supply voltage	4.5	5	5.5	4.5	5	5.5	V
V_{IH}	High-level input voltage	2			2			V
V_{IL}	Low-level input voltage			$0.8^‡$			0.8	V
				$0.7^§$				
I_{OH}	High-level output current			−0.4			−0.4	mA
I_{OL}	Low-level output current			4			8	mA
T_A	Operating free-air temperature	−55		125	0		70	°C

‡ Applies over temperature range −55°C to 70°C
§ Applies over temperature range 70°C to 125°C

electrical characteristics over recommended operating free-air temperature range (unless otherwise noted)

PARAMETER	TEST CONDITIONS		SN54ALS08			SN74ALS08			UNIT
			MIN	TYP¶	MAX	MIN	TYP¶	MAX	
V_{IK}	$V_{CC} = 4.5$ V,	$I_I = −18$ mA			−1.5			−1.5	V
V_{OH}	$V_{CC} = 4.5$ V to 5.5 V,	$I_{OH} = −0.4$ mA	$V_{CC} − 2$			$V_{CC} − 2$			V
V_{OL}	$V_{CC} = 4.5$ V	$I_{OL} = 4$ mA		0.25	0.4		0.25	0.4	V
		$I_{OL} = 8$ mA					0.35	0.5	
I_I	$V_{CC} = 5.5$ V,	$V_I = 7$ V			0.1			0.1	mA
I_{IH}	$V_{CC} = 5.5$ V,	$V_I = 2.7$ V			20			20	μA
I_{IL}	$V_{CC} = 5.5$ V,	$V_I = 0.4$ V			−0.1			−0.1	mA
$I_O^\#$	$V_{CC} = 5.5$ V,	$V_O = 2.25$ V	−20		−112	−30		−112	mA
I_{CCH}	$V_{CC} = 5.5$ V,	$V_I = 4.5$ V		1.3	2.4		1.3	2.4	mA
I_{CCL}	$V_{CC} = 5.5$ V,	$V_I = 0$		2.2	4		2.2	4	mA

¶ All typical values are at $V_{CC} = 5$ V, $T_A = 25$°C.
# The output conditions have been chosen to produce a current that closely approximates one half of the true short-circuit output current, I_{OS}.

TEXAS
INSTRUMENTS
POST OFFICE BOX 655303 ● DALLAS, TEXAS 75265

2

(b)

Figure 2.3 *An extract from a semiconductor datasheet*

Diecast Boxes

IP65 Sealed/Painted

A range of high-quality diecast alumnium boxes with an optional grey epoxy paint finish to RAL7001. The lid features an integral synthetic rubber sealing gasket and captive stainless steel fixing screws. Mounting holes and lid fixing screws are outside the seal, giving the enclosure protection to IP 65.

Standard supply multiple = 1   Delivery normally ex-stock

| Size | | | | | | | Price each | | |
L	W	H	T	Finish	Manufacturer's ref:	Stock code	1-9	10-24	25+
90	45	30	3.0	none	1770-1541-21	DB65-01	£4.52	£3.95	£3.50
90	45	30	3.0	grey	1770-1542-21	DB65-01P	£5.40	£4.90	£4.45
110	50	30	4.5	none	1770-1543-22	DB65-02	£5.25	£4.50	£4.15
110	50	30	4.5	grey	1770-1544-22	DB65-02P	£6.42	£5.37	£4.95
125	85	35	5.0	none	1770-1545-23	DB65-03	£6.15	£5.17	£4.71
125	85	35	5.0	grey	1770-1546-23	DB65-03P	£7.10	£6.05	£5.65

Figure 2.4 *An extract from a short-form catalogue*

Test your knowledge 2.4

Your company requires 15 diecast boxes suitable for enclosing a printed circuit board of thickness 3 mm measuring 80 mm × 35 mm. The tallest component stands 15 mm above the board and a minimum clearance of 5 mm is to be allowed all round the board. The enclosure is to be supplied ready for mounting the printed circuit board and should not need any further finishing other than drilling. Prepare a fax message to Dragon Components (the hardware supplier whose short-form catalogue extract appears in Figure 2.4) giving all the information required to fulfil your order.

	Application note	Data sheet	Shortform catalogue	Technical report
Summary of the precautions to be observed when handling a chemical etching fluid				
Cost of die-cast boxes supplied in various quantities, from 1 to 100				
Maximum working temperature for a power transistor				
Recommended printed circuit board layout for an audio amplifier				
Comparison of different types of surface finish for the interior of a domestic microwave oven				
Physical dimensions of a marine radar for fitting to a small boat				
Description of tests applied to an off-road vehicle				
Performance specification for a satellite TV aerial				

Copy the table into a word-processing package and then print out your work.

Databases

A database is simply an organized collection of data. This data is usually organized into a number of records each of which contains a number of fields. Because of their size and complexity and the need to be able to quickly and easily search for information, a database is usually stored within a computer and a special program – a *database manager* or *database management system* (DBMS) – provides an interface between users and the data itself. The DBMS keeps track of where the information is stored and provides an index so that users can quickly and easily locate the information they require (see Figure 2.5).

The database manager will also allow users to search for related items. For example, a particular component may be used in a number of different products. The database will allow you to quickly identify

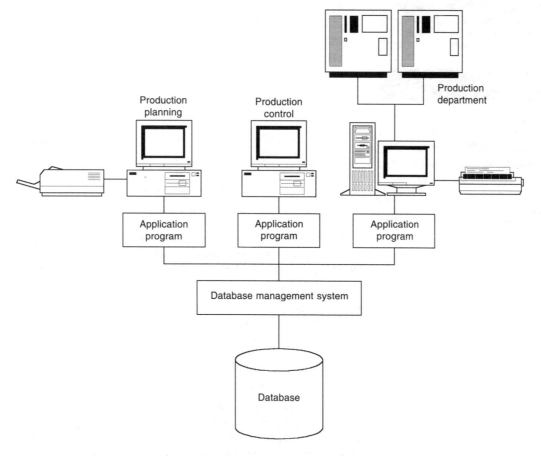

Figure 2.5 *A database management system (DBMS)*

each product that uses the component as well as the materials and processes that are used to produce it.

The structure of a simple database is shown in Figure 2.6. The database consists of a number of *records* arranged in the form of one or more *tables*. Each record is divided into a number of *fields*. The fields contain different information but they all relate to a particular component. The fields are organized as follows:

Field 1	Key (or index number)
Field 2	Part number
Field 3	Type of part
Field 4	Description or finish of the part

The *key* (in this case Field 1) allows us to uniquely identify the data stored in the record. Keys may be meaningful data (such as a product name) or they may be an index number that we have invented for the purpose. The important feature of a key is that it must be *unique*.

Key point

A database consists of *records* arranged into one or more *tables*. Records are uniquely identified by their *key*. Each record is divided into a number of *fields*, one of which is the key.

Test your knowledge 2.6

A college uses a database to hold information on its students. In this database application, explain why a student's last name is not suitable for use as the key to a record.

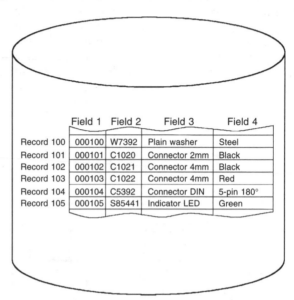

	Field 1	Field 2	Field 3	Field 4
Record 100	000100	W7392	Plain washer	Steel
Record 101	000101	C1020	Connector 2mm	Black
Record 102	000102	C1021	Connector 4mm	Black
Record 103	000103	C1022	Connector 4mm	Red
Record 104	000104	C5392	Connector DIN	5-pin 180°
Record 105	000105	S85441	Indicator LED	Green

Figure 2.6 *The structure of a simple parts database*

The database management system is used to build and maintain the database. It also provides the interface between the user and the information stored in the database. Tasks performed by the DBMS include:

- adding new records
- deleting unwanted records
- amending records
- linking or cross-referencing records
- searching and sorting the database records
- printing reports of selected records.

Most engineering companies use several specialized databases in order to manage different functions. To help you understand this, consider the case of an engineering company that manufactures fork-lift trucks. The company might use the following databases to help it organize the different aspects of its operation:

- a *product database* containing records of each vehicle manufactured and any modifications fitted
- a *manufacturing database* containing records of all components and materials used during manufacture
- a *customer database* containing records of all customers and the vehicles supplied to them
- a *spare parts database* containing records of all spare parts held showing where they are stored and the quantity held.

Figure 2.7 shows how these four databases relate to three different departments within the company. Since these three functions cannot operate in isolation there is a need to exchange data between the three databases. To give you some idea of how this might work, let's assume that the sales department has been given the task of marketing a new fork-lift truck that has just become available. The manager of the sales department has been given a target that requires him to sell ten fork-lift trucks by the end of the next quarterly period. He has been allocated a fixed budget specifically to meet the costs associated with this sales campaign.

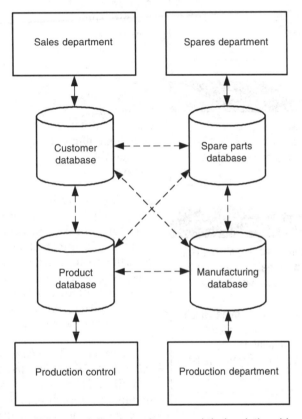

Figure 2.7 *Four specialized databases and their relationships with different company functions*

The sales manager decides to produce a brochure giving details of the new fork-lift truck and offering a substantial discount to any previous customers who may wish to 'trade in' their existing fleet of fork-lift trucks. The brochure will be mailed to all UK customers that have purchased fork-lift trucks in the last ten years. Sales staff will then follow this up with a telephone call to each named customer contact.

The information required to produce the new fork-lift truck brochure will be drawn from the company's product database. The mailing list, contact names and telephone numbers will be taken from a report generated from the customer database.

The structure and content of a typical record in the product database will include the following information:

Product reference number: FLT1022
Product name: Challenger

Vehicle chassis type:	BCX077
Quantity in stock:	7
Scheduled production (current period):	10
Scheduled production (next period):	12

The structure and content of a typical record in the customer database will include the following information:

Customer reference number:	13871
Company name:	Enterprise Air Freight
Address (line 1):	Unit 8
Address (line 2):	Bath Road Industrial Estate
Town/city:	Feltham
County or state:	Middlesex
Post code or zip code:	UB10 3BY
Country:	UK
Contact name:	David Evans
Contact title:	Purchasing manager
Contact salutation:	Dear Dave
Telephone number:	020-8979-7756
Fax number:	020-8979-7757
e-mail:	devans@enterprise.co.uk

Nowadays, there is a trend towards integrating many of the databases within an engineering company into one large database. This database becomes central to all of the functions within the company. In effect, it becomes the 'glue' that holds all of the departments together. The concept of a centralized manufacturing database is a very sound one because it ensures that every function within the company has access to the same data. By using a single database, all departments become aware of changes and modifications at the same time and there is less danger of data becoming out of date.

In a *relational database*, you are able to create views that display only selected fields in a data file or that combine fields from multiple data files. There are several good reasons for breaking a large amount of information into a number of smaller data files. In some cases small data files will process more quickly than using one very large data file. Better security can be maintained by breaking a large data file into several smaller data files. By controlling access to each data file you can also control what users can see! In any event, when designing a database it is essential to take time planning the individual data files that will be needed and, in particular, considering the relative association that the records contained in the data files may have to each other.

Figure 2.8 shows how information can be taken from separate data tables and combined into a single report. In this case, the matching field is the employee number and the report lists those employees earning a salary of more than £30 000 together with their National Insurance (NI) numbers.

Test your knowledge 2.7

Explain the following terms when applied to a database:

(a) field
(b) record
(c) key
(d) relational.

Test your knowledge 2.8

Give two examples of information that would be stored in a manufacturing database.

Test your knowledge 2.9

In setting up the mailshot (see text), what fields within the customer database will be:

(a) used to determine which customers are included in the mailing list
(b) used to generate the label that will be attached to the information pack.

Test your knowledge 2.10

There is a need to ensure that information contained in a database is kept up to date. In the case of the customer database (see text), which records and field may need updating as a result of the mailshot and telephone sales campaign?

The World Wide Web

The Internet is the name given to a huge network of computers all over the world that communicate with each other. The Internet

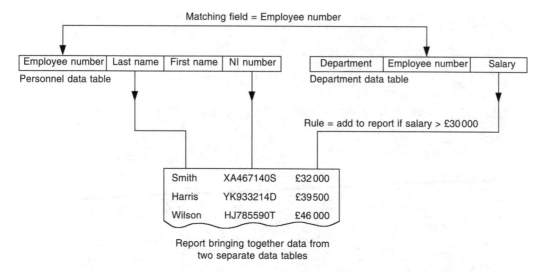

Figure 2.8 *Matching data tables to produce sophisticated reports*

makes it possible for you to access the World Wide Web, a vast collection of on-line information kept on numerous web servers throughout the world. The World Wide Web began in March 1989, when Tim Berners-Lee of the European Particle Physics Laboratory at CERN (the European centre for nuclear research) proposed the project as a means to better communicate research ideas among members of a widespread organization.

If you have an Internet connection to your PC, the Web is accessed via a program called a *browser* that helps you to navigate between the *pages* of information held on the Web. The browser available within the college is Microsoft's *Internet Explorer*.

Information on the Web is stored as pages of electronic text with *hyperlinks*. Clicking on a hyperlink will allow you to jump from one page to another, or from one place in a page to another place on the same page. Hyperlinks often appear in a different colour text and are usually underlined. They can also take the form of pictures and graphics that you can click on. A hyperlink can give you access to anything that can be stored electronically: text, graphics, video, sound, or a file that you can *download* to your PC. A typical web page is shown in Figure 2.9.

Web pages are written using This language contains embedded commands. The important distinction between a web page and a document produced by a word processor is that, in the case of the web page, the instructions for displaying the document are imbedded in the document itself (they are called *tags*). The web browser reads the tags and then uses them to construct the page. The tags are defined by a language called *HyperText Markup Language* (HTML). Figure 2.10 shows the source HTML code for the page shown in Figure 2.9. See if you can make sense of it!

Every page on the Web has its own unique address called a *Uniform Resource Locator* (URL). The URL specifies the protocol used to retrieve the data – usually either *HyperText Transfer Protocol* (HTTP) or *File Transfer Protocol* (FTP), the address of the machine, the path and filename of the page, e.g. http://www.brooklands.ac.uk where:

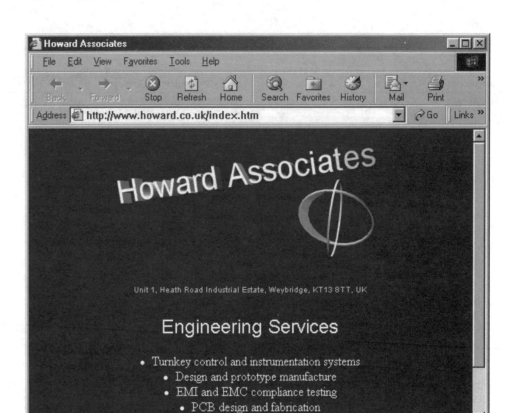

Figure 2.9 *A typical engineering company's website displayed in a web browser*

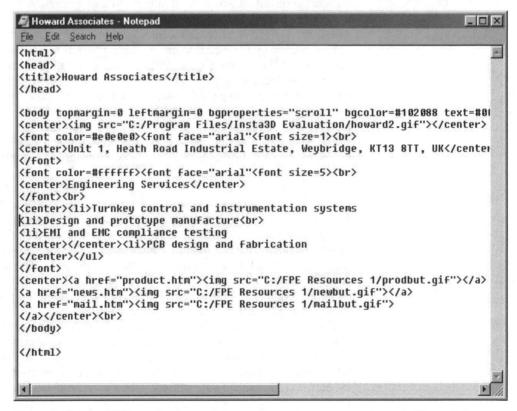

Figure 2.10 *The HTML code responsible for generating the page shown in Figure 2.9*

http://	specifies the hypertext transfer protocol
www	specifies a World Wide Web (WWW) site accessed via a program called a *web browser*
brooklands.ac.uk	is Brooklands College's *domain name* (the .ac.uk tells you that the domain is that of a UK-based academic institution.

The last part of a URL gives you some information about the type of site you are visiting as well as where the owner's business is located. For example, a .com site is a commercial site (often, but not exclusively, located in the USA). A .co.uk site is a commercial site located in the UK. Similarly, a site with a URL ending in .co.fr is likely to be a commercial site located in France. Different rules apply to different countries and registration authorities; however, the following is worth noting.

UK registrations

co.uk is for companies and general use
org.uk is for organizations and non-profit-making companies. However, anyone can register a org.uk domain
net.uk is for Internet Service Providers (ISP). The rules for .net.uk are very strict; the governing body for UK domain names will not allow a non-ISP to register these domains
ltd.uk is for UK limited companies. You must provide your company registration number
plc.uk is for UK public limited companies.

International registrations

com is for companies and general use
net is for ISP. The international rules are more relaxed than the UK in that anyone can register a .net domain
org is for organizations and non-profit-making companies. Once again anyone can register a org domain.

Search engines

One of the most useful features of the Web is the availability of search engines that will help you to locate the information that you require. Search engines are just large computers that contain cross-referenced lists of URLs. To use a search engine, you simply enter the name of the search engine that you wish to use in your browser. Once the search engine's home page has loaded you can enter the text that you wish to search for (see Figure 2.11). To get the best out of a search engine you need to give some thought as to what it is you are searching for and how best to describe it unambiguously. If you give the search engine too little to go on, say just one word, you may be rewarded with a huge number of references to sites that contain references to the word that you have used. On the other hand, using words that are very specific may restrict the search to too few sites. In any event, it's always worth experimenting with a search engine. With a little trial and error you will soon get to know how to get the best out of it!

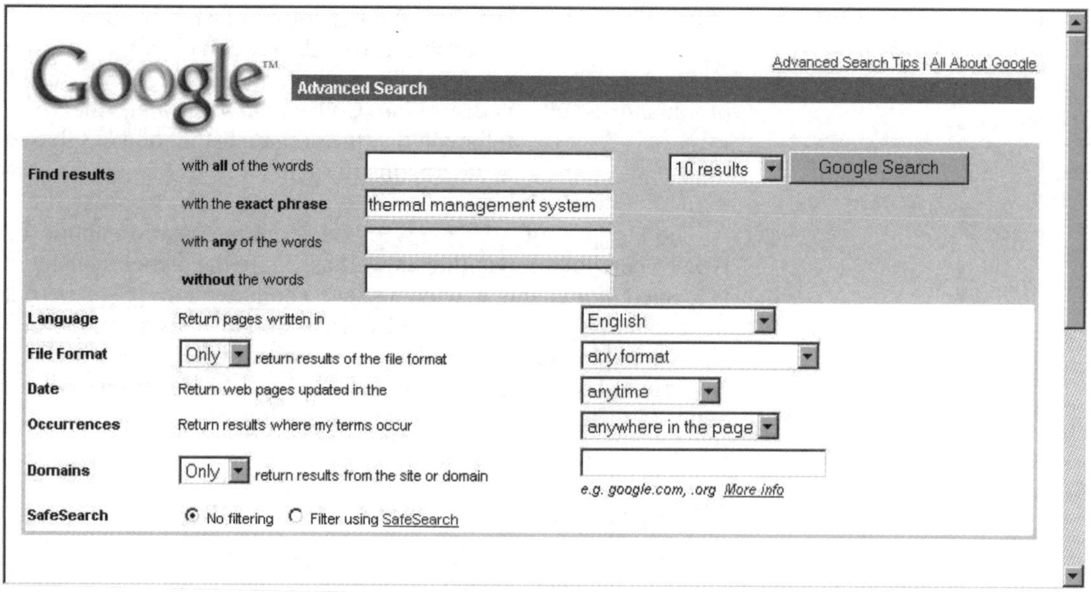

Figure 2.11 *Using a search engine (here we are using Google's Advanced Search facility)*

On-line services

On-line services are services that add value to the World Wide Web. Originally, these services built and maintained trunk networks that could be used by their customers. They also added their own *content* (such as news and weather reports, software libraries, etc.) to the Web. Users who are prepared to pay for the service can access this material. Customers still pay for some of the on-line services but the trend, in recent years, has been to make added value services free. Currently the most popular on-line services are America Online (AOL), CompuServe, Prodigy and the Microsoft Network (MSN). All of these services provide access to e-mail, support libraries and on-line communities where people with similar interests can communicate for business or pleasure.

Intranets

Intranets work like the web (with browsers, web servers, and web sites) but companies and other organizations use them internally. Companies use them because they let employees share corporate data, but they're cheaper and easier to manage than most private networks because nobody needs any software more complicated or more expensive than a web browser, for instance. They also have the added benefit of giving employees access to the web. Intranets are closed off from the rest of the Net by *firewall* software, which lets employees surf the web but keeps all the data on internal web servers hidden from those outside the company.

Test your knowledge 2.11

List TWO advantages and TWO disadvantages of a catalogue supplied on CD-ROM compared with one that is published on the World Wide Web.

Extranets

One of the most recent developments has been that of the *extranet*. Extranets are several intranets linked together so that businesses can share information with their customers and suppliers. Consider, for example, the production of a European aircraft by four major aerospace companies located in different European countries. They might connect their individual company intranets (or parts of their intranets) to each other, using private leased lines or even using the public Internet. The companies may also decide to set up a set of private newsgroups so that employees from different companies can exchange ideas and share information.

E-mail

Like ordinary mail, e-mail consists of a message, an address and a carrier that has the task of conveying the message from one place to another. The big difference is that e-mail messages (together with any attached files) are broken down into small chunks of data (called *packets*) that travel independently to their destination along with innumerable other packets travelling to different destinations. The packets that correspond to a particular e-mail message may travel by several different routes and may arrive out of order and at different times. Once all the packets have arrived, they are recombined into their original form. This may all sound rather complicated but it is nevertheless efficient because it prevents large messages hogging all of the available bandwidth. To put this into context, a simple page of A4 text can be transferred half-way round the world in less than a minute! Figure 2.12 shows a typical e-mail message written using Microsoft Outlook Express.

Test your knowledge 2.12

Briefly explain what is meant by a URL. Give an example of a URL for a UK-based engineering company. What is the URL of the web page shown in Figure 2.9?

Test your knowledge 2.13

The URLs of three web sites are:

1. www.gloscat.ac.uk
2. www.sony.co.uk
3. www.btec.org.uk

What does this information tell you about each of the three sites?

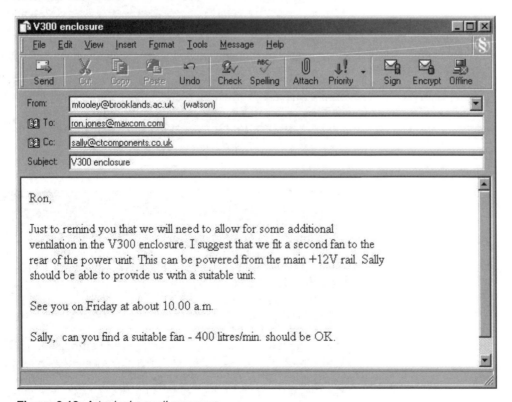

Figure 2.12 *A typical e-mail message*

Activity 2.6

Investigate the use of databases in your school or college. Working as part of a group, begin by finding out the name of the person who has overall responsibility for collecting and processing student data. Interview him or her and find out what information is held in the database and how it is organized. Also find out about the reports that are generated by the database and who has access to the information. Prepare a brief presentation to the rest of your class using appropriate handouts and visual aids.

Activity 2.7

Visit the Texas Instruments web site at www.ti.com

(a) Use the search facility to search for information on '74ALS08'.
(b) When the device has been located, open the 'Product Folder' to view more information.
(c) Click to download the full data sheet for the device (this is provided in the form of an Adobe Acrobat (.PDF) file).
(d) View and print the full data sheet.
(e) Given that the device employs 'totem-pole' outputs, sketch the circuit that should be used for measuring the switching characteristics of the device and include component values.
(f) What are typical values for the 'low state' and 'high state' output voltages?

Present your work in the form of a word-processed 'fact sheet'. Insert your circuit diagram into this document (either by scanning it or cutting and pasting it from a drawing package). Attach your 'fact sheet' to the printed data sheet.

Activity 2.8

One important type of data sheet is known as a Material Safety Data Sheet (MSDS). Use the World Wide Web to locate information on ferric chloride solution (printed circuit board etching fluid). To locate this information you will probably need to make use of one of several on-line MSDS databases. These include MSDSonline, MSDS-Search, MSDS Solutions and Oxford University. You should locate these (and other) sources using a popular search engine. Once you have identified several information sources use them to answer the following questions:

1. Is this material flammable?
2. Is this material corrosive?
3. What is the specific gravity of this material?
4. Is this material hazardous when in contact with the skin?
5. What does the material look like?
6. What precautions should be observed when handling and storing this material?

Present your work in the form of a word-processed 'fact sheet'.

Activity 2.9

Visit the web site of TAG McLaren Audio, a manufacturer of high-quality audio-visual equipment. Investigate the company and view some of its products. The URL for TAG McLaren Audio is:

http://www.tagmclarenaudio.com

Search the site for information on the Aphrodite Music System (see Figure 2.13) and obtain the full technical specification for the product. From the specification, determine:

(a) the frequency response of the amplifier
(b) the power output produced by the amplifier

Figure 2.13 *The TagMclaren Aphrodite audio system*

(c) the total harmonic distortion produced by the amplifier
(d) the dimensions of the integrated amplifier/tuner/CD player
(e) the weight of the integrated amplifier/tuner/CD player.

Write a brief word-processed report describing THREE innovative features of this system. Include a picture of the system in your report.

Activity 2.10

Use a search engine (such as Lycos or AltaVista) to locate information on electric drills. Visit the first four sites displayed as a result of your search and note down the URL of each of these sites. Summarize the contents of each of the sites by writing a paragraph describing each site. Then rate each site on a scale of 1 to 10 on the basis of content, presentation and ease of use. Summarize your results in a table.

Now repeat the activity using a web directory (such as Yahoo or Excite). Use the directory to navigate to four different sites giving details of electric drills. Once again, note down the URL of each site, summarize its contents and rate it on a scale of 1 to 10 (again presenting your results in the form of a table). Compare these two search methods.

Present your findings in the form of a word-processed 'fact sheet'.

Activity 2.11

If you have not already done so, set up a web-based e-mail account in your own name. Note down all of the steps that you took to open the account including details of any electronic forms that you had to complete.

Present your findings in the form of a brief word-processed article for your local paper showing how easy it is to open and make use of an e-mail account. You should assume that the reader is non-technical.

Presentation techniques

Engineers rely heavily upon graphical methods of communication. Drawings and charts produced to international standards and using international symbols and conventions suffer no language barriers. They are not liable to be misinterpreted by translation errors. Graphical communication does not replace spoken and written communication but instead is used to simplify, reinforce and complement other means of communication.

Having now established the need for communicating engineering information, let's look at the various methods of graphical communication available. We can broadly divide engineering information into two categories. That which is mathematically based and that which is technically based. We will start by looking at ways of representing mathematical data.

Graphs

Just as engineering drawings are used as a clear and convenient way of describing complex components and assemblies, so can graphs be used to give a clear and convenient picture of the mathematical relationships between engineering and scientific quantities. Figure 2.14(a) shows a graph of the relationship between distance s and time t for the mathematical expression $s = \frac{1}{2}at^2$ where the acceleration, $a = 10$ m s^{-1}.

In this instance it is correct to use a continuous flowing curve to connect the points plotted. Not only do these points lie on the curve, but every corresponding value of s and t between the points plotted also lie on the curve.

However, this is not true for every type of line graph. Figure 2.14(b) shows a graph relating speed and distance for a journey. From A to B the vehicle is accelerating. From B to C the vehicle

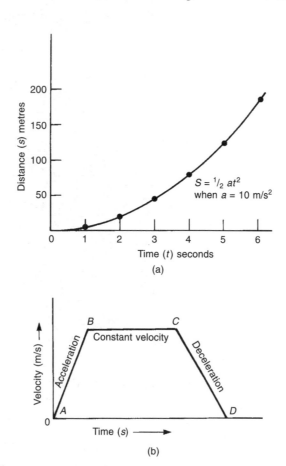

Figure 2.14 *Line graphs*

is travelling at a constant speed. From C to D the vehicle is decelerating (slowing down). In this example it is correct to join the points by straight lines. This is because each stage of the journey is represented by a linear mathematical expression which is unrelated both to the previous and following stages of the journey.

Activity 2.12

Draw a graph showing how the following voltage varies with time:

Time, t (s)	0	0.1	0.2	0.3	0.4	0.5	0.6	0.7
Voltage, v (V)	0	1.95	2.95	3.46	3.72	3.86	3.93	3.96

What can you infer from the shape of the graph?
Present your findings in the form of a hand-drawn graph on graph paper. Label your drawing clearly.

Activity 2.13

Given the formula $N = \dfrac{1000S}{\pi d}$

where: N = spindle speed in rev/min
S = cutting speed = 33 m/min
d = drill diameter in mm
π = 3.142

Use the formula to complete the table below, then enter the data into a spreadsheet package and use it to produce a graph relating drill diameter, d, and spindle speed, N. From the graph determine the spindle speed of a drill having a diameter of 5 mm.

Drill diameter, d (mm)	2	4	6	8	10
Spindle speed, N (rev/min)					

Present your work in the form of a printed spreadsheet and graph.

Histograms

Histograms are used for plotting information where the change is discrete rather than continuous. To put this into context consider the number of National Diplomas awarded by a college over a six-year period. The award of a National Diploma happens at the end of each academic year when a particular number of students achieve the qualification. This varies from year to year but, since it does not change on a continuous basis, it is inappropriate to illustrate the relationship using a line. Instead, we use a series of bars representing the number of National Diplomas awarded. Despite the fact that there is no line joining the bars together we can still clearly see a trend.

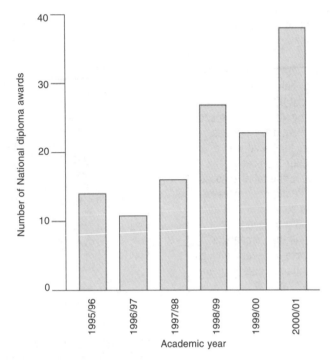

Figure 2.15 *A histogram*

Bar charts

Like histograms, bar charts are also used for displaying statistical data, but are usually plotted horizontally. They are also often made to look more attractive by using 3-D drawing, as shown in Figure 2.16.

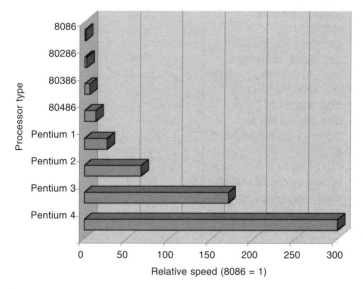

Figure 2.16 *A 3-D bar chart*

Ideographs (pictograms)

Ideographs (or pictograms) are frequently used to simplify statistical data so that it can be made meaningful to the general public. A typical example is the number of cars produced by a car manufacturer, over a four-year period, as shown in Figure 2.17. In this example each symbol represents 1000 cars. Therefore in 1998, 4000 cars were produced (four symbols each representing 1000 cars makes a total of 4000 cars).

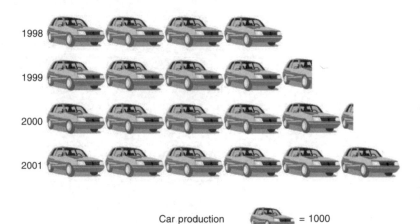

Figure 2.17 *An ideograph*

Some ideographs use the size of a symbol (rather than a number of symbols of identical size) to represent quantities. An example of this is shown in Figure 2.18.

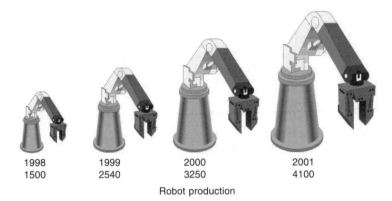

| 1998 | 1999 | 2000 | 2001 |
| 1500 | 2540 | 3250 | 4100 |

Robot production

Figure 2.18 *An alternative form of ideograph*

Activity 2.15

Use an ideograph to compare the production of lager and bitter for a local brewery over a six-week period (each symbol is to represent 1000 pints):

Week	Lager production (pints)	Bitter production (pints)
1	1750	2500
2	2500	4000
3	3500	6000
4	4000	6500
5	4250	3500
6	3500	5000

Use a simple drawing or art package to produce the ideograph and present the printed result.

Pie-charts

Pie-charts are used to show how a total quantity is divided up into its individual parts. Take a look at Figure 2.19(a). Since a complete circle is 360°, we can represent 25 per cent of a complete circle as $\left(360° \times \frac{25}{100}\right) = 90°$. Figure 2.19(b) shows how the annual expenditure of an engineering company can be represented by a pie-chart.

As with bar charts, pie-charts can be presented in 3-D in order to make them more attractive. Figure 2.20 shows a 3-D pie-chart showing how the costs of a PC system unit are divided between the main components.

Test your knowledge 2.16

Use Figure 2.20 to answer the following questions:

1. What is the most expensive component in a PC system unit?
2. Which is the least expensive component in a PC system unit?
3. Roughly what proportion of the cost is attributable to the case?
4. What is the proportion of cost of the motherboard and processor taken together?
5. If the total cost of a system unit is £250, determine the approximate cost of the power supply.

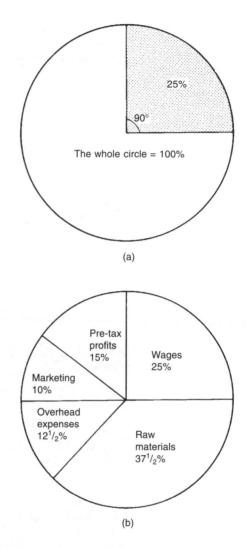

Figure 2.19 *Pie-charts*

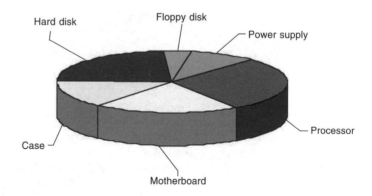

Figure 2.20 *A 3-D pie-chart*

Scatter diagrams

Scatter diagrams are used when we need to show that a change in one value is likely to result in a change in another. Scatter diagrams are usually produced by plotting corresponding pairs of values on an X–Y chart. This is usually done for a large number of observations (much larger than the number of values taken to plot a conventional line graph). The resulting grouping of dots on the diagram helps us to understand the *correlation* between the two values and whether there are any significant trends.

Figure 2.21 shows how the bond strength for an adhesive is

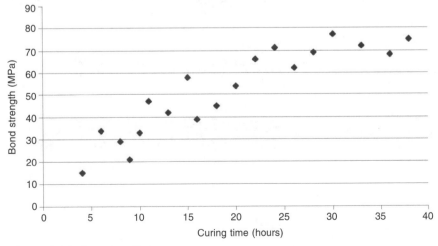

Figure 2.21 *A scatter diagram*

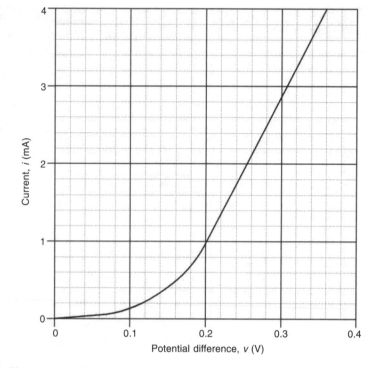

Figure 2.22 *See Test your knowledge 2.18*

related to the curing time. Here, the performance of 19 individual samples has been plotted and a clear trend can be seen.

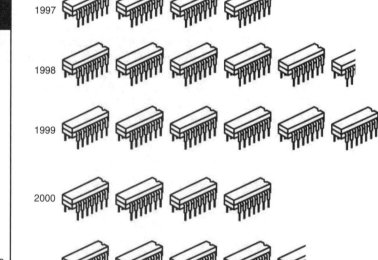

TTL logic devices

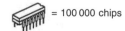 = 100 000 chips

Figure 2.23 *See Test your knowledge 2.19*

Key point

When selecting a method of graphical communication in order to convey information it is important to consider what sort of audience it is intended for. This means that you will often need to simplify the representation of the data so that it can be more easily grasped by the user. In every case, it is essential to avoid any possibility of misunder-standing the information that you are trying to represent.

Activity 2.17

Complete the table below placing a tick in the column showing the most appropriate method of communicating the information in each situation listed:

Situation	Line graph	Histogram	Pictogram	Pie-chart
Number of passengers carried by an airline each year in a six-year period				
Variation of temperature in a room over a 24-hour period				
Monthly production of TV sets by a consumer electronics manufacturer				
Average starting salary for a newly employed graduate over a ten-year period				
Market share of five companies that produce cars				

Copy the table into a word-processing package and then print out your work.

Activity 2.18

The following data relates to a 12-month operating period for a small engineering company:

Total income:	£7 200 000
Material costs:	£1 450 000
Labour costs:	£2 900 000
Overhead costs:	£1 150 000

Construct a 3-D pie-chart (using an appropriate spreadsheet package) to show the relationship between the costs and the profit.

Activity 2.19

The following data was obtained during a number of observations of the working life of an optical projection unit when used in different ambient temperatures. Plot this data using a scatter diagram. What can you infer from these results?

Ambient temperature (°C)	Time to failure (hours)
15	2505
25	1950
20	2807
32	850
27	1365
19	2219
35	606
37	722
18	2605
12	3503
22	3083
21	2430
29	1971
22	2004
24	2696
20	3431
18	3122
16	3909

Use a spreadsheet to process the data and to create an X–Y scatter diagram. Then insert the scatter diagram into a word-processed document with your comments.

Design sketches

Engineers and designers frequently use quick hand-drawn sketches to illustrate design concepts as well as features associated with a particular product or process. In fact, being able to use sketches to illustrate your ideas is an essential part of becoming an engineer! Sketches are also an excellent way of communicating your ideas to other engineers and designers without having to resort to words or formal engineering drawings. When producing a design sketch there are no 'hard-and-fast' rules other than ensuring that the sketch is clear and unambiguous (i.e. not liable to be misinterpreted) and that it is adequately labelled.

Key point

Design sketches are initial design ideas, requirements, calculations and concepts. Sketches are often used to convey design parameters to other designers who are responsible for producing more detailed drawings.

Activity 2.20

Produce a design sketch for an entry-level digital camera. The camera is to incorporate an in-built flash unit and an LCD screen. The battery is to be powered by two AA batteries and is to have a USB connection for downloading images to a computer. Make sure that all of these features are identified in your sketch. Present your work in the form of hand-drawn sketches and hand-written notes.

Technical drawings

Like the graphs that we have just considered, there are many different ways of representing and communicating technical information. To avoid confusion, such information should make use of nationally and internationally recognized symbols, conventions and abbreviations. These are listed and their use explained in the appropriate British Standards. Such standards are lengthy and costly but a summary is available for educational use. This document is entitled 'Engineering drawing practice for schools and colleges' and it has the British Standards reference number PP 8888. The document is abridged from the earlier British Standard BS 308.

Block diagrams

These show the relationship between the various elements of a system. Figure 2.24 shows the block diagram for a simple radio receiver. This sort of diagram is used in the initial stages of conceptualizing a design or to provide an overview of the way in which an engineering system operates.

Activity 2.21

Construct a block diagram showing the arrangement of the braking system of a car. Label your drawing clearly.
Present your results in the form of a printout from a simple drawing or technical illustration package.

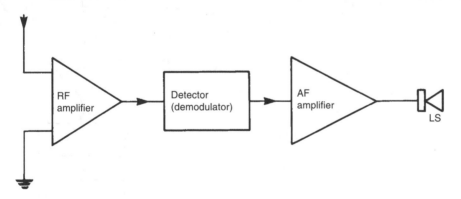

Figure 2.24 *A block diagram*

Flow diagrams

Flow diagrams are used to illustrate a sequence of events. They are used in a wide variety of applications including the planning of engineering processes and the design of computer software. Figure 2.25 shows a flowchart for the process of drilling a hole. The shape of the symbols used in this flowchart have particular meanings as shown in Figure 2.26. For the complete set of symbols and their meanings you should refer to the appropriate British Standard BS 4058.

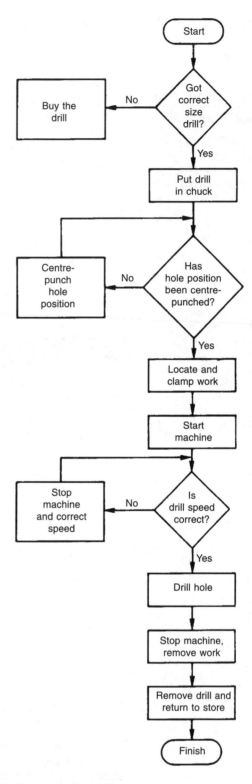

Figure 2.25 *Flow chart for drilling a hole*

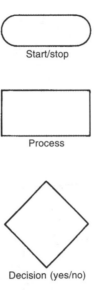

Figure 2.26 *Some common flow chart symbols*

Your bicycle tyre is flat and may have a puncture or may simply need reinflating. Draw a flow chart for checking the tyre and, if necessary, replacing the wheel. Figure 2.27 will provide you with a starting point. Present your results in the form of a printout from a simple drawing or technical illustration package.

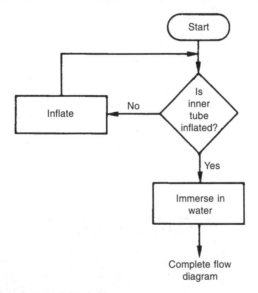

Figure 2.27 *See Activity 2.22*

Circuit and related diagrams

Circuit diagrams are used to show the functional relationships between the components in an electric or electronic circuit. The components are represented by symbols and the electrical connections between the components drawn using straight lines. It is important to note that the position of a component in a circuit diagram does not represent its actual physical position in the final assembly. Circuit diagrams are sometimes also referred to as schematic diagrams or schematic circuits.

Figure 2.28(a) shows the circuit for an electronic filter unit using standard component symbols. Figure 2.28(b) shows the

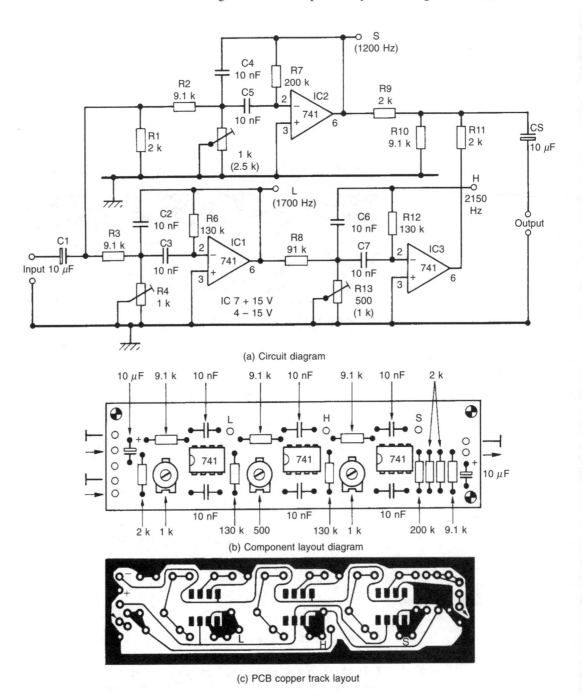

(a) Circuit diagram

(b) Component layout diagram

(c) PCB copper track layout

Figure 2.28 *A typical electronic circuit diagram with corresponding layout diagram and PCB copper track layout*

corresponding physical layout diagram with the components positioned on the upper (component side) of a printed circuit board (PCB). Finally, Figure 2.28(c) shows the copper track layout for the PCB. This layout is developed photographically as an etch-resistant pattern on the copper surface of a copper-clad board.

The term 'wiring diagram' is usually taken to refer to a diagram that shows the physical interconnections between electrical and electronic components. Typical applications for wiring diagrams include the wiring layout of control desks, control cubicles and power supplies. Wiring diagrams are directly related to circuit schematics (circuit diagrams). As an example, architects use circuit schematics to show the electrical wiring and components inside a building or plant. They will also provide installation drawings to show where the components are to be sited. In addition, they may also provide a wiring diagram to show how the wires and cables are to be routed to and between the components. The symbols used in architectural installation drawings and wiring diagrams are not the same as those used in circuit diagrams.

Schematic circuit diagrams are also used to represent pneumatic (compressed air) circuits and hydraulic circuits. Pneumatic circuits and hydraulic circuits share the same symbols. You can tell which circuit is which because pneumatic circuits should have open arrow heads, while hydraulic circuits should have solid arrowheads. Also, pneumatic circuits exhaust to the atmosphere, while hydraulic circuits have to have a return path to the oil reservoir. Figure 2.29 shows a typical hydraulic circuit.

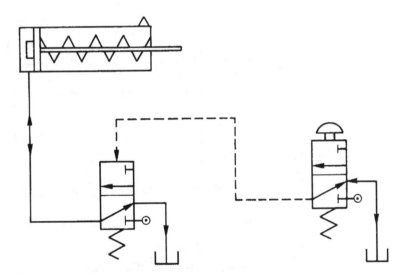

Figure 2.29 *A typical hydraulic circuit*

Just as electrical circuit diagrams may have corresponding installation and wiring diagrams, so do hydraulic, pneumatic and plumbing circuits. Only this time the wiring diagram becomes a pipework diagram. A plumbing example is shown in Figure 2.30. As you may not be familiar with the symbols, we have named them for you. Normally this is not necessary and the symbols are recognized by their shapes.

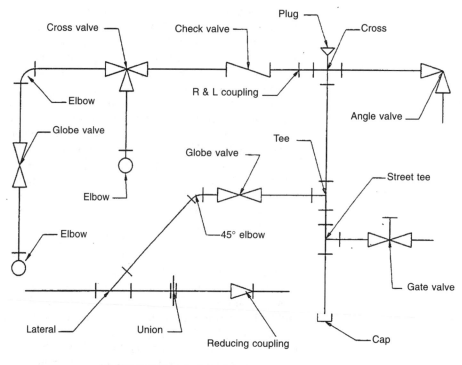

(a) Circuit diagram (schematic)

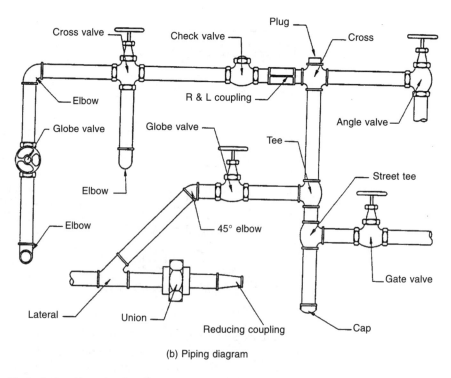

(b) Piping diagram

Figure 2.30 *Typical plumbing circuit with corresponding piping diagram*

General arrangement drawings

Figure 2.31 shows the layout of a typical drawing sheet. To save time these are printed to a standardized layout for a particular company, ready for the draftsperson to add the drawing and complete the boxes and tables.

The basic information found on most drawing sheets consists of:

- drawing number and name of the company
- title and issue details
- scale
- method of projection (first or third angle)
- initials of persons responsible for: drawing, checking, approving, tracing, etc., together with the appropriate dates
- unit(s) of measurement (inches or millimetres) and general tolerances
- material and finish
- copyright and standards reference
- guidance notes such as 'do not scale'
- reference grids so that 'zones' on the drawing sheet can be quickly found
- modifications table for alterations which are related to the issue number on the drawing and identified by the means of the reference grid.

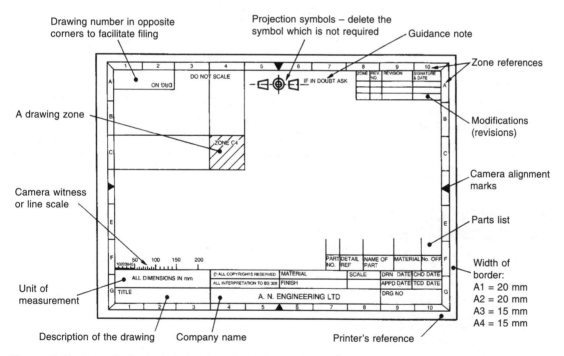

Figure 2.31 *Layout of a typical drawing sheet*

The following additional information may also be included:

- fold marks
- centre marks for camera alignment when microfilming
- line scale, so that the true size is not lost when enlarging or reducing copies
- trim marks
- orientation marks.

Figure 2.32 shows a typical general arrangement (GA) drawing. This shows as many of the features listed above as are appropriate for this drawing. It shows all the components correctly assembled together. Dimensions are not usually given on GA drawings although, sometimes, overall dimensions will be given for reference when the GA drawing is of a large assembly drawn to a reduced scale.

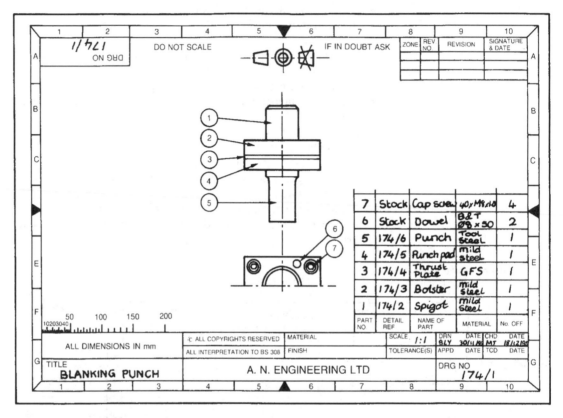

Figure 2.32 *A typical general arrangement (GA) drawing*

The GA drawing shows all the parts used in an assembly. These are listed in a table together with the quantities required. Manufacturers' catalogue references are also given for any components that are not being manufactured. The parts are usually 'bought-in' as 'off-the-shelf' parts from other suppliers. The detail drawing numbers are also included for components that have to be manufactured as special items.

Detail drawings

As the name implies, detail drawings provide all the details required to make the component shown on the drawing. Referring back to Figure 2.32 we can see from the table that the detail drawing for the punch has the reference number 174/6. Figure 2.33 shows this detail drawing. In this instance, the drawing provides the following information:

• the shape of the punch

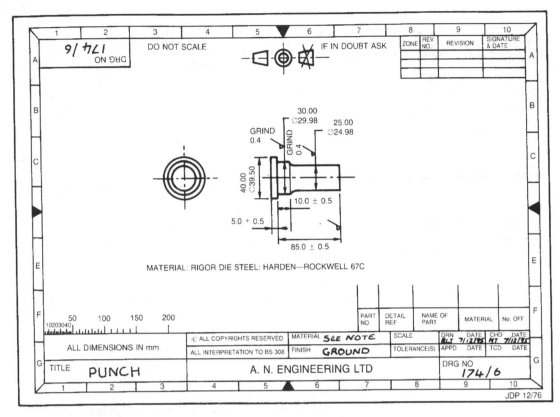

Figure 2.33 *A typical detail drawing*

- the dimensions of the punch and the manufacturing tolerances
- the material from which the punch is to be made and its subsequent heat treatment
- the unit of measurement (millimetre)
- the projection (first angle)
- the finish
- the guidance note 'Do not scale drawing'
- the name of the company
- the name of the draftsperson
- the name of the person checking the drawing.

It should go without saying that the amount of information given will depend upon the nature of the job. Drawings for a critical aircraft component, for example, will be much more fully detailed than those for a wheelbarrow component!

Production and storage of technical drawings

Computer aided design and manufacture (CAD/CAM) has now largely replaced manual methods used for engineering drawing and data storage. CAD software is used in conjunction with a computer and the drawing produced on the computer screen is saved in a computer file on disk. Networked CAD/CAM and computer aided engineering (CAE) systems have made it possible

Key point

Computer aided design (CAD) refers to the use of a computer to design a part and to produce technical drawings. Two-dimensional (2-D) CAD is confined to the layout and graphic representation of parts using traditional standard industry conventions. Drawings are representations of the product plotted on paper. Three-dimensional (3-D) CAD is normally the starting point for a product design.

to share data and drawings over a network. This allows many people to have access to the same data (usually stored on a powerful network server). This is now the most efficient and economical method of data storage. Very many complex drawings can be saved onto a hard disk or optical drive on a single file server. When required, the file can be recalled for immediate viewing on a computer screen and hard copy can be printed out to any desired scale at the touch of a key.

Despite the fact that computerized methods of data storage are now commonplace, it is important to know something about the older, traditional methods of storage. These are described briefly below:

- *Tracing linen* This was the traditional material making technical drawings. It was strong and durable and stood up well to the effects of the ultraviolet arc lamps used for making 'blue-print' copies. These got their name from the fact that the print appeared as white lines on a blue background.
- *Tracing paper* This is widely used in conjunction with manual drawing techniques. It is cheap and readily available. It is also easy to draw on. Unfortunately the paper becomes brittle with age and requires careful handling. Therefore it is not suitable where print copies have to be made frequently.
- *Tracing film* This is tough plastic film that is shiny on one side and matt on the other. You draw on the matt surface. No special techniques are required in its use and it stands up to repeated handling without deterioration. It is more expensive than tracing paper.
- *Microfilm* The storage of full size 'negatives', as the tracings are called, when produced on linen, paper or film takes up a lot of room. These large drawings can be reduced photographically onto 16 mm or 35 mm film stock for storage. This saves considerable space. The microfilm copies can be projection printed (enlarged) full size when required for issue.
- *Microfiche* Libraries, offices and stores use microfiche systems. Data is stored photographically in a grid of frames on a large rectangle of film. A desktop viewer is used to select and enlarge a single frame. The frame is then projected onto a rear projection screen for easy reading. This system is more likely to be used for storing literal and numerical data than for drawings.

Activity 2.23

(a) Compare the printed drawing sheets provided by your tutor with the model sheet shown in Figure 2.31. List any differences.
(b) Examine the general arrangement (GA) drawings provided by your tutor and compare them with the model GA shown in Figure 2.32. List any differences.
(c) Examine the detail drawings provided by your tutor and compare them with the model detail drawing shown in Figure 2.33. List any differences.

Present your results in the form of a set of brief handwritten notes.

Engineering drawing techniques

Engineering drawings can be produced using a variety of different techniques. The choice of technique is dependent upon a number of factors such as:

Speed How much time can be allowed for producing the drawing. How soon the drawing can be commenced.

Media The choice will depend upon the equipment available (e.g. CAD or conventional drawing board and instruments) and the skill of the person producing the drawing.

Complexity The amount of detail required and the anticipated amount and frequency of development modifications.

Cost Engineering drawings are not works of art and have no intrinsic value. They are only a means to an end and should be produced as cheaply as possible. Both initial and ongoing costs must be considered.

Presentation This will depend upon who will see/use the drawings. Non-technical people can visualize pictorial representations better than orthographic drawings.

Nowadays technical drawings are increasingly produced using computer aided drawing (CAD) techniques. Developments in software and personal computers have reduced the cost of CAD and made it more powerful. At the same time, it has become more 'user friendly'. Computer aided drawing does not require the high physical skill required for manual drawing, which takes years of practise to achieve. It also has a number of other advantages over manual drawing. Let's consider some of these advantages:

Accuracy Dimensional control does not depend upon the draftsperson's eyesight.

Radii Radii can be made to blend with straight lines automatically.

Repetitive features For example, holes round a pitch circle do not have to be individually drawn but can be easily produced automatically by 'mirror imaging'. Again, some repeated, complex features need only be drawn once and saved as a matrix. They can then be called up from the computer memory at each point in the drawing where they appear at the touch of a key.

Editing Every time you erase and alter a manually produced drawing on tracing paper or plastic film the surface of the drawing is increasingly damaged. On a computer you can delete and redraw as often as you like with no ill effects.

Storage No matter how large and complex the drawing, it can be stored digitally on floppy disk. Copies can be taken and transmitted between factories without errors or deterioration.

Prints Hard copy can be produced accurately and easily on laser printers, flat bed or drum plotters and to any scale. Colour prints can also be made.

Key point

Modelling is use to describe the design stage of a 3-D model or part. A model can be created physically or by using a 3-D solid modelling package on a computer.

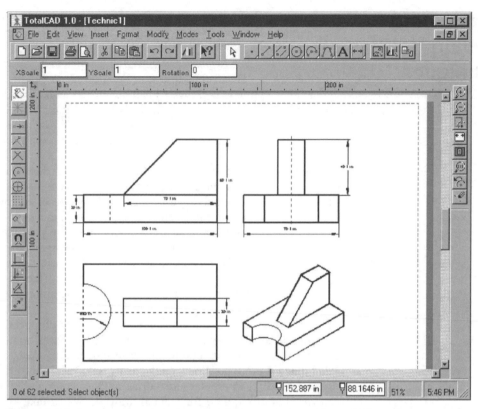

Figure 2.34 *A conventional engineering drawing produced by a CAD package*

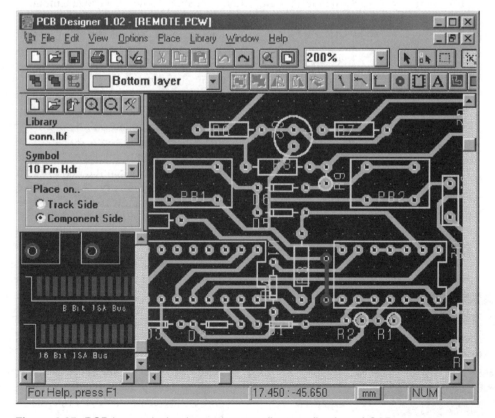

Figure 2.35 *PCB layout design is another excellent application of CAD*

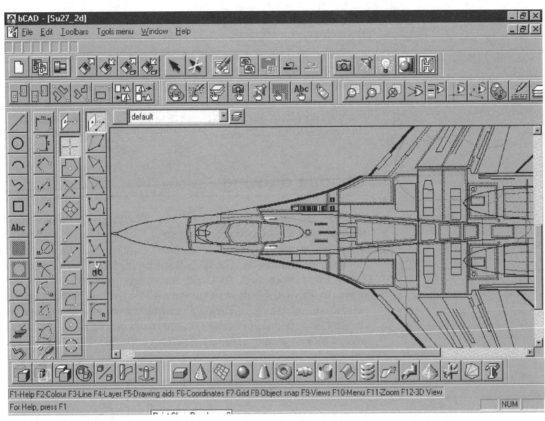

Figure 2.36 *Complex drawings can be produced prior to generating solid 3-D views*

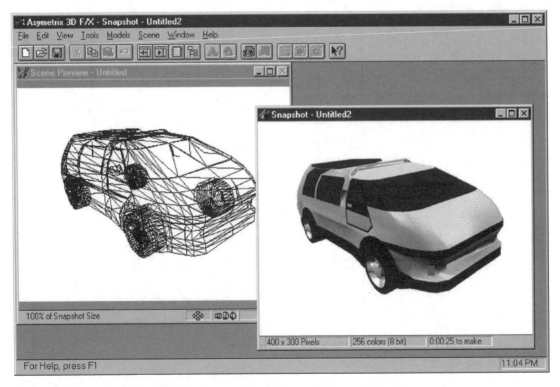

Figure 2.37 *A wire frame 3-D drawing and its corresponding rendered view*

Pictorial techniques

Engineering drawings such as general arrangement drawings and detail drawings are produced by a technique called orthographic drawing using the conventions set out in BS 308. Since we will be asking you to make orthographic drawings from more easily recognized pictorial drawings, we will start by introducing you to the two pictorial techniques widely used by draftspersons.

Oblique drawing

Figure 2.38 shows a simple oblique drawing. The front view (elevation) is drawn true shape and size. Therefore this view should be chosen so as to include any circles or arcs so that these can be drawn with compasses. The lines forming the side views appear to travel away from you, so these are called 'receders'. They are drawn at 45° to the horizontal using a 45° set-square. They may be drawn full length as in cavalier oblique drawing or they may be drawn half-length as in cabinet oblique drawing. This latter method gives a more realistic representation, and is the one we will be using.

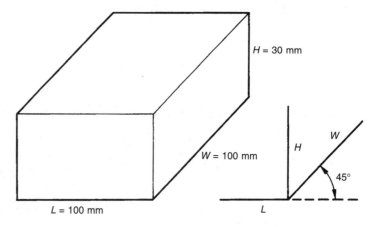

(a) Cavalier oblique projection

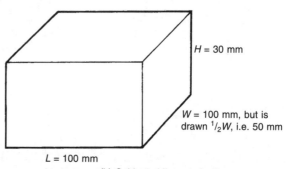

(b) Cabinet oblique projection

Figure 2.38 *A simple oblique drawing*

Isometric drawing

Figure 2.39(a) shows an isometric drawing of our previous box. To be strictly accurate, the vertical lines should be drawn true length and the receders should be drawn to a special isometric scale. However, this sort of accuracy is rarely required and, for all practical purposes, we draw all the lines full size. As you can see, the receders are drawn at 30° to the horizontal for both the elevation and the end view.

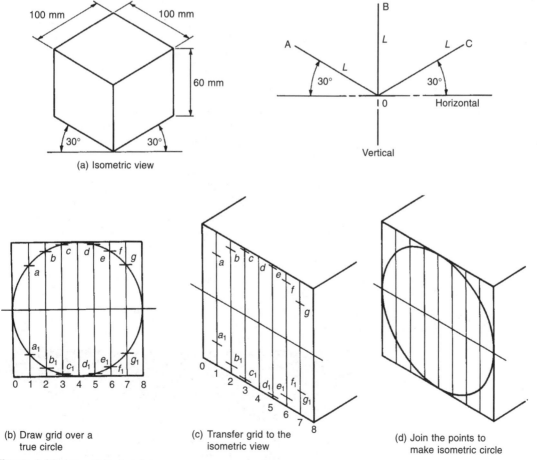

(a) Isometric view

(b) Draw grid over a true circle

(c) Transfer grid to the isometric view

(d) Join the points to make isometric circle

Figure 2.39 *Isometric drawing*

Although an isometric drawing is more pleasing to the eye, it has the disadvantage that all circles and arcs have to be constructed. They cannot be drawn with compasses. Figures 2.39(b), (c) and (d) show you how to construct an isometric curve. You could have used this technique in Activity 2.24 to draw the circle on the side of the box drawn in oblique projection.

First we draw the required circle. Then we draw a grid over it as shown in Figure 2.39(b). Next number or letter the points where the circle cuts the grid as shown. Now draw the grid on the side elevation of the box and step off the points where the circle cuts the grid with your compasses as shown in Figure 2.39(c). All that remains is to join up the dots and you have an isometric circle as shown in Figure 2.39(d).

Activity 2.25

(a) Draw, full size, an isometric view of the box shown in Figure 2.39. Isometric ruled paper will be of great assistance if you can obtain some.
(b) Draw a 50 mm diameter isometric circle on the TOP face of the box (remember that Figure 2.39 shows it on the side of the box).

Present your results in the form of a hand-constructed drawing with hand-written notes.

Another way of drawing isometric circles and curves is the 'four-arcs' method. This does not produce true curves but they are near enough for all practical purposes and quicker and easier than the previous method for constructing true curves. The steps are shown in Figure 2.40.

1. Join points B and E as shown in Figure 2.40(b). The line BE cuts the line GC at the point J. The point J is the centre of the first arc. With radius BJ set your compass to strike the first arc as shown.
2. Join the points A and F as shown in Figure 2.40(c). The line AF cuts the line GC at the point K. The point K is the centre of the second arc. With radius KF set your compasses to strike the second arc as shown. If your drawing is accurate both arcs should have the same radius.
3. With centre A and radius AF or AD strike the third arc as shown in Figure 2.40(d).
4. With centre E and radius EH or EB strike the fourth and final arc as shown in Figure 2.40(e).
5. If your drawing is accurate, arcs 3 and 4 should have the same radius.

Activity 2.26

Use the technique just described to draw a 40 mm diameter circle on the side face of our box. Start off by drawing a 40 mm isometric square in the middle of the side face. Present your results in the form of a hand-constructed drawing with hand-written notes.

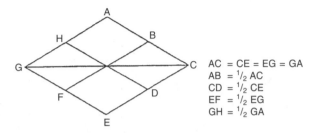

AC = CE = EG = GA
AB = ¹/₂ AC
CD = ¹/₂ CE
EF = ¹/₂ EG
GH = ¹/₂ GA

(a) Draw an isometric grid of appropriate size

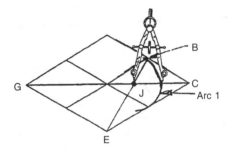

(b) Construct the 1st arc using a compass located as shown

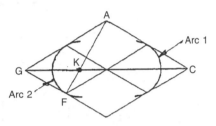

(c) Draw the 2nd arc using the construction process shown

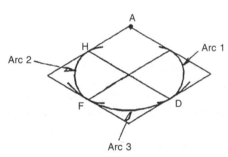

(d) Draw the 3rd arc through the points shown

(e) Complete the process drawing the 4th arc from the opposite corner

Figure 2.40 *The 'four arcs' method*

Activity 2.27

(a) Figure 2.41(a) shows some further examples of isometric drawings. Redraw them as cabinet oblique drawings.

(b) Figure 2.41(b) shows some further examples of cabinet oblique drawings. Redraw them as isometric drawings. Any circles and arcs on the vertical surfaces should be drawn using the grid construction method. Any arcs and circles on the horizontal (plan) surfaces should be drawn using the 'four-arcs method'.

Present your results in the form of a hand-constructed drawing with hand-written notes.

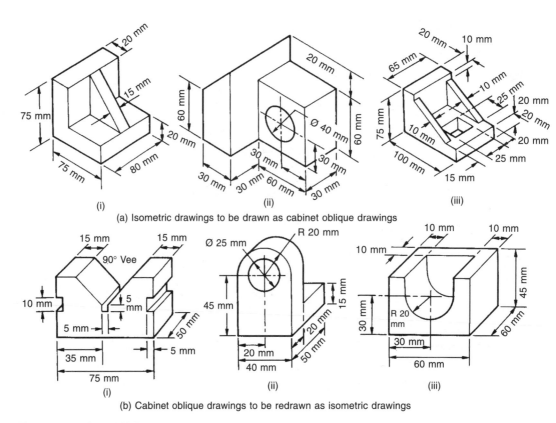

(a) Isometric drawings to be drawn as cabinet oblique drawings

(b) Cabinet oblique drawings to be redrawn as isometric drawings

Figure 2.41 *See Activity 2.27*

Orthographic drawing

GA and detail drawings are produced by the use of a drawing technique called orthographic projection. This is used to represent three-dimensional solids on the two-dimensional surface of a sheet of drawing paper so that all the dimensions are true length and all the surfaces are true shape. To achieve this when surfaces are inclined to the vertical or the horizontal we have to use auxiliary views, but more about these later. Let's keep things simple for the moment.

First angle projection

Figure 2.42(a) shows a simple component drawn in isometric projection. Figure 2.42(b) shows the same component as an orthographic drawing. This time we make no attempt to represent the component pictorially. Each view of each face is drawn separately either full size or to the same scale. What is important is how we position the various views as this determines how we 'read' the drawing.

Engineers use two orthographic drawing techniques, either first angle or third angle projection. The former is called 'English projection' and the latter is called 'American projection'. The drawing in Figure 2.42 is in first angle projection. The views are arranged as follows.

Elevation This is the main view from which all the other views are positioned. You look directly at the side of the component and draw what you see.

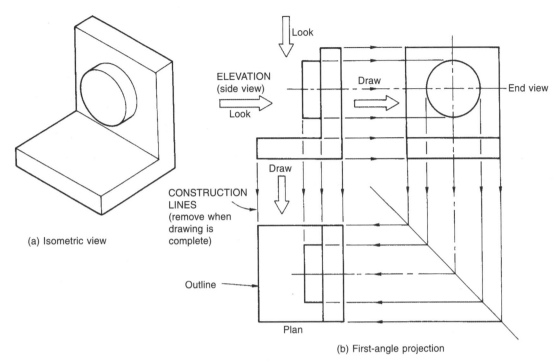

Figure 2.42 *An isometric view and its corresponding first angle projection*

Plan   To draw this, you look directly down on the top of the component and draw what you see below the elevation.

End view   This is sometimes called an 'end elevation'. To draw this you look directly at the end of the component and draw what you see at the opposite end of the elevation. There may be two end views, one at each end of the elevation, or there may be only one end view if this is all that is required to completely depict the component. Figure 2.42 requires only one end view. When there is only one end view this can be placed at either end of the elevation depending upon which position gives the greater clarity and ease of interpretation. Whichever end is chosen the rules for drawing this view must be obeyed.

Use feint construction lines to produce the drawing as shown in Figure 2.42(b). When these are complete, 'line-in' the outline more heavily. Carefully remove the construction lines to leave the drawing uncluttered, thus improving the clarity. Figure 2.43 shows the finished drawing.

Activity 2.28

Figure 2.43 showed some components using pictorial projections. We now want you to redraw these components in first-angle orthographic projection. To start you off we have drawn the first one for you. This is shown in Figure 2.44. Note how we have positioned the end view this time so that you can see the web. Present your results in the form of a hand-constructed drawing with hand-written notes.

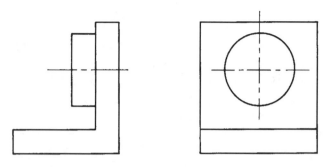

First angle projection

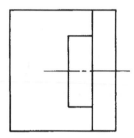

Figure 2.43 *Completed first angle projection*

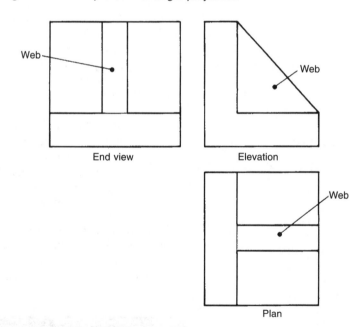

Figure 2.44 *See Activity 2.28*

Third angle projection

Figure 2.45 shows the same component, but this time we have drawn it in third angle projection for you.

Elevation Again we have started with the elevation or side view of the component and, as you can see, there is no difference.

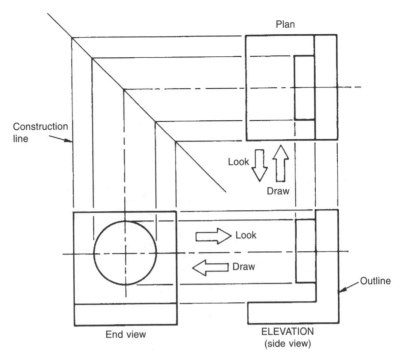

Figure 2.45 *Third angle projection*

Plan   Again we look down on top of the component to see what the plan view looks like. However, this time we draw the plan view above the elevation. That is, in third angle projection we draw all the views from where we look.

End view   Note how the position of the end view is reversed compared with first angle projection. This is because, like the plan view, we draw the end views at the same end from which we look at the component.

Again use feint construction lines to produce the drawing as shown in Figure 2.45. Then 'line-in' the outline more heavily and carefully remove the construction lines for clarity, unless you have been instructed otherwise. Figure 2.46 shows the finished drawing in third angle projection.

Test your knowledge 2.20

(a)   Figure 2.48 shows some components drawn in first angle projection and some in third angle projection. We have not necessarily drawn all the views each time. Instead, we have only drawn as many of the views as are needed. State which is first angle and which is third angle.

(b)   Two of the drawings are standard symbols for indicating whether drawing is in first angle or whether it is in third angle. Which drawings do you think are these symbols?

Activity 2.29

Figure 2.41 showed some components using pictorial projections. We now want you to redraw these components in third angle orthographic projection. To start you off, we have again drawn the first one for you. This is shown in Figure 2.47. Once again, note how we have positioned the end view so that you can see the web. Present your results in the form of a hand-constructed drawing with hand-written notes.

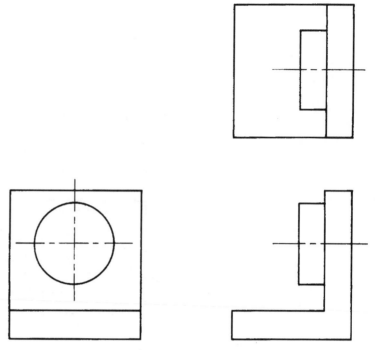

Figure 2.46 *Completed third angle projection*

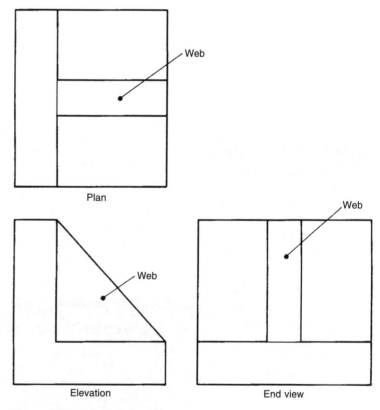

Plan

Elevation

End view

Figure 2.47 *See Activity 2.29*

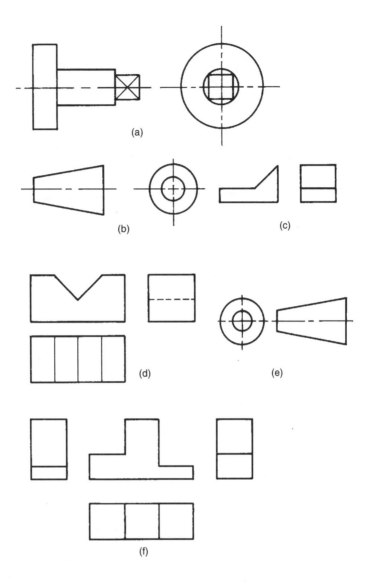

Figure 2.48 *See Test your knowledge 2.20*

Activity 2.30

Figure 2.49 shows pictorial views of some more solid objects.

(a) Use a 2-D CAD package to draw these objects in first angle orthographic projection and label the views.
(b) Use a 2-D CAD package to draw these objects in third angle orthographic projection and label the views.

Present your results in the form of printed drawings using appropriate drawing sheets (these will usually be supplied in the form of a template that is approved for use in your school or college).

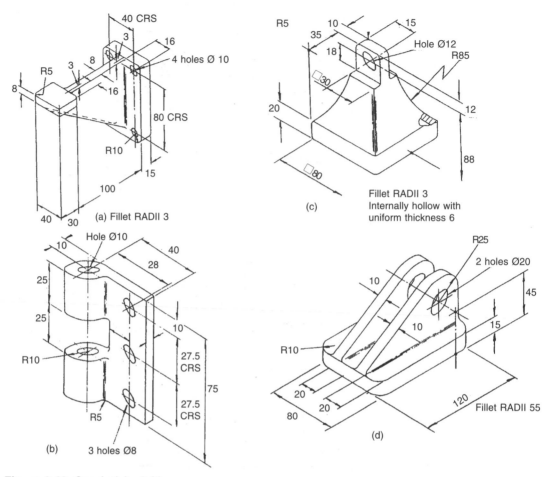

Figure 2.49 *See Activity 2.30*

Auxiliary views

In addition to the main views on which we have just been working, we sometimes have to use auxiliary views. We use auxiliary views when we cannot show the true outline of the component or a feature of the component in one of the main views. For example, when a surface of the component is inclined as shown in Figure 2.50.

Activity 2.31

Use a CAD package to draw the component shown in Figure 2.51 in isometric projection given that each square has a side length of 10 mm.
Also draw the component in:

(a) first angle orthographic projection (only two views are required)
(b) third angle orthographic projection (only two views are required)
(c) cabinet oblique projection.

Present your work in the form of a portfolio of printed drawings. Clearly mark each drawing with the projection used.

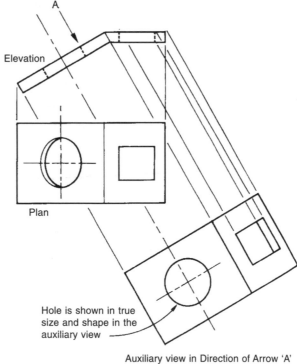

Figure 2.50 *An auxiliary view*

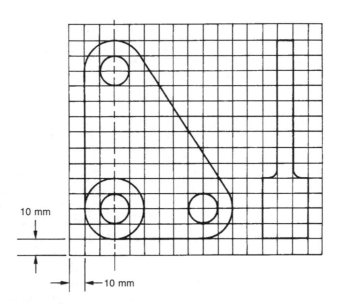

Figure 2.51 *See Activity 2.31*

Production of engineering drawings

Standard conventions are used in order to avoid having to draw, in detail, common features in frequent use. Figure 2.52 shows a typical dimensioned engineering drawing. Some conventions can help us save a great deal of time and effort. For example, Figure 2.53(a) shows a pictorial representation of a screw thread while Figure 2.53(b) shows the standard convention for a screw thread. Clearly the latter is much quicker and easier to draw!

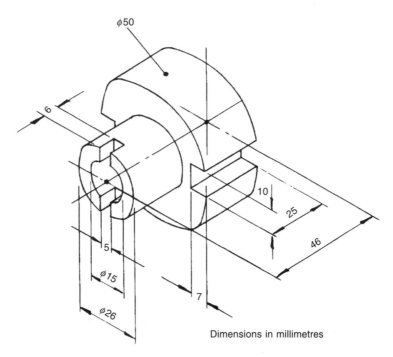

Dimensions in millimetres

Figure 2.52 *A dimensioned drawing*

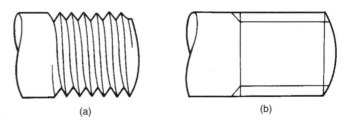

(a)   (b)

Figure 2.53 *Screw threads*

All engineering drawings should be produced using appropriate drawing standards and conventions for the following reasons.

Time It speeds up the drawing process by making life easier for the draftsman as indicated above. This reduces costs and also reduces the 'lead time' required to get a new product into production.

Appearance It makes your drawings look more professional and improves the 'image' of yourself and your company. Badly presented drawings can send out the wrong messages and can call your competence into question.

Portability Drawings produced to international standards and conventions can be read and correctly interpreted by any trained engineer, anywhere in this country or abroad. This avoids misunderstandings that could lead to expensive and complex components and assemblies being scrapped and dangerous situations arising. The only difficulties will arise from written notes that are language dependent.

Drawing conventions used by engineers in the UK are specified in BS 308. This is produced in three parts:

- Part 1   General principles.
- Part 2   Dimensioning and tolerancing of size.
- Part 3   Geometrical tolerancing.

These are all 'harmonized' with their appropriate ISO (International Standards Organization) counterparts.

As has been stated earlier, you will need to locate a copy of the British Standards Institution's publication PP 8888: Engineering Drawing Practice for Schools and Colleges. Also useful is PP 7307: Graphical Symbols for use in Schools and Colleges. This latter standard contains symbols for use in electrical, electronic, pneumatic and hydraulic schematic circuit diagrams. It also contains many other useful symbols.

Other British Standards of importance to engineering draftspersons are:

BS 4500   ISO Limits and Fits (these are used by mechanical and production engineers).

BS 3939   Graphical symbols for electrical power, telecommunications and electronics diagrams.

BS 2197   Specifications for graphical symbols used in diagrams for fluid power systems and components.

Planning the drawing

Before we start the drawing and lay pencil to paper we should plan what we are going to do. This saves having to alter the drawing later on. We have to decide whether the drawing is to be pictorial, orthographic or schematic. If orthographic we have to decide on the projection we are going to use. We also have to decide whether we need a formal drawing or whether a freehand sketch is all that is required. If a formal drawing is needed then we have to decide whether to use manual techniques or CAD.

Paper size

When you start to plan your drawing you have to decide on the paper size. Engineering drawings are usually produced on 'A' size paper. Paper size A0 is approximately one square metre in area and is the basis of the system. Size A1 is half the area of size A0, size A2 is half the area of size A1 and so on down to size A4. Smaller sizes are available but they are not used for drawing. All the 'A' size sheets have their sides in the ratio of 1:12. This gives the following paper sizes:

A0   841 mm × 1189 mm
A1   594 mm × 841 mm
A2   420 mm × 549 mm
A3   297 mm × 420 mm
A4   210 mm × 297 mm

These relationships are shown diagrammatically in Figure 2.54.

The paper size you choose will depend upon the size of the drawing and the number of views required. Be generous, nothing

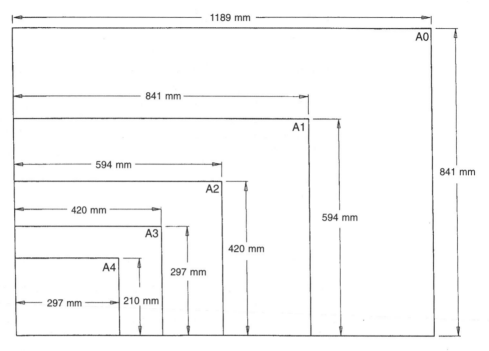

Figure 2.54 *Paper sizes*

looks worse than a cramped drawing and overcrowded dimensions. It is also false economy since overcrowding invariably leads to reading errors. As you will already have seen from some of the previous examples, the drawing should always have a border and a title block. This restricts the blank area available to draw on. Figures 2.55 and 2.56 show how the views should be positioned. These

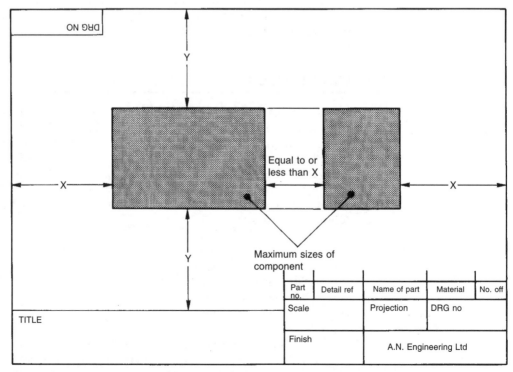

Figure 2.55 *Positioning the drawing*

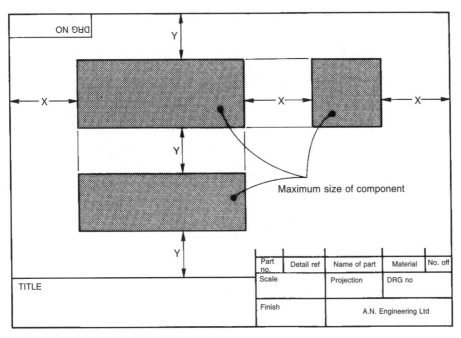

Figure 2.56 *Positioning the drawing*

layouts are only a guide but they offer a good starting point until you gain more experience. If only one view is required then it is centred in the drawing space available.

Title block

A typical title block was shown in Figure 2.32. If you refer back to this figure you will see that it is expandable vertically and horizontally to accommodate any written information that is required. The title block should contain:

- The drawing number (which should be repeated in the top left-hand corner of the drawing).
- The drawing name (title).
- The drawing scale.
- The projection used (standard symbol).
- The name and signature of the draftsperson together with the date on which the drawing was signed.
- The name and signature of the person who checks and/or approves the drawing, together with the date of signing.
- The issue number and its release date.
- Any other information as dictated by company policy.

Scale

The scale should be stated on the drawing as a ratio. The recommended scales are as follows:

- Full size = 1:1

• Reduced scales (smaller than full size) are:

1:2	1:5	1:10
1:20	1:50	1:100
1:200	1:500	1:1000

(NEVER use the words full size, half size, quarter size, etc.)

• Enlarged scales (larger than full size) are:

2:1	5:1	10:1
20:1	50:1	100:1

Lines and linework

The lines of a drawing should be uniformly black, dense and bold. On any one drawing they should all be in pencil or in black ink. Pencil is quicker to use but ink prints more clearly. Lines should be thick or thin as recommended below. Thick lines should be twice as thick as thin lines. Figure 2.57 shows the type of lines

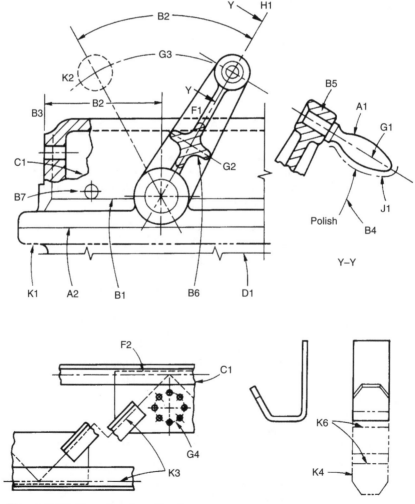

Figure 2.57 *Use of various line types*

Table 2.1 *Types of line (see Figure 2.57)*

A	**Continuous thick**
A1	Visible outline
A2	Visible edge
B	**Continuous thin**
B1	Imaginary line of intersection
B2	Dimension line
B3	Projection line
B4	Leader line
B5	Hatching
B6	Outline of a revolved section
B7	Short centre line
C	**Continuous thin irregular**
C1	Limit of partial view
D	**Continuous thin zig-zag**
D1	Limit of partial view
E	**Dashed thick**
E1	Hidden outline
E2	Hidden edge
F	**Dashed thin**
F1	Hidden outline
F2	Hidden edge
G	**Chain thin**
G1	Centre line
G2	Line of symmetry
G3	Trajectory and locus line
G4	Pitch line and pitch circle
H	**Chain thin, thick at ends**
H1	Cutting plane
J	**Chain thick**
J1	Indicates special requirement
K	**Chain thin double dashed**
K1	Outline of adjacent part
K2	Outline of extreme position
K3	Centroid line
K4	Initial outline
K5	Part in front of cutting plane
K6	Bend line

---

recommended in BS 308 for use in engineering drawing and how the lines should be used. This is reinforced by Table 2.1.

Sometimes the lines overlap in different views. When this happens, as shown in Figure 2.58, the following order of priority should be observed.

- Visible outlines and edges (type A) take priority over all other lines.
- Next in importance are hidden outlines and edges (type E).
- Then cutting planes (type G).
- Next come centre lines (types F and B).
- Outlines and edges of adjacent parts, etc. (type H).
- Finally, projection lines and shading lines (type B).

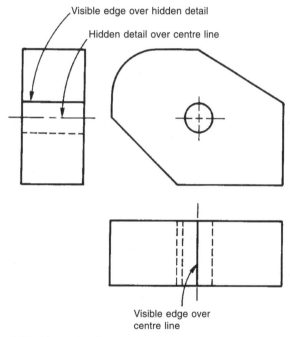

Figure 2.58 *Line priorities*

Leader lines

Leader lines, as their name implies, lead written information or dimensions to the points where they apply. Leader lines are thin lines (type B) and they end in an arrowhead or in a dot as shown in Figure 2.60(a). Arrowheads touch and stop on a line, while dots should always be used within an outline.

- When an arrowed leader line is applied to an arc it should be in line with the centre of the arc as shown in Figure 2.60(b).
- When an arrowed leader line is applied to a flat surface, it should be nearly normal to the lines representing that surface as shown in Figure 2.60(c).
- Long and intersecting leader lines should not be used, even if this means repeating dimensions and/or notes as shown in Figure 2.60(d).
- Leader lines must not pass through the points where other lines intersect.

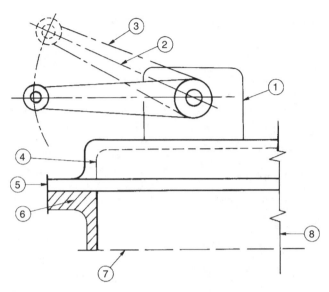

Figure 2.59 *See Test your knowledge 2.21*

Test your knowledge 2.22

Figure 2.61 shows some applications of leader lines with arrowheads and leader lines with dots. List the numbers and state whether the application is correct or incorrect. If incorrect explain (with sketches if required) how the application should be corrected.

- Arrowheads should be triangular with their length some three times larger than the maximum width. They should be formed from straight lines and the arrowheads should be filled in. The arrowhead should be symmetrical about the leader line, dimension line or stem. It is recommended that arrowheads on dimension and leader lines should be some 3 mm to 5 mm long.
- Arrowheads showing direction of movement or direction of viewing should be some 7 mm to 10 mm long. The stem should be the same length as the arrowhead or slightly greater. It must never be shorter.

Letters and numerals

Style	The style should be clear and free from embellishments. In general, capital letters should be used. A suitable style could be:

ABCDEFGHIJKLMNOPQRSTUVWXYZ
1 2 3 4 5 6 7 8 9 0

Size	The characters used for dimensions and notes on drawings should not be less than 3 mm tall. Title and drawing numbers should be at least twice as big.
Direction of lettering	Notes and captions should be positioned so that they can be read in the same direction as the information in the title block. Dimensions have special rules and will be dealt with later.
Location of notes	General notes should all be grouped together and not scattered about the drawing. Notes relating to a specific feature should be placed adjacent to that feature.

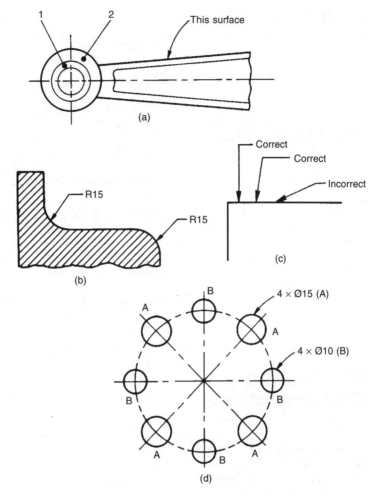

Figure 2.60 *Examples of the use of leader lines*

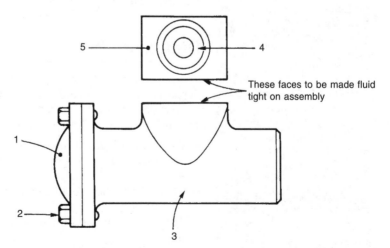

Figure 2.61 *See Test your knowledge 2.22*

Emphasis

Characters, words and/or notes should not be emphasized by underlining. Where emphasis is required the characters should be enlarged.

Test your knowledge 2.23

With reference to appropriate standards, complete Table 2.2.

Test your knowledge 2.24

With reference to appropriate standards, complete Figure 2.62. You must take care to use the same types of line as shown in the standard or the conventions become meaningless. This applies particularly to line thickness.

Table 2.2 *Types of line (see Test your knowledge 2.23)*

Abbreviation or symbol	Term
AF	
ASSAY	
CHAM	
	Countersink
CYL	
	Diameter (in a note)
	Figure
	Hexagon
MATL	
	Maximum
	Minimum
	Radius (in a note)
	Radius (preceding a dimension)
REQD	
RH	
SPEC	
Ø	

Key point

A *dimension* is a numeric value expressed in appropriate units of measure and indicated on a drawing and in other documents along with lines, symbols and notes to define the size or geometric characteristic, or both, of a part or part feature.

A *reference dimension* (usually without tolerance) is used for information only. It is considered auxiliary information and does not govern production or inspection operations. A reference dimension repeats a dimension or size already given or derived from other values shown on the drawing or related drawing. Reference dimensions are enclosed in brackets, for example (23.50).

Tolerance is the total amount by which a specific dimension is permitted to vary. The tolerance is the difference between the maximum and minimum limits.

Symbols and abbreviations

If all the information on a drawing were written out in full, the drawing would become very cluttered. Therefore symbols and abbreviations are used to shorten written notes. Those recommended for use on engineering drawings are listed in BS 308, and in the corresponding student version PP 7308.

Conventions

These are a form of 'shorthand' used to speed up the drawing of common features in regular use. The full range of conventions and examples of their use can be found in appropriate standards so we will not waste space by listing them here. However, by completing the next exercise you will use some of the more common conventions and this will help you to become familiar with them.

Dimensioning

When a component is being dimensioned, the dimension lines and the projection lines should be thin full lines (type B). Where possible dimensions should be placed outside the outline of the object as shown in Figure 2.63(a). The rules are:

- Outline of object to be dimensioned in thick lines (type A).
- Dimension and projection lines should be half the thickness of the outline (type B).
- There should be a small gap between the projection line and the outline.
- The projection line should extend to just beyond the dimension line.
- Dimension lines end in an arrowhead that should touch the projection line to which it refers.
- All dimensions should be placed in such a way that they can be read from the bottom right-hand corner of the drawing.

The purpose of these rules is to allow the outline of the object to stand out prominently from all the other lines and to prevent confusion.

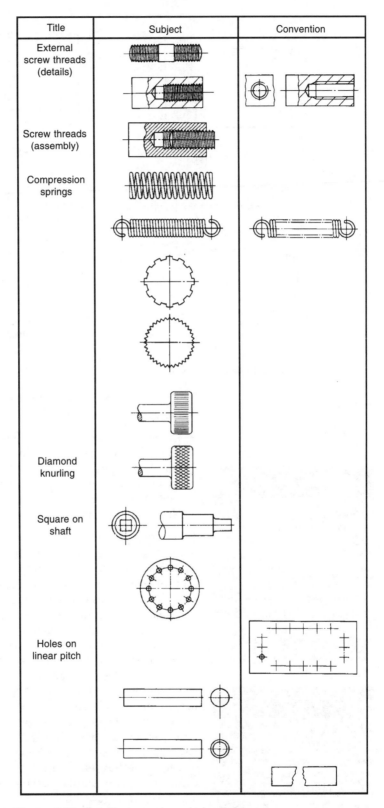

Title	Subject	Convention
External screw threads (details)		
Screw threads (assembly)		
Compression springs		
Diamond knurling		
Square on shaft		
Holes on linear pitch		

Figure 2.62 *See Test your knowledge 2.24*

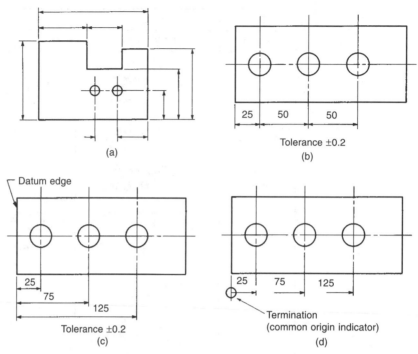

Figure 2.63 *Dimensioning*

There are three ways in which a component can be dimensioned. These are:

- Chain dimensioning as shown in Figure 2.63(b).
- Absolute dimensioning (dimensioning from a datum) using parallel dimension lines as shown in Figure 2.63(c).
- Absolute dimensioning (dimensioning from a datum using superimposed running dimensions as shown in Figure 2.63(d). Note the common origin (termination) symbol.

It is neither possible to manufacture an object to an exact size nor to measure an exact size. Therefore important dimensions have to be *toleranced*. That is, the dimension is given two sizes: an upper limit of size and a lower limit of size. Providing the component is made so that it lies between these limits it will function correctly. Information on limits and fits can be found in BS 4500.

The method of dimensioning can also affect the accuracy of a component and produce some unexpected effects. Figure 2.63(b) shows the effect of chain dimensioning on a series of holes or other features. The designer specifies a common tolerance of ±0.2 mm. However, since this tolerance is applied to each and every dimension, the cumulative tolerance becomes ±0.6 mm by the time you reach the final, right-hand hole. Not what was intended. Therefore absolute dimensioning as shown in Figure 2.63(c) and Figure 2.63(d) is to be preferred in this example. With absolute dimensioning, the position of each hole lies within a tolerance of ±0.2 mm and there is no cumulative error. Further examples of dimensioning techniques are shown in Figure 2.64.

It is sometimes necessary to indicate machining processes and surface finish. The machining symbol together with examples of process notes and the surface finishes in micro-metres (μm) is shown in Figure 2.65.

Key point

The *origin* or *datum point* is the name given to the point from where the location or geometric characteristics of a part are established. The correct identification of datums on a component and the related dimensioning can be vitally important in the manufacturing process (e.g. CNC machining).

Key point

A *feature* is a general term applied to a physical portion of a part, e.g. a surface, hole or slot.
A *datum feature* is a geometric feature of a part that is used to establish a datum. For example, a point, line, surface, hole, etc.

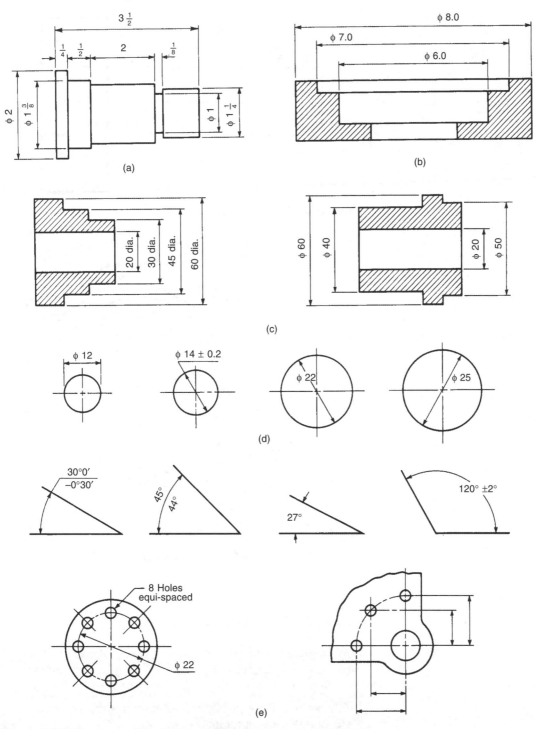

Figure 2.64 *More examples of dimensioning*

Activity 2.32

Figure 2.66 shows a component drawn in isometric projection. Use a CAD package to redraw it in first angle orthographic projection and add the dimensions using the following techniques:

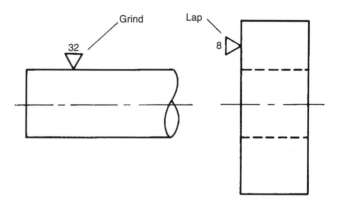

Figure 2.65 *Indicating surface finishes*

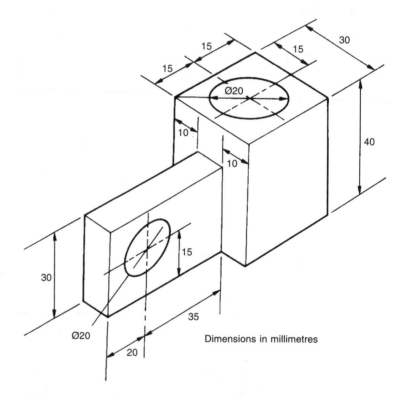

Figure 2.66 *See Activity 2.32*

(a) Absolute dimensioning using parallel dimension lines.
(b) Absolute dimensioning using superimposed running dimensions.

Present your work in the form of printed drawings using appropriate drawing sheets.

Sectioning

Sectioning is used to show the hidden detail inside hollow objects more clearly than can be achieved using dashed thin (type E) lines. Figure 2.67(a) shows an example of a simple sectioned drawing. The cutting plane is the line A–A. In your imagination you remove

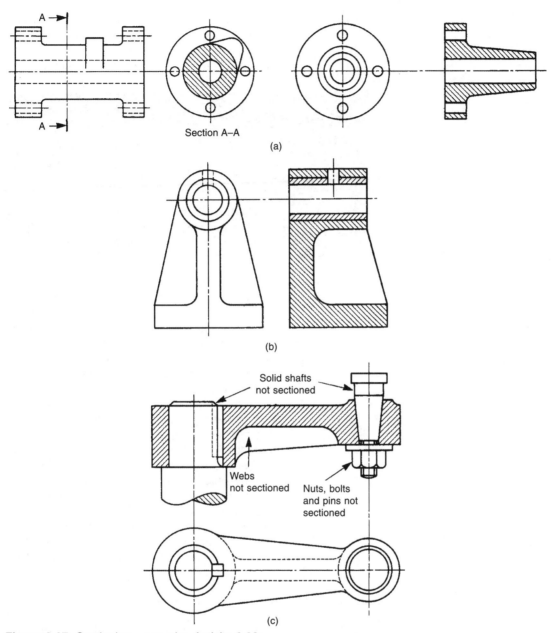

Figure 2.67 *Sectioning – see also Activity 2.33*

everything to the left of the cutting plane, so that you see only what remains to the right of the cutting plane looking in the direction of the arrowheads. Another example is shown in Figure 2.67(b).

Figure 2.67(c) shows how to section an assembly. Note how solid shafts and the key are not sectioned. Also note that thin webs that lie on the section plane are not sectioned. When interpreting sectioned drawings, some care is required. It is easy to confuse the terms sectional view and section.

Sectional view   In a sectional view you see the outline of the object at the cutting plane. You also see all the visible outlines seen beyond the cutting plane in the direction of viewing. Therefore Figure 2.67(a) is a sectional view.

Section A section shows only the outline of the object at the cutting plane. Visible outlines beyond the cutting plane in the direction of viewing are not shown. Therefore a section has no thickness.

Activity 2.33

(a) Use a CAD package to redraw Figure 2.67(a) as a section. (Remember that my drawing is a sectional view.)
(b) Explain why Figure 2.67(b) can be a section or a sectional view.

Present your work in the form of a printed drawing (using an appropriate drawing sheet) together with a brief hand-written explanation.

Cutting planes You have already been introduced to cutting planes in the previous examples. They consist of type G lines. That is a thin chain line that is thick at the ends and at changes of direction. The direction of viewing is shown by arrows with large heads. The points of the arrowheads touch the thick portion of the cutting plane. The cutting plane is labelled by placing a capital letter close to the stems of the arrows. The same letters are used to identify the corresponding section or sectional view.

Hatching You will have noticed that the shading of sections and sectional views consists of sloping, thin (type B) lines. This is called hatching. The lines are equally spaced, slope at 45° and are not usually less than 4 mm apart. However, when hatching very small areas the hatching can be reduced, but by never less than 1 mm. The drawings in this book may look as though they do not obey these rules. Remember that they have been reduced from much bigger drawings to fit onto the pages.

Figure 2.68 shows the basic rules of hatching. The hatching of separated areas is shown in Figure 2.68(a). Separate sectioned areas of the same component should be hatched in the same direction and with the same spacing.

Figure 2.68(b) shows how to hatch assembled parts. Where the different parts meet on assembly drawings, the direction of hatching should be reversed. The hatching lines should also be staggered. The spacing may also be changed.

Figure 2.68(c) shows how to hatch large areas. This saves time and avoids clutter. The hatching is limited to that part of the area that touches adjacent hatched parts or just to the outline of a large parts.

Figure 2.68(d) shows how sections through thin materials can be blocked in solid rather than hatched. There should be a gap of not less than 1 mm between adjacent parts even when these are a tight fit in practice.

Finally we have included some further examples of sectioning in Figure 2.69. These include assemblies, half-sections, part sections

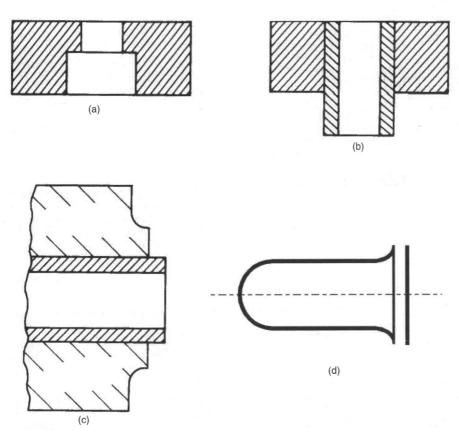

Figure 2.68 *Hatching*

and revolved sections. Then it will be your turn to produce some engineering drawings including some or all of the features outlined in this section.

Fluid power schematic diagrams

These diagrams cover both pneumatic and hydraulic circuits. The symbols that we shall use do not illustrate the physical make-up, construction or shape of the components. Neither are the symbols to scale or orientated in any particular position. They are only intended to show the 'function' of the component they portray, the connections and the fluid flow path.

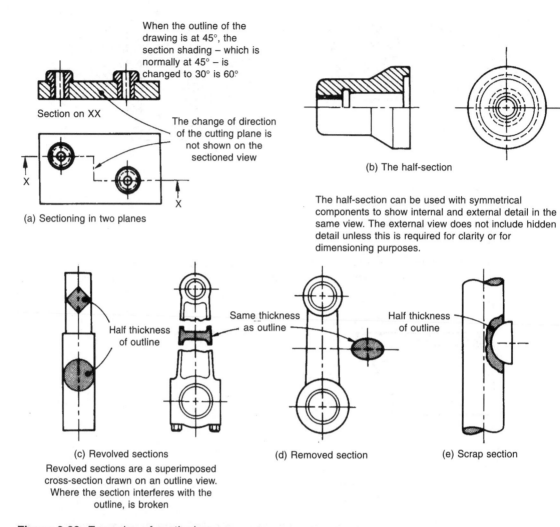

When the outline of the drawing is at 45°, the section shading – which is normally at 45° – is changed to 30° is 60°

Section on XX

The change of direction of the cutting plane is not shown on the sectioned view

(a) Sectioning in two planes

(b) The half-section

The half-section can be used with symmetrical components to show internal and external detail in the same view. The external view does not include hidden detail unless this is required for clarity or for dimensioning purposes.

Half thickness of outline

Same thickness as outline

Half thickness of outline

(c) Revolved sections

Revolved sections are a superimposed cross-section drawn on an outline view. Where the section interferes with the outline, is broken

(d) Removed section

(e) Scrap section

Figure 2.69 *Examples of sectioning*

Complete symbols are made up from one or more basic symbols and from one or more functional symbols. Examples of some basic symbols are shown in Figure 2.72 and some functional symbols are shown in Figure 2.73.

Energy converters

Let's now see how we can combine some of these basic and functional symbols to produce a complete symbol representing a component. For example, let's start with a motor. The complete symbol is shown in Figure 2.74.

The large circle indicates that we have an energy conversion unit such as a motor or pump. Notice that the fluid flow is into the device and that it is pneumatic. The direction of the arrowhead indicates the direction of flow. The fact that the arrowhead is clear (open) indicates that the fluid is air. Therefore the device must be a motor. If it were a pump the fluid flow would be out of the circle. The single line at the bottom of the circle is the outlet (exhaust)

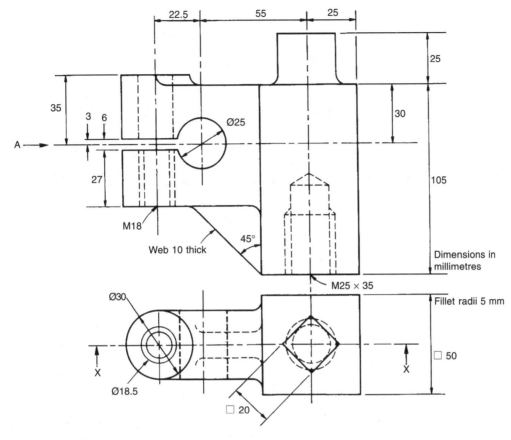

Figure 2.70 *See Activity 2.34*

from the motor and the double line is the mechanical output from the motor.

Now let's analyse the symbol shown in Figure 2.75.

- The circle tells us that it is an energy conversion unit.
- The arrowheads show that the flow is from the unit so it must be a pump.
- The arrowheads are solid so it must be a hydraulic pump.
- The arrowheads point in opposite directions so the pump can deliver the hydraulic fluid in either direction depending upon its direction of rotation.
- The arrow slanting across the pump is the variability symbol, so the pump has variable displacement.
- The double lines indicate the mechanical input to the pump from some engine or motor.

Summing up, we have a variable displacement, hydraulic pump that is bi-directional.

Directional control valves

The function of a directional control valve is to open or close flow lines in a system. Control valve symbols are always drawn in square

---

Test your knowledge 2.25

Draw the symbol for:

(a) a unidirectional, fixed displacement pneumatic pump (compressor)
(b) a fixed capacity hydraulic motor.

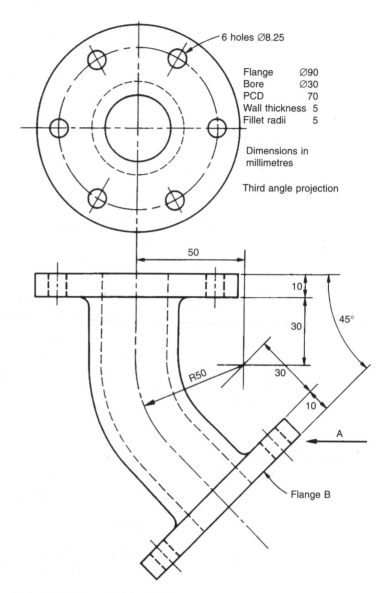

Figure 2.71 *See Activity 2.34*

A valve symbol is shown in Figure 2.77.

(a) State the numerical code that describes the valve.
(b) Describe the flow path drawn.
(c) Sketch and describe the flow path when the valve is in its alternative position.

boxes or groups of square boxes to form a rectangle. This is how you recognize them. Each box indicates a discrete position for the control valve. Flow paths through a valve are known as 'ways'. Thus a 4-way valve has four flow paths through the valve. This will be the same as the number of connections. We can, therefore, use a number code to describe the function of a valve. Figure 2.76 shows a 4/2 directional control valve. This valve has four flow paths, ports or connections and two positions. The two boxes indicate the two positions. The appropriate box is shunted from side to side so that, in your imagination, the internal flow paths line up with the connections. Connections are shown by the lines that extend 'outside' the perimeters of the boxes.

As drawn, the fluid can flow into port 1 and out of port 2. Fluid can also flow into port 3 and out of port 4. In the second position, the fluid flows into port 3 and out of port 1. Fluid can also flow into port 4 and out of port 2.

Description	Symbol
Flow lines Continuous: working line return line feed line	 ——————
Long dashes: pilot control lines	 – – – – –
Short dashes: drawn lines	 – – – – – –
Long chain enclosure line	 —— – – ——
Flow line connections	
Mechanical link, roller, etc.	○
Semi-rotary actuator	
As a rule, control valves (valve) except for non-return valves	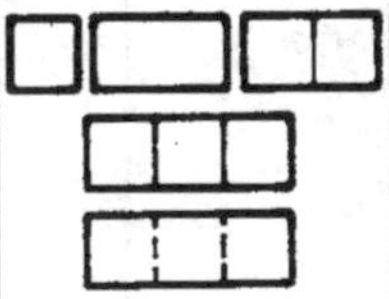
Conditioning apparatus (filter, separator, lubricator, heat exchanger)	 ◇

Description	Symbol
Spring	
Restriction: affected by viscosity unaffected by viscosity	
As a rule, energy conversion units (pump, compressor motor)	 ○
Measuring instruments	 ○
Non-return valve, rotary connection, etc.	○

Figure 2.72 *Basic symbols used in fluid power diagrams*

Valve control methods

Before we look at other examples of directional control valves, let's see how we can control the positions of a valve. There are four basic methods of control, these are:

- Manual control of the valve position.
- Mechanical control of the valve position.
- Electromagnetic control of the valve position.
- Pressure control of the valve positions (direct and indirect).
- Combined control methods.

The methods of control are shown in Figure 2.78. With simple electrical or pressure control, it is possible only to move the valve to one, two or three discrete positions. The valve spool may be located in such positions by a spring loaded detent.

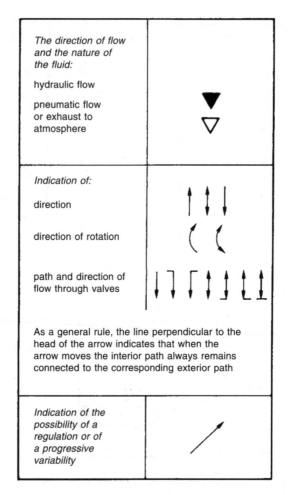

The direction of flow and the nature of the fluid: hydraulic flow pneumatic flow or exhaust to atmosphere	
Indication of: direction direction of rotation path and direction of flow through valves As a general rule, the line perpendicular to the head of the arrow indicates that when the arrow moves the interior path always remains connected to the corresponding exterior path	
Indication of the possibility of a regulation or of a progressive variability	

Figure 2.73 *Functional symbols used in fluid power diagrams*

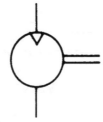

Figure 2.74 *Basic symbol for a motor*

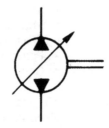

Figure 2.75 *Energy converter symbol*

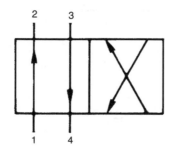

Figure 2.76 *4/2 directional control valve*

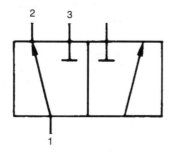

Figure 2.77 *See Test your knowledge 2.26*

Combinations of the above control methods are possible. For example, a single solenoid with spring return for a two position valve. Let's now look at some further directional control valves (DCVs).

- Figure 2.79(a) shows a 4/2 DCV controlled by a single solenoid with a spring return.
- Figure 2.79(b) shows a 4/3 DCV. That is, a directional control valve with four ports (connections) and three positions. It is operated manually by a lever with spring return to the centre. The service ports are isolated in the centre position. An application of this valve will be shown later.
- Figure 2.79(c) shows a 4/2 DCV controlled by pneumatic pressure by means of a pilot valve. The pilot valve is actuated by a single solenoid and a return spring.

Test your knowledge 2.27

Describe the DCV whose symbol is shown in Figure 2.80.

Linear actuators

A linear actuator is a device for converting fluid pressure into a mechanical force capable of doing useful work and combining this force with limited linear movement. Put more simply, a piston in a cylinder. The symbols for linear actuators (also known as 'jacks' and 'rams') are simple to understand and some examples are shown in Figure 2.81.

- Figure 2.81(a) shows a single-ended, double-acting actuator. That is, the piston is connected by a piston rod to some external mechanism through one end of the cylinder only. It is double acting because fluid pressure can be applied to either side of the piston.

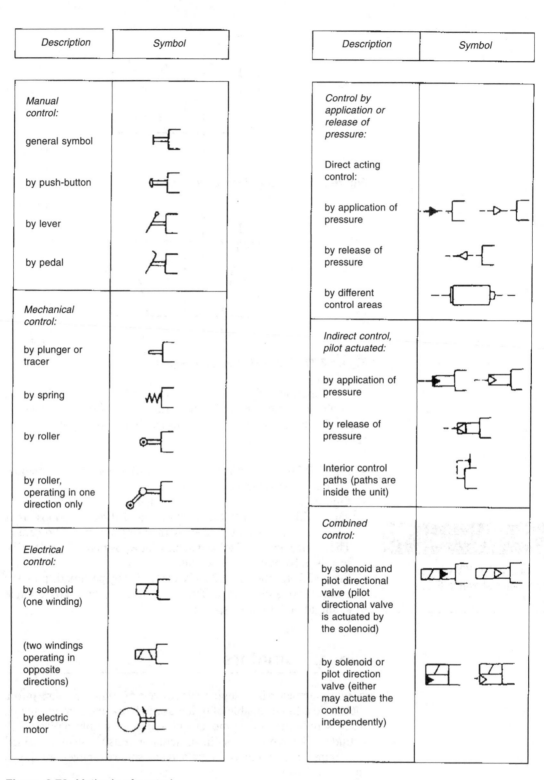

Figure 2.78 *Methods of control*

- Figure 2.81(b) shows a single-ended, single-acting actuator with spring return. Here the fluid pressure is applied only to one side of the piston. Note the pneumatic exhaust to atmosphere so that the air behind the piston will not cause a fluid lock.

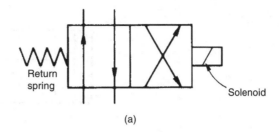

(a)

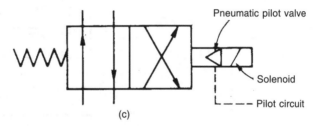

(b)

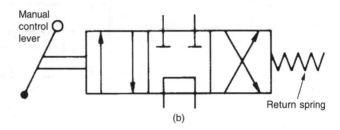

(c)

Figure 2.79 *Various types of DCV*

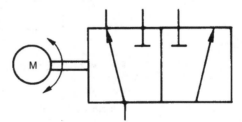

Figure 2.80 *See Test your knowledge 2.27*

- Figure 2.81(c) shows a single-ended, single-acting actuator, with double variable cushion damping. The cushion damping prevents the piston impacting on the ends of the cylinder and causing damage.
- Figure 2.81(d) shows a double-ended, double-acting actuator fitted with single, fixed cushion damping.

We are now in a position to use the previous component symbols to produce some simple fluid power circuits.

Figure 2.82 shows a single-ended, double-acting actuator controlled by a 4/3 tandem centre, manually operated DCV. Note that in the neutral position both sides of the actuator piston are blocked off, forming a hydraulic lock. In this position the pump flow is being returned directly to the tank. Note the tank symbol.

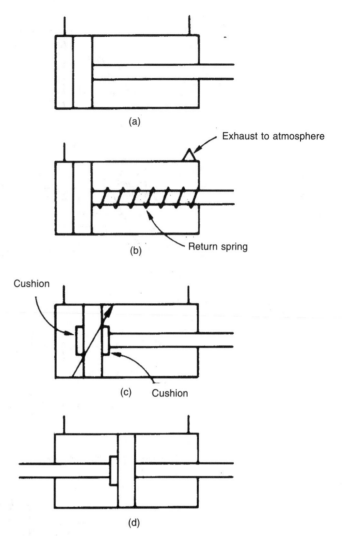

Figure 2.81 *Various types of linear actuator*

This system is being supplied by a single direction fixed displacement hydraulic pump.

Figure 2.83 shows a simple pneumatic hoist capable of raising a load. The circuit uses two 2-port manually operated push-button valves connected to a single-ended, single-acting actuator. Supply pressure is indicated by the circular symbol with a black dot in its centre. Valve 'b' has a threaded exhaust port indicated by the extended arrow. When valve 'a' is operated, compressed air from the air line is admitted to the underside of the piston in the cylinder. This causes the piston to rise and to raise the load. Any air above the piston is exhausted to the atmosphere through the threaded exhaust port at the top of the cylinder. Again this is indicated by a long arrow. When valve 'b' is operated, it connects the cylinder to the exhaust and the actuator is vented to the atmosphere. The load is lowered by gravity.

Both these circuits are functional, but they do not have protection against overpressurization, neither do they have any other safety devices fitted. Therefore, we need to increase our vocabulary of components before we can design a safe, practical circuit. We will now consider the function and use of pressure and flow control valves.

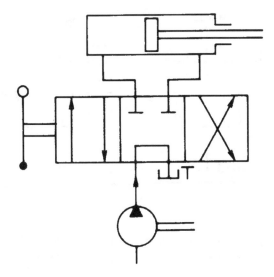

Figure 2.82 *Actuator controlled by a DCV*

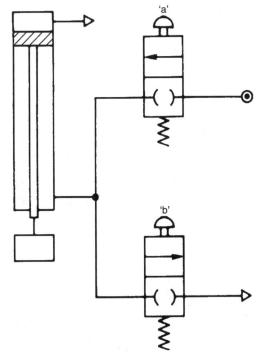

Figure 2.83 *A simple pneumatic hoist*

Pressure relief and sequence valves

Figure 2.84 shows an example of a pressure relief (safety) valve. In Figure 2.84(a) the valve is being used in a hydraulic circuit. Pressure is controlled by opening the exhaust port to the reservoir tank against an opposing force such as a spring. In Figure 2.84(b) the valve is being used in a pneumatic circuit so it exhausts to the atmosphere.

Figures 2.84(c) and 2.84(d) show the same valves except that this time the relief pressure is variable, as indicated by the arrow

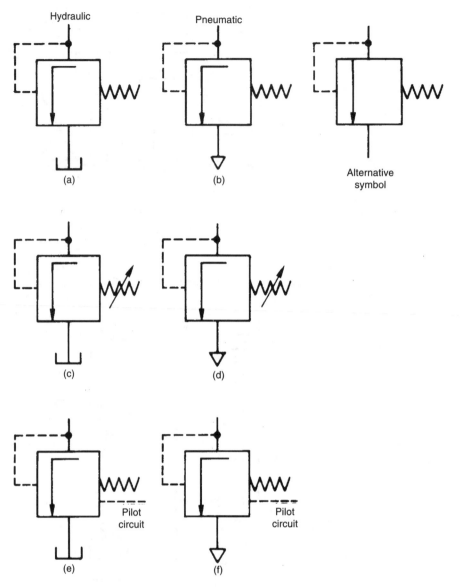

Figure 2.84 *Use of a pressure release valve*

drawn across the spring. If the relief valve setting is used to control the normal system pressure as well as acting as an emergency safety valve, the adjustment mechanism for the valve must be designed so that the maximum safe working pressure for the circuit cannot be exceeded.

Figures 2.84(e) and 2.84(f) show the same valves with the addition of pilot control. This time the pressure at the inlet port is not only limited by the spring but also by the pressure of the pilot circuit superimposed on the spring. The spring offers a minimum pressure setting and this can be increased by increasing the pilot circuit pressure up to some predetermined safe maximum. Sometimes the spring is omitted and only pilot pressure is used to control the valve.

Sequence valves are closely related to relief valves in both design and function and are represented by very similar symbols. They permit the hydraulic fluid to flow into a subcircuit, instead of back to the reservoir, when the main circuit pressure reaches the setting of the sequence valve. You can see that Figure 2.85 is very similar to a pressure relief valve (PRV) except that, when it opens, the fluid is directed to the next circuit in the sequence instead of being exhausted to the reservoir tank or allowed to escape to the atmosphere.

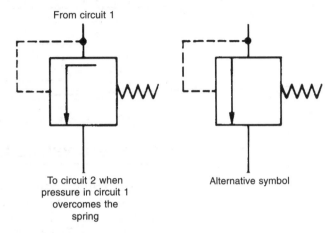

Figure 2.85 *Sequence valve*

Flow control valves

Flow control valves, as their name implies, are used in systems to control the rate of flow of fluid from one part of the system to another. The simplest valve is merely a fixed restrictor. For operational reasons this type of flow control valve is inefficient, so the restriction is made variable as shown in Figure 2.86(a). This is a throttling valve. The full symbol is shown in Figure 2.86(b). In this example the valve setting is being adjusted mechanically. The valve rod ends in a roller follower in contact with a cam plate.

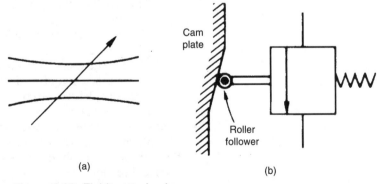

Figure 2.86 *Fluid control valves*

Sometimes it is necessary to ensure that the variation in inlet pressure to the valve does not affect the flow rate from the valve. Under these circumstances we use a pressure compensated flow control valve (PCFCV). The symbol for this type of valve is shown in Figure 2.87. This symbol suggests that the valve is a combination of a variable restrictor and a pilot operated relief valve. The enclosing box is drawn using a long-chain line. This signifies that the components making up the valve are assembled as a single unit.

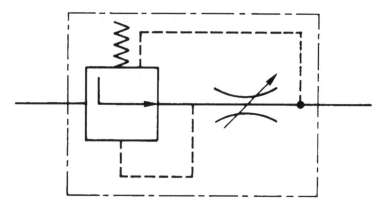

Figure 2.87 *Pressure compensated flow control valve*

Non-return valves and shuttle valves

The non-return valve (NRV), or check valve as it is sometimes known, is a special type of directional control valve. It allows the fluid to flow in one direction only and it blocks the flow in the reverse direction. These valves may be operated directly or by a pilot circuit. Some examples are shown in Figure 2.88.

- Figure 2.88(a) shows a valve that opens (is free) when the inlet pressure is higher than the outlet pressure (back-pressure).
- Figure 2.88(b) shows a spring-loaded valve that only opens when the inlet pressure can overcome the combined effects of the outlet pressure and the force exerted by the spring.
- Figure 2.88(c) shows a pilot controlled NRV. It opens only if the inlet pressure is greater than the outlet pressure. However, these pressures can be augmented by the pilot circuit pressure.
  - (i)   The pilot pressure is applied to the inlet side of the NRV. We now have the combined pressures of the main (primary) circuit and the pilot circuit acting against the outlet pressure. This enables the valve to open at a lower main circuit pressure than would normally be possible.
  - (ii) The pilot pressure is applied to the outlet side of the NRV. This assists the outlet or back-pressure in holding the valve closed. Therefore it requires a greater main circuit pressure to open the valve. By adjusting the pilot pressure in these two examples we can control the circumstances under which the NRV opens.
- Figure 2.88(d) shows a valve that allows normal full flow in the forward direction but restricted flow in the reverse direction. The valves previously discussed did not allow any flow in the reverse direction.
- Figure 2.88(e) shows a simple shuttle valve. As its name implies, the valve is able to shuttle backwards and forwards. There are two inlet ports and one outlet port. Imagine that inlet port A has the higher pressure. This pressure overcomes the inlet pressure at B and moves the shuttle valve to the right. The valve closes inlet port B and connects inlet port A to the outlet port. If the pressure at inlet port B rises, or that at A falls, the shuttle will move back to the left. This will close inlet port A and connect inlet port B to the outlet. Thus, the inlet port with the higher pressure is automatically connected to the outlet port.

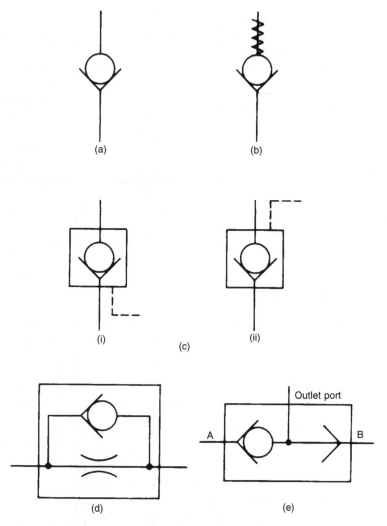

Figure 2.88 *Some examples of non-return valves*

Conditioning equipment

The working fluid, be it oil or air, has to operate in a variety of environments and it can become overheated and/or contaminated. As its name implies, conditioning equipment is used to maintain the fluid in its most efficient operating condition. A selection of conditioning equipment symbols is shown in Figure 2.89. Note that all conditioning device symbols are diamond shaped.

Filters and *strainers* have the same symbol. They are normally identified within the system by their position. The filter element (dashed line) is always positioned at 90° to the fluid path.

Water traps are easily distinguished from filters since they have a drain connection and an indication of trapped water. Water traps are particularly important in pneumatic systems because of the humidity of the air being compressed.

Lubricators are particularly important in pneumatic systems. Hydraulic systems using oil are self-lubricating. Pneumatic systems use air, which has no lubricating properties so oil, in the form of a mist, has to be added to the compressed air line.

Filters, water traps, lubricators and miscellaneous apparatus

Description	Symbol
Filter or strainer	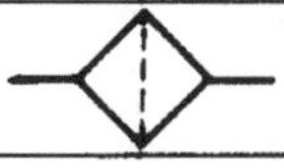
Water trap: with manual control	
automatically drained	
Filter with water trap: with manual control	
automatically drained	
Air dryer	
Lubricator	
Conditioning unit detailed symbol	
simplified symbol	

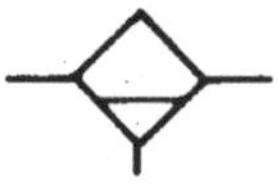

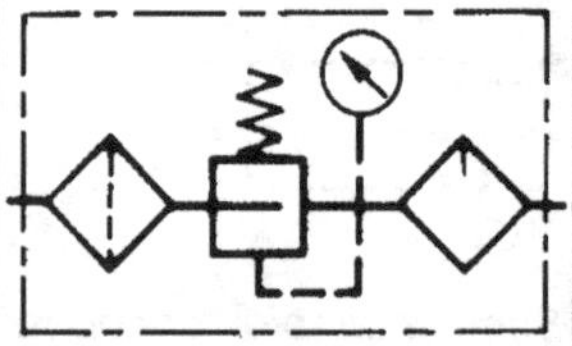

Heat exchangers

Description	Symbol
Temperature controller (arrows indicate that heat may be either introduced or dissipated)	
Cooler (arrows indicate the extraction of heat) Without representation of the flow lines of the coolant	
With representation of the flow lines of the coolant	
Heater (arrows indicate the introduction of heat)	

Figure 2.89 *Symbols for air conditioning devices*

Heat exchangers can be either heaters or coolers. If the hydraulic oil becomes too cool it becomes thicker (more viscous) and the system becomes sluggish. If the oil becomes too hot it will become too thin (less viscous) and not function properly. The direction of the arrows in the symbol indicates whether heat energy is taken from the fluid (cooler) or given to the fluid (heater). Notice that the cooler can show the flow lines of the coolant.

There is one final matter to be considered before you can try your hand at designing a circuit, and that is the pipework circuit to connect the various components together. The correct way of representing pipelines is shown in Figure 2.91.

- Figure 2.91(a) shows pipelines that are crossing each other but are not connected.

Test your knowledge 2.28

Figure 2.90 shows a selection of fluid circuit symbols. Name the symbols and briefly explain what they do.

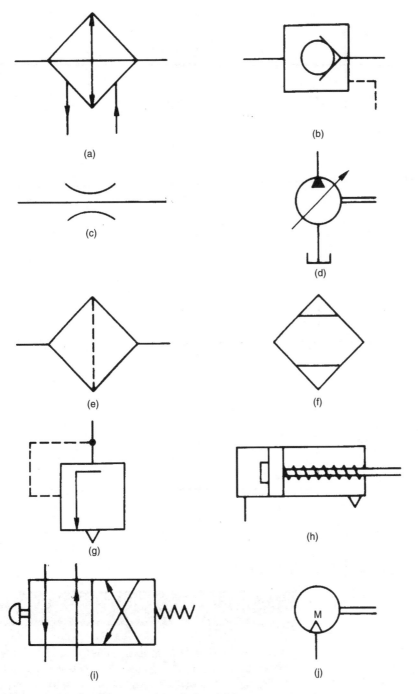

Figure 2.90 *See Test your knowledge 2.28*

- Figure 2.91(b) shows three pipes connected at a junction. The junction (connection) is indicated by the solid circle (or large dot, if you prefer).
- Figure 2.91(c) shows four pipes connected at a junction. On no account can the connection be drawn as shown in Figure 2.91(d). This is because there is always a chance of the ink running where lines cross on a drawing. The resulting 'blob' could then be misinterpreted as a connection symbol with disastrous results.

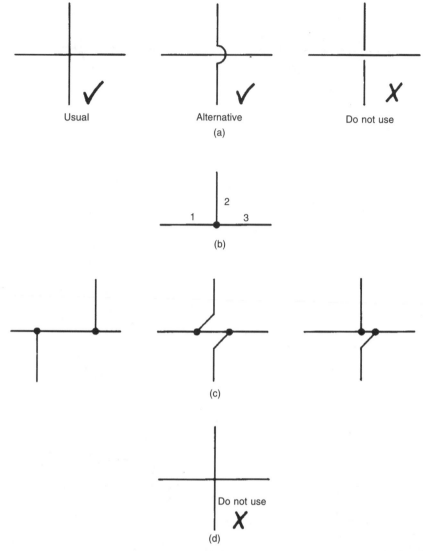

Figure 2.91 *Representing pipelines*

Activity 2.35

Figure 2.92 shows the general principles for the hydraulic drive to the ram of a shaping machine. The ram is moved backwards and forwards by a double-acting single-ended hydraulic actuator. The drawing was made many years ago and it uses outdated symbols. Use CAD or a technical illustration package to draw a schematic hydraulic diagram for the machine using current symbols and practices (as set out in BS PP 7307).

Present your work in the form of a printed diagram.

Electrical and electronic circuit schematics

Electrical and electronic circuits can also be drawn using schematic symbols to represent the various components. The full range of

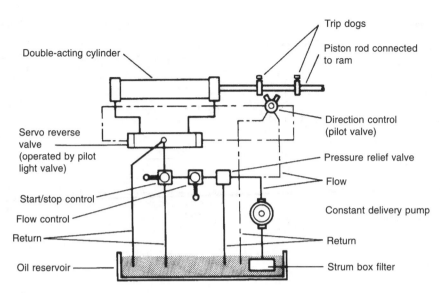

Figure 2.92 *See Activity 2.35*

symbols and their usage can be found in BS 3939. This is a very extensive standard and well beyond the needs of this book. For our immediate requirements you should refer to PP 7307 Graphical Symbols for Use in Schools and Colleges. Figure 2.93 shows a selection of symbols that will be used in the following examples.

- A *cell* is a source of direct current (DC) electrical energy. Primary cells have a nominal potential of 1.5 volts each. They cannot be recharged and are disposable. Secondary cells are rechargeable. Lead–acid cells have a nominal potential of 2 volts and nickel cadmium (NiCd) cells have a nominal potential of 1.2 volts. Cells are often connected in series to form a battery.

- *Batteries* consist of a number of cells connected in series to increase the overall potential. A 12 volt car battery consists of six lead–acid secondary cells of 2 volts each.

- *Fuses* protect the circuit in which they are connected from excess current flow. This can result from a fault in the circuit, from a fault in an appliance connected to the circuit or from too many appliances being connected to the same circuit. The current flowing in the circuit tends to heat up the fuse wire. When the current reaches some predetermined value the fuse wire melts and breaks the circuit so the current can no longer flow. Without a fuse the circuit wiring could overheat and cause a fire.

- *Resistors* are used to control the magnitude of the current flowing in a circuit. The resistance value of the resistor may be fixed or it may be variable. Variable resistors may be preset or they may be adjustable by the user. The electric current does work in flowing through the resistor and this heats up the resistor. The resistor must be chosen so that it can withstand this heating effect and sited so that it has adequate ventilation.

- *Capacitors*, like resistors, may be fixed in value or they may be preset or variable. Capacitors store electrical energy but, unlike secondary cells, they may be charged or discharged almost instantaneously. The stored charge is much smaller than the charge stored by a secondary cell. Large value capacitors are

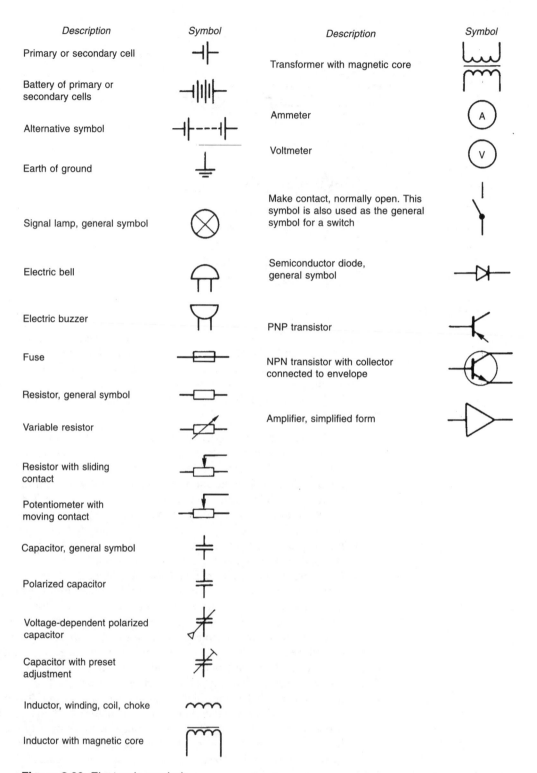

Figure 2.93 *Electronic symbols*

used to smooth the residual ripple from the rectifier in a power pack. Medium value capacitors are used for coupling and decoupling the stages of audio frequency amplifiers. Small value capacitors are used for coupling and decoupling radio frequency signals and they are also used in tuned (resonant) circuits.

- *Inductors* act like electrical 'flywheels'. They limit the build-up of current in a circuit and try to keep the circuit running by putting energy back into it when the supply is turned off. They are used as current limiting devices in fluorescent lamp units, as chokes in telecommunications equipment and, together with capacitors, to make up resonant (tuned) circuits in telecommunications equipment.
- *Transformers* are used to raise or lower the voltage of alternating currents. Inductors and transformers cannot be used in direct current circuits. You can't get something for nothing, so if you increase the voltage you decrease the current accordingly so that (neglecting losses), $V \times I = k$ where k is a constant for the primary and secondary circuits of any given transformer.
- *Ammeters* measure the current flowing in a circuit. They are always wired in series with the circuit so that the current being measured can flow through the meter.
- *Voltmeters* measure the potential difference (voltage) between two points in a circuit. To do this they are always wired in parallel across that part of the circuit where the potential is to be measured.
- *Switches* are used to control the flow of current in a circuit. They can only open or close the circuit. So the current either flows or it doesn't.
- *Diodes* are like the non-return valves in hydraulic circuits. They allow the current to flow in one direction only as indicated by the arrowhead of the symbol. They are used to rectify alternating current (AC) and convert it into DC.
- *Transistors* are used in high speed switching circuits and to magnify radio and audio frequency signals.
- *Integrated circuits* consist of all the components necessary to produce amplifiers, oscillators, central processor units, computer memories and a host of other devices fabricated onto a single slice of silicon; each chip being housed in a single compact package.

Let's look at some examples of schematic circuit diagrams using these symbols. All electric circuits consist of:

- A source of electrical energy (e.g. a battery or a generator).
- A means of controlling the flow of electric current (e.g. a switch or a variable resistor).
- An appliance to convert the electrical energy into useful work (e.g. a heater, a lamp, or a motor).
- Except for low power battery operated circuits, an overcurrent protection device (fuse or circuit breaker).
- Conductors (wires) to connect these various circuit elements together. Note that the rules for drawing conductors that are connected and conductors that are crossing but not connected are the same as for drawing pipework as previously described in Figure 2.74.

Figure 2.94 shows a very simple circuit that satisfies the above requirements. In Figure 2.94(a) the switch is 'closed' therefore the circuit as a whole is also a closed loop. This enables the electrons that make up the electric current to flow from the source of electrical energy through the appliance (lamp) and back to the source of energy ready to circulate again – rather like the fluid in our earlier

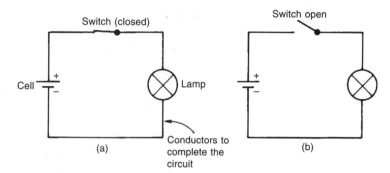

Figure 2.94 *A simple electronic circuit*

hydraulic circuits. In Figure 2.94(b) the switch is 'open' and the circuit is no longer a closed loop. The circuit is broken. The electrons can no longer circulate. The circuit ceases to function. We normally draw our circuits with the switches in the 'open' position so that the circuit is not functioning and is 'safe'.

Figure 2.95 shows a simple battery operated circuit for determining the resistance of a fixed value resistor. The resistance value is obtained by substituting the values of current and potential into the formula, $R = V/I$. The current in amperes is read from the ammeter and the potential in volts is read from the voltmeter. Note that the ammeter is wired in series with the resistor so that the current can flow through it. The voltmeter is wired in parallel with the resistor so that the potential can be read across it. This is always the way these instruments are connected.

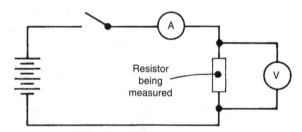

Figure 2.95 *Circuit for determining resistance*

Figure 2.96 shows a circuit for operating the light over the stairs in a house. The light can be operated either by the switch at the bottom of the stairs or by the switch at the top of the stairs. Can you work out how this is achieved? The switches are of a type called 'two-way, single-pole'. The circuit is connected to the mains supply. It is protected by a fuse in the 'consumer unit'. This unit

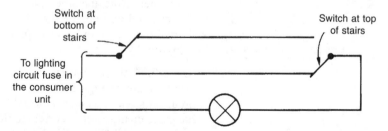

Figure 2.96 *Two-way lighting switch*

contains the main switch and all the fuses for the house and is situated adjacent to the supply company's meter and main fuse.

Figure 2.97 shows a two-stage transistorized amplifier. It also shows a suitable power supply. Table 2.3 lists and names the components.

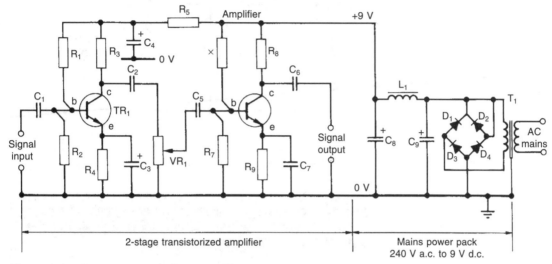

Figure 2.97 *A two-stage transistor amplifier*

Table 2.3 *Components used in a two-stage transistorized amplifier*

Component	Description
R_1–R_9	Fixed resistors
VR_1	Variable resistor
C_1–C_9	Capacitors
D_1–D_4	Diodes
TR_1, TR_2	Transistors
T_1	Mains transformer
L_1	Inductor (choke)

Figure 2.98 shows a similar amplifier using a single chip. Such an amplifier would have the same performance but fewer components are required. Therefore it is cheaper and quicker to make.

Activity 2.36

Draw a schematic circuit diagram for a battery charger having the following features:

- the primary circuit of the transformer (i.e. the side that is connected to the a.c. mains supply) is to have an on/off switch, a fuse and an indicator lamp
- the secondary circuit of the transformer is to have a bridge rectifier, a variable resistor to control the charging current, a fuse and an ammeter to indicate the charging current.

Test your knowledge 2.29

Figure 2.99 shows a selection of electrical and electronic symbols. Name the symbols and briefly explain what they do.

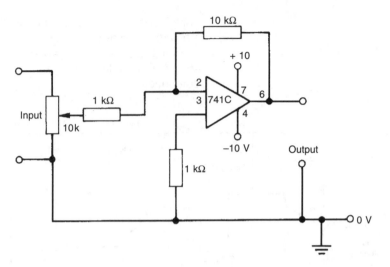

Figure 2.98 *A single-chip amplifier circuit*

Use CAD or a technical illustration package to produce the schematic circuit diagram and present your work in the form of a printed diagram.

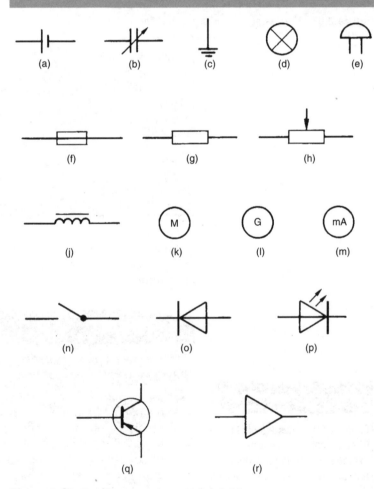

Figure 2.99 *See Test your knowledge 2.29*

Activity 2.37

Figure 2.100 shows an electronic circuit.

(a) Draw up a component list that numbers and names each of the components (include values where given).
(b) Suggest what the circuit might be used for. (Hint: The circuit only has an output!)

Present your work in the form of a single A4 printed page of word-processed text.

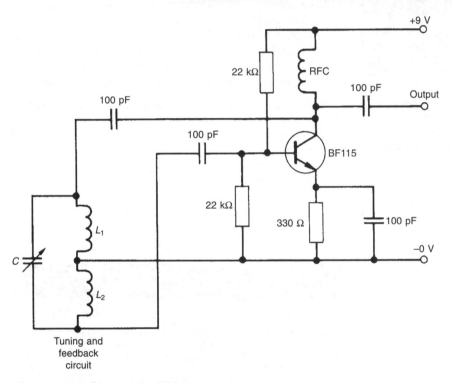

Figure 2.100 *See Activity 2.37*

Activity 2.38

Your tutor will supply you with a selection of (at least five) engineering components (mechanical, hydraulic, pneumatic, electrical and electronic). Identify each component and make a freehand sketch of it.
Present your work in the form of a portfolio of sketches.

Verbal and written communication

Verbal communication (i.e. speaking and listening) is widely used in everyday situations, including:

• Informal discussions either on the telephone or face to face.
• Formal presentations to groups of persons who all require the same information.

Where a group of persons all require the same information, a formal presentation must be used. On no account should information be 'passed down the line' from person to person because errors are bound to creep in. There is a story that during World War I the message 'send reinforcements, we are going to advance' arrived at headquarters, by word of mouth, as 'send three and four pence (old money), we are going to a dance'. We will let you decide on the truth behind this story, but we feel it makes the point.

In any event, it is important to remember that the spoken word is easily forgotten and oral communication should be reinforced by:

Key point

When speaking to people or giving a verbal presentation it is important to check that everybody has understood what you have been saying. There are various ways to do this including, in a small group, asking each person if they have any questions and if necessary, testing *their* understanding by asking *them* questions. Your tutor should be doing this all the time!

- Notes taken at the time.
- Tape recording the conversation.
- A written summary. For example, the published 'proceedings' of formal lectures and presentations. Another example is a 'press release' that is provided to journalists and reporters in order to ensure the factual accuracy of information intended for the public.

Oral communication must be presented in a manner appropriate to the audience. It must be brief and to the point. The key facts must be emphasized so that they can be easily remembered. The presentation must be interesting so that the attention of the audience does not wander.

When communicating by the spoken word, it is as equally important to be a good listener as it is to be a good speaker. This applies to conversations between two or three people as well as to formal presentations.

Activity 2.39

Use presentation software to prepare a five-minute presentation to the rest of the class (using appropriate visual aids) on any one of the following topics:

- How to choose a digital camera.
- How to connect to the Internet.
- What to look for when purchasing a second-hand car.

You should prepare a set of brief printed notes summarizing the key points for your audience. Also include printed copies of any screens or overhead projector transparencies that you use. At the end of your talk you should invite questions from your audience and provide appropriate answers.

Activity 2.40

Conduct a brief interview (lasting no longer than 15 minutes) with another student and take notes to summarize the outcome. Don't forget to allow time for questions at the end of the interview. Your interview should be based around the following questions:

- Why did you decide to take a course in Engineering?
- Why did you choose the BTEC National Diploma course?
- What made you choose this school/college?
- What subjects/topics have you enjoyed the most?
- What subjects/topics have you enjoyed the least?
- What plans have you got for the future?
- Where would you hope to be and what would you hope to be doing in ten years' time?

You should add further questions to clarify the above. Don't forget to thank your interviewee! Present your findings in the form of hand-written interview notes.

Written communication

This is a more reliable method of communication since it usually provides a permanent written record of the key information. The same information is available for all those who require it.

Anyone who has ever marked an English comprehension test will know that the same written passage can mean very different things to different people. Therefore, care must be taken in preparing written information. To avoid confusion, the normal conventions of grammar and punctuation must be used. Words must be correctly spelt. Use a dictionary if you are uncertain. If you are using a word-processing package use the spell-checker. However, take care, many software packages originate in the USA and the spell-checker may reflect this.

Never use jargon terms and acronyms unless you are sure that those reading the message are as equally familiar with them as is the writer.

An engineer often has to write notes, memoranda and reports. He/she often has to maintain logbooks and complete service sheets. An engineer may also have to communicate with other engineers, suppliers and customers by letter. Being able to express yourself clearly and concisely is of great importance.

Key point

It is important to use correct grammar and punctuation when writing. Since you won't always get the opportunity to check that everybody has understood what you have written it's important to ensure that everything is clear and concise. If possible, get someone to check what you have written *before* you distribute it to its intended audience!

Activity 2.41

Prepare a brief article for the local press (using not more than 1000 words) on any one of the following topics:

- A sporting event that you took part in.
- A recent school or college activity.
- A newly available product or technology.

Include contact or other details for further information. Present your work in word-processed form and include relevant photographs, diagrams or sketches.

Activity 2.42

Prepare:

(a) a word-processed letter
(b) an e-mail message

to an engineering supplier requesting details of a product or service. This may simply take the form of a request for a short-form catalogue or for the supply of a data sheet or application note. Present your work in the form of printed copies of correspondence and e-mail messages.

Problems

2.1 Describe the main features of:

(a) a detail drawing
(b) a general arrangement drawing
(c) an assembly drawing.

2.2 Explain how a search engine is used to locate information on the World Wide Web.

2.3 What information is typically held in each of the following types of database?

(a) a product database
(b) a manufacturing database
(c) a customer database
(d) a spare parts database.

2.4 List the main headings used in a technical report.

2.5 Draw a graph showing how the following current varies with time:

Time, t (s)	0	0.1	0.2	0.3	0.4	0.5	0.6	0.7
Current, i (V)	0	0.22	0.45	0.65	0.89	1.09	1.32	1.54

What does the shape of the graph suggest? Use the graph to determine the value of i when $t = 0.25$ s.

2.6 The address of a website is: http://www.daviesengineering.co.uk. Explain how this address is constructed.

2.7 The following data relates to a 12-month operating period for an engineering company:

Total income: £19.2 million
Material costs: £6.8 million
Overhead costs: £2.2 million
Labour costs: £5.9 million

Use a pie-chart to illustrate the income, expenditure and profit made by the company.

2.8 Explain, with the aid of an example, how a scatter diagram is used.

2.9 The following data refers to the number of TV sets produced by a particular manufacturer:

1997   2500
1998   4000
1999   5500
2000   7250
2001   4500

Illustrate this information using an ideograph.

2.10 Sketch typical line styles used to illustrate:

(a) a centre line
(b) the limit of a partial or interrupted view.

2.11 How many A2 drawing sheets can be cut from an A0 drawing sheet? Explain your answer with a sketch.

2.12 State FOUR items that should be included within the title block of an engineering drawing.

2.13 Sketch engineering drawing symbols that are used to indicate the following components:

(a) a 4/2 directional control valve
(b) a non-return valve
(c) a battery
(d) a variable resistor
(e) a semiconductor diode
(f) an iron-cored transformer.

2.14 Draw, using appropriate symbols, a two-way lighting circuit. Label your drawing clearly.

2.15 List FOUR advantages of using CAD in the preparation of engineering drawings compared with purely manual methods.

2.16 Identify the projection used in Figures 2.101 and 2.102.

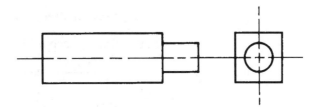

Figure 2.101 *See Problem 2.16*

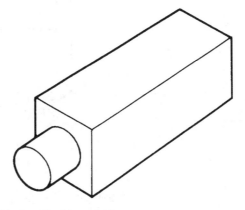

Figure 2.102 *See Problem 2.16*

2.17 Figure 2.103 shows a pie-chart. If the labour costs amount to £2.5 million, what profit is made by the company?

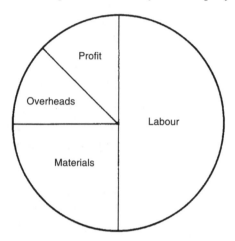

Figure 2.103 *See Problem 2.17*

2.18 Convert the pie-chart in question 2.17 into a bar chart.
2.19 Identify the lines marked A and B in Figure 2.104.

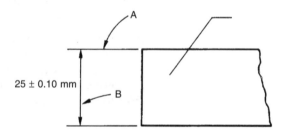

Figure 2.104 *See Problem 2.19*

2.20 Identify each of the components shown in Figure 2.105.

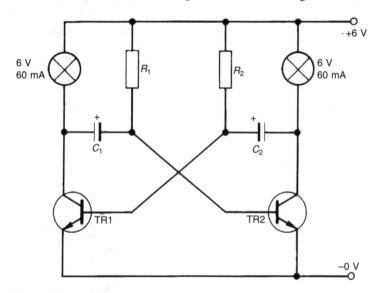

Figure 2.105 *See Problem 2.20*

Unit 3 Science for technicians

This unit aims to provide you with a foundation in scientific principles, which will enable you to tackle and solve a variety of unique engineering problems at the technician level. After successfully working your way through the subject matter presented in this unit, you will also find that you are adequately prepared for further study of your chosen specialist engineering science units, in addition to having the necessary scientific base for progression onto higher education engineering programmes.

You will be introduced to the *outcomes* that make up the unit, that is: elementary mechanics, the nature of energy, electrical principles and a brief introduction to engineering systems. In the first outcome on mechanics, we consider the way in which *forces* act on rigid bodies, this area of mechanics is known as *statics*. We then study bodies in motion or *dynamics*, where in particular we look at the *nature of the forces* that cause such motion.

The concept of *energy* is fundamental to our understanding of all science. We therefore, consider the major forms of energy, work and power which are particularly applicable to our study of engineering. We look in particular at the various forms of mechanical energy and then at heat energy, leaving electrical energy until we study the outcome on electrical principles.

In the outcome on electrical principles, as already mentioned, we start by considering the nature of *electrical energy*. We then look at *DC circuits* and *magnetism* where, in the latter topic, we investigate the principles of electrical machines.

In the final outcome of the unit we will briefly study engineering systems, in particular we look at the nature of *closed-loop engineering systems* and their engineering uses. We then apply what we have learnt to three simple examples involving *electrical, hydraulic* and *pneumatic engineering systems*.

No study of science would be complete without considering the units of measurement that underpin the subject. For this reason a brief introduction to the *Système International d'unités* (SI) is included as part of the following section on fundamentals.

Fundamentals

Before embarking on our study of the unit outcomes, we first consider some fundamental scientific properties such as mass, matter, weight, etc., in addition to an initial look at the SI system of units. This section may be treated as revision for those who have already successfully studied Physics or Engineering Science at GCSE or First Diploma level. For those new to the subject, this section will provide a useful glossary of essential fundamental concepts, which underpin all that follows.

Units

The SI system of units is now the accepted international standard for all units of scientific measure. It has been legally accepted in America even though it has not been fully adopted – the American preference still being for the old English or imperial system of measurement. All scientific concepts and science teaching in the UK, as well as throughout Europe and many other areas of the world, has legally adopted the SI system and so these units will be used throughout this book for scientific and technological applications.

The International System of Units (SI) consists of three main groups: seven base units, two supplementary units and a number of derived units. Table 3.1 shows the seven base units by *dimension*, *SI name* and *symbol*. Do not worry too much at this stage about *dimension*, you will meet this concept in your later studies. Suffice to say at this stage that, for example: *length* has the *dimension L, no matter what units of length are being considered*. So it does not matter whether we measure the length in miles, metres, kilometres or knots, they all have the same dimension, that of length. The full scientific definitions of these units, as defined by the International Committee for Weights and Measures (CIPM), are given below. You might find these definitions rather strange. They are, however, true and accurate definitions of these quantities we know so well! Do not worry too much if you cannot understand these definitions at this time. Treat this section on units, *as a source of reference*, that is always here for you when needed.

Table 3.1 *SI base units*

Basic quantity/dimension		SI unit name	SI unit symbol
Mass	M	Kilogram	kg
Length	L	Metre	m
Time	T	Second	s
Electric current	I	Ampere	A
Temperature	Θ	Kelvin	K
Amount of substance	N	Mole	mol
Luminous intensity	J	Candela	cd

Kilogram

The kilogram or kilogramme is the unit of mass; it is equal to the mass of the international prototype of the kilogram, as defined by CIPM.

Metre

The metre is the length of the path travelled by light in a vacuum during the time interval of $1/299\,792\,458$ seconds.

Second

The second is the duration of 9 192 631 770 periods of radiation corresponding to the transition between the two hyperfine levels of the ground state of the cesium 133 atom.

Ampere

The ampere is that constant current which if maintained in two straight parallel conductors of infinite length, of negligible circular cross-section, and placed 1 metre apart in a vacuum, would produce between these conductors a force equal to 2×10^{-7} newton per metre length.

Kelvin

The kelvin, unit of thermodynamic temperature, is the fraction 1/273.16 of the thermodynamic temperature of the triple point of water.

Mole

The mole is the amount of substance of a system which contains as many elementary particles as there are atoms in 0.012 kg of carbon 12. When the mole is used, the elementary entities must be specified and may be atoms, molecules, ions, electrons, or other particles, or specified groups of such particles.

Candela

The candela is the luminious intensity, in a given direction, of a source that emits monochromatic radiation of frequency 540×10^{12} Hz and that has a radiant intensity in that direction of 1/683 watt per steradian (see below).

In addition to the seven base units given above, as mentioned before there are two supplementary units, the radian for plane angles (which you will meet later) and the steradian for solid three-dimensional angles (Table 3.2). Both of these units are non-dimensional, that is they are ratios and have a dimension of unity (1). In fact *all ratios are non-dimensional*, for example if we divide a distance with dimension L by another distant with dimension L, we have dimension $L/L = 1$. *Ratios* also *have no units*, for the same reason as above, e.g. metres/metres = 1. Again, do not worry too much at this stage, it may become clearer later, when we look at radian measures in our study of dynamics.

Table 3.2 *SI supplementary units*

Supplementary unit	SI unit name	SI unit symbol
Plane angle	Radian	rad
Solid angle	Steradian	srad

The SI derived units are defined by simple equations relating two or more base units. The names and symbols of some of the derived units may be substituted by special names and symbols. Some of the derived units, which you may be familiar with, are listed in Table 3.3 with their special names as appropriate.

Table 3.3 *SI derived units*

Name	Symbol	Physical quantity	Dimension	Equivalent in SI base units
Coulomb	C	Quantity of electricity, electric charge	IT	$1\,C = 1\,A.s$
Farad	F	Electric capacitance	$M^{-1}L^{-2}T^4I^2$	$1\,F = 1\,kg^{-1}.m^{-2}.s^4.A^2$
Henry	H	Electrical inductance	$ML^2T^{-2}I^{-2}$	$1\,H = 1\,kg.m^2.s^2.A^{-2}$
Hertz	Hz	Frequency	T^{-1}	$1\,H = 1\,s^{-1}$
Joule	J	Energy, work, heat	ML^2T^{-2}	$1\,J = 1\,kg.m^2.s^{-2}$
Lux	Lx	Illuminance	$J\Omega L^{-2}$	$1\,lx = 1\,cd.sr.m^{-2}$
Newton	N	Force, weight	MLT^{-2}	$1\,N = 1\,kg.ms^{-2}$
Ohm	Ω	Electrical resistance	$ML^2T^{-3}I^{-2}$	$1\,\Omega = 1\,kg.m^2.s^{-3}.A^{-2}$
Pascal	Pa	Pressure, stress	$ML^{-1}T^{-2}$	$1\,Pa = 1\,kg.m^{-1}.s^{-2}$
Siemen	S	Electrical conductance	$M^{-1}L^{-2}T^3I^2$	$1\,S = 1\,kg^{-1}.m^{-2}.s^3.A^2$
Tesla	T	Induction field, magnetic flux density	$MT^{-2}I^{-1}$	$1\,T = 1\,kg.A^{-1}.s^{-2}$
Volt	V	Electric potential, electromotive force	$ML^2T^{-3}I^{-1}$	$1\,V = 1\,kg.m^2.s^{-3}.A^{-1}$
Watt	W	Power, radiant flux	ML^2T^{-3}	$1\,W = 1\,kg.m^2.s^{-3}$
Weber	Wb	Induction magnetic flux	$ML^2T^{-2}I^{-1}$	$1\,Wb = 1\,kg.m^2.s^{-2}.A^{-1}$

You will be introduced to many of these units as you progress through this chapter and you will be asked to use them when solving problems. The SI is a decimal system of units where fractions have been eliminated, so multiples and submultiples are formed from a series of prefixes. That is, we multiple the original unit by powers of ten for units greater than one and by decimal fractions of ten for numbers less than one. Some of these multiples (with which I am sure you are familiar) are detailed in Table 3.4.

Table 3.4 *SI prefixes*

Prefix	Symbol	Multiply by
peta	P	10^{15}
tera	T	10^{12}
giga	G	10^9
mega	M	10^6
kilo	k	10^3
hecto	h	10^2
deca	da	10^1
deci	d	10^{-1}
centi	c	10^{-2}
milli	m	10^{-3}
micro	μ	10^{-6}
nano	n	10^{-9}
pico	p	10^{-12}
femto	f	10^{-15}

So, for example: 1 millimetre = 1 mm = 10^{-3} m, 1 cm^3 = $(10^{-2}\,m)^3 = 10^{-6}\,m^3$ and 1 mm = 10^{-6} m. Note the way in which powers of ten are used. The above examples show us the correct way for representing multiples and submultiples of units.

Finally, before we leave our short study on units you should be aware of some commonly used, legally accepted, *non-SI units*. These are detailed in Table 3.5.

Table 3.5 *Non-SI units*

Name	Symbol	Physical quantity	Dimension	Equivalent in SI base units
Ampere-hour	Ah	Electric charge	IT	1 Ah = 3600 C
Day	d	Time, period	T	1 d = 86 400 s
Degree	°	Plane angle	α	$1° = \pi/180$ rad
Electronvolt	eV	Electric potential	$ML^2T^{-3}I^{-1}$	1 eV = (e/C) J
Kilometre per hour	kph	Velocity	LT^{-1}	1 kph = (1/3.6) m s^{-1}
Hour	h	Time, period	T	1 h = 3600 s
Litre	L, l	Capacity, volume	L^3	1 L = 10^{-3} m^3
Minute	Min	Time, period	T	1 min = 60 s
Metric tonne	t	Mass	M	1 t = 10^3 kg

Note that velocity in kilometres per hour is a *derived unit*, that is, made up from the units of length (distance) and time.

Having briefly introduced the idea of units of measurement, we are now going to consider some fundamental quantities such as mass, force, weight, density, pressure, temperature, the nature of matter and the concept of energy, which plays a vital role in our understanding of science in general. Knowledge of these fundamental physical parameters will be required when we look in detail at the outcomes of the unit.

Mass, weight and gravity

Mass

The *mass* of a body is a measure of the *quantity of matter* in the body. The amount of matter in a body does not change when the position of the body changes so, *the mass of a body does not change with position.*

As can be seen from Table 3.1, the SI unit of mass is the kilogram (kg). The standard kilogram is the mass of a block of platinum alloy kept at the Office of Weights and Measures in Sèvres near Paris.

Weight

The weight of a body is the gravitational *force* of attraction between the mass of the earth and the mass of a body. The weight of a body decreases as the body is moved away from the earth's centre. It obeys the inverse square law, which states that if the distance of the body is doubled, the weight is reduced to a quarter of its previous value. The SI unit of weight is the newton (N).

Using mathematical symbols this law may be written as:

weight $(W) \propto 1/d^2$

where d = distance and $\propto$ is the symbol for proportionality.

Key point

The mass of a body is un-affected by its position.

Key point

In the SI system, weight is measured in newtons (N).

Key point

At sea level, the acceleration due to gravity, *g*, is approximately 9.81 m/s^2.

So, for example, consider a body of weight (*W*) at an initial distance of 50 m from the gravitational source, then: $W \propto (1/50)^2 = 4 \times 10^{-4}$.

Now if we double this distance then the weight $(W) \propto (1/100)^2 = 1 \times 10^{-4}$, which clearly shows that if the distance is doubled the weight is reduced to a quarter of its original value.

Gravitational acceleration

When a body is allowed to fall it moves towards the earth's centre with an acceleration caused by the weight of the body. If air resistance is ignored, then at the same altitude all bodies fall with the same gravitational acceleration. Although heavier bodies have more weight, at the same altitude they fall with the same gravitational acceleration because of their greater resistance to acceleration. The concept of resistance to acceleration will be explained more fully when we deal with Newton's laws of motion.

Like weight, gravitational acceleration depends on distance from the earth's centre. At sea level *gravitational acceleration* (*g*) has an accepted standard value of **9.80665 m s^{-2}**. For the purpose of calculations in this unit, we will use the approximation *g* = 9.81 m s^{-2}.

The mass–weight relationship

From what has already been said, we may define the weight of a body as the product of its mass and the value of gravitational acceleration at the position of the body. This is expressed in symbols as:

$$W = mg$$

where in the SI system the weight (*W*) is in newtons (N) the mass is in kilograms (kg) and the acceleration due to gravity is taken as 9.81 m s^{-2} unless specified differently.

We now know that the mass of a body does not change with changes in altitude but its weight and gravitational acceleration do. However, for bodies that do not move outside the earth's atmosphere, the changes in gravitational acceleration (and therefore weight) are small enough to be ignored for most practical purposes. We may therefore assume our approximation for *g* = 9.81 m s^{-2} to be reasonably accurate, unless told otherwise.

To clarify the mass–weight relationship let us consider an example calculation, using standard SI units.

Example 3.1

A missile having a mass of 25 000 kg is launched from sea level on a course for the moon. If the gravitational acceleration of the moon is one-sixth that on earth, determine:

(a)   the weight of the rocket at launch
(b)   the mass of the rocket on reaching the moon
(c)   the weight of the rocket on reaching the moon.

(a) Using the relationship $W = mg$, then the weight on earth:

$W = (25\,000 \times 9.81) = 245\,250$ N or **245.25 kN**

(b) We know from our definition of mass, that it does not change with change in position therefore the mass on the moon remains the same as on earth, i.e. **25 000 kg**.

(c) We know that the gravitational acceleration on the moon is approximately one-sixth that on earth.

So $g_m = 9.81/6$ m s^{-2} = 1.635 m s^{-2} and again from $W = mg_m$ then weight of rocket on the moon = $(25\,000 \times 1.635)$ = $40\,875$ N = **40.875 kN**

Note that a much easier method of solution for part (c) would have been to divide the weight on earth by 6.

Density and relative density

Density

The density (ρ) of a body is defined as its mass per unit volume. Combining the SI units for mass and volume gives the unit of density as kg m^{-3}. Using symbols the formula for density is given as:

$$\rho = \frac{m}{V}$$

where again the mass is in kilograms (kg) and the volume is in m³.

Relative density

The relative density of a body is the *ratio* of the density of the body with that of the density of pure water measured at 4°C. The density of water under these conditions is 1000 kg m^{-3}. Since relative density is a ratio it has *no units*. The old name for relative density was *specific gravity* (SG) and this is something you need to be aware of in case you meet this terminology in the future.

The density of some of the more common engineering elements and materials is laid out in Table 3.6. To find the relative density of any element or material divide its density by 1000 kg m^{-3}.

Example 3.2

A mild steel component has a mass of 240 grams. Using the density of mild steel given in Table 3.6, calculate the volume of the component in cm³.

From the table mild steel has a density of 7850 kg m^{-3}, therefore using our definition for density:

Since $\rho = \dfrac{m}{V}$, then $V = \dfrac{m}{\rho} = \dfrac{240 \times 10^{-3}}{7850} = 3.057 \times 10^{-3}\ m^3$

Thus, volume of component = **30.57 cm³**

Table 3.6 *Density of some engineering elements/materials*

Element/Material	Density (kg/m^3)	Element/Material	Density (kg/m^3)
Acrylic	1200	Nickel	8900
Aluminium	2700	Nitrogen	0.125
Boron	2340	Nylon	1150
Brass	8400–8600	Oxygen	0.143
Cadmium	8650	Platinum	21450
Cast iron	7350	Polycarbonate	914–960
Chromium	7190	Polyethylene	1300–1500
Concrete	2400	Rubber	860–2000
Copper	8960	Sodium	971
Glass	2400–2800	Stainless steel	7905
Gold	19 320	Tin	7300
Hydrogen	0.09	Titanium	4507
Iron	7870	Tungsten	1900
Lead	11 340	UPVC	19 300
Magnesium	1740	Vanadium	6100
Manganese	7430	Wood (Douglas fir)	608
Mercury	13 600	Wood (oak)	690
Mild steel	7850	Zinc	7130

Note that to obtain the standard unit for mass, the 240 grams was converted to kilograms using the multiplier 10^{-3} and, multiplying m^3 by 10^6 converts them into cm^3, as required. Be careful with your conversion factors when dealing with squared or cubic measure!

Example 3.3

An engineering component made from an aluminium alloy weighs 16 N and has a volume of 600 cm^3, determine the relative density of the alloy.

We need to use the mass–weight relationship $m = \dfrac{W}{g}$ to find the mass of the component,

i.e. mass, $m = \dfrac{16}{9.81} = 1.631$ kg

Then, density $= \dfrac{m}{V} = \dfrac{1.631}{600 \times 10^{-6}} = 2.718$ kg m^{-3}

The relative density (RD) is then given by RD $= \dfrac{2718 \text{ kg m}^{-3}}{1000 \text{ kg m}^{-3}}$

$= \mathbf{2.718}$

Force

In its simplest sense a force is a push or pull exerted by one object on another. In a member in a static structure, a push causes compression and a pull causes tension. Members subject to compressive and tensile forces have special names. A member of a structure that is in *compression* is known as a *strut* and a member in *tension* is called a *tie*.

Only rigid members of a structure have the capacity to act as both a strut and tie. Flexible members, such as ropes, wires or chains, can act only as ties.

Force cannot exist without opposition, as you will see later when you study Newton's laws. An applied force is called an *action* and the opposing force it produces is called *reaction*.

The effects of any force depend on its three characteristics, illustrated in Figure 3.1.

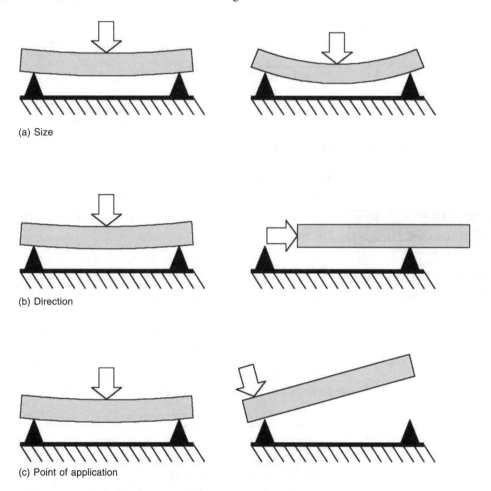

(a) Size

(b) Direction

(c) Point of application

Figure 3.1 *Characteristics of a force*

In general force (F) = mass (m) × acceleration (a) is used as the measure of force.

$$F = ma$$

The SI unit of force is the newton (N). Note that weight force mentioned earlier is a special case where the acceleration acting on the mass is that due to gravity, so *weight force* may be defined as $F = mg$, as mentioned earlier. The newton is thus defined as follows:

1 newton is the force that gives a mass of 1 kg an acceleration of 1 m s^{-2}.

It can be seen from Figure 3.1 that a force has size (magnitude), direction and a point of application. A force is thus a *vector quantity*, that is, it has magnitude and direction. A *scalar* quantity has only magnitude, for example mass. A force may therefore be represented graphically in two dimensions by drawing an arrow to scale with its length representing the magnitude of a force and the head of the arrow indicating the direction in relation to a set of previously defined axes. Figure 3.2 illustrates the graphical representation of a force.

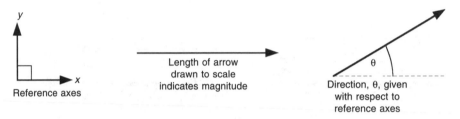

Figure 3.2 *Graphical representation of a force*

Pressure

Pressure, due to the application of force or load, is defined as force per unit area.

$$pressure = \frac{force \text{ or load applied perpendicular to a surface}}{area \text{ over which the force or thrust acts}}$$

The units of pressure in the SI system are normally given as: N m^{-2}, N mm^{-2}, MN m^{-2} or pascal (Pa) where 1 Pa = 1 N m^{-2}. Also pressures in fluid systems are often quoted in bar, where 1 bar = 10^5 Pa, or 100 000 N m^{-2}.

The bar should not be taken as the value for standard atmospheric pressure at sea level. The value quoted in bar for *standard atmospheric pressure* is 1.0132 bar or 101 320 N m^{-2} or 101.32 kPa.

Example 3.4

The area of ground surface contained by the skirt of a hovercraft is 240 m^2. The unladen weight of the craft is 480 kN and total laden weight is 840 kN. Determine the minimum air pressure needed in the skirt to support the craft when unladen and when fully loaded.

When unloaded: pressure = force area = $\dfrac{480 \text{ kN}}{240 \text{ m}^2}$ = **2 kN m^{-2}**

When fully loaded: pressure = $\dfrac{840 \text{ kN}}{240 \text{ m}^2}$ = **3.5 kN m^{-2}**

In practice the skirt would be inflated to the higher of these two pressures and the craft (when static) would rest in the water, at the appropriate level.

Temperature

Temperature is a measure of the quantity of energy possessed by a
body or substance, it is a measure of the molecular vibrations
within the body. The more energetic these vibrations become, the
hotter will be the body or substance. For this reason, in its simplest
sense, temperature may be regarded as the 'degree of hotness of a
body'. We will look more closely at the concept of temperature
when we study heat energy.

Problems 3.1

1. A rocket launched into the earth's atmosphere is subject to an
   acceleration due to gravity of 5.2 m s^{-2}. If the rocket has a
   mass of 120 000 kg, determine:

   (a) the weight of the rocket on earth
   (b) the weight of the rocket in orbit.

2. A solid rectangular body measures 1.5 m × 20 cm × 3 cm and
   has a mass of 54 kg. Calculate:

   (a) its volume in m^3
   (b) its density in Pa
   (c) its relative density.

3. A fuel storage depot consists of four fuel tanks. Two tanks
   each have a volume of 20 m^3 and the other two tanks each have
   a volume of 30 m^3. The fuel used in the depot has a relative
   density (RD) of 0.85. Determine the weight of fuel when all
   the tanks are full.

4. A body has a weight of 550 N on the surface of the earth.

   (a) What force is required to give it an acceleration of 6 m s^{-2}?
   (b) What will be the inertia reaction of the body when given
       this acceleration?

Matter

We have already defined *mass* as the amount of matter in a body
but what is the nature of this matter?

All *matter* or *material* is made up from elementary building
blocks that we know as atoms and molecules. The *atom* may be
further subdivided into *protons*, *neutrons* and *electrons*. Physicists
have discovered many more elementary subatomic particles that,
for the purposes of this discussion, we do not need to consider.

A *molecule* consists of a collection of two or more atoms, which
are joined chemically, in a certain way, to give the material its
macroscopic properties. The act of joining atoms and/or joining
molecules to form parent material is known as *chemical bonding*.

The driving force that encourages atoms and molecules to combine in certain ways is *energy*. Like everything else in nature, matter or material is formed as a consequence of the atoms and/or molecules combining in such a way that, once formed, attain their lowest energy state. We may define *energy* as the capacity to do work. Like nature, we measure our efficiency with respect to work, in terms of the least amount of energy we expend.

The states of matter

Matter exists in *three states*; each state is due to the struggle between the interatomic or intermolecular binding forces and the motion that these atoms and/or molecules have because of their own *internal energy*.

Solids

When matter is chemically bonded together there are *forces of attraction* which bind the atoms and molecules together. Also at very short distances from the centre of the adjacent atoms there are *repulsion forces*. Now, whether or not the force of attraction or repulsion dominates, depends on the atomic distance between the atoms/molecules when combined. It has been shown that at distances greater than one atomic diameter the forces of attraction dominate, while at very small separation distances, the reverse is true. In solids the forces of attraction and repulsion tend to balance each other out. Thus in a solid, if the atoms are brought closer by compression, they will repel each other and if pulled further apart they attract.

Liquids

As temperature increases the *amplitude* (size) of the internal vibration energy of the atoms increases, until they are able to partly overcome the interatomic bonding forces of their immediate neighbours. For short spells they are within range of forces exerted by other atoms which are not quite so near. There is less order and so the solid liquefies. Although the atoms and molecules of a liquid are not much further apart than in a solid, they have greater speeds due to increased temperature and so move randomly in the liquid, while continuing to vibrate. However, the primary differences between liquids and solids may be attributed to *differences in structure*, rather than distance between the atoms. It is these differences in the forces between the molecules which give the liquid its flow characteristics while at the same time holding it sufficiently together, to exhibit shape, within a containing vessel.

Gases

In a gas the atoms and molecules move randomly with high speeds and take up all the space in the containing vessel. Gas molecules are therefore relatively far apart when compared with solids and liquids. Because of the relatively large distances involved, molecular interaction occurs only for those brief spells when molecules collide and large repulsive forces operate between them.

The idea of a gas filling the vessel in which it is contained has its origins in Newton's first law of motion. Each molecule will, in

Key point

Matter is generally considered to exist in solid, liquid and gaseous forms.

Key point

The atoms within solids tend to combine in such a manner that the interatomic binding forces are balanced by the very short range repulsion forces.

Key point

Gases always fill the available space of the vessel into which they are introduced.

Test your knowledge 3.4

1. Explain the essential difference between solids and liquids.
2. Over what sort of distances do the atomic repulsion forces act?
3. How is the *internal energy* within matter defined?

accordance with this law, travel in a straight line until it collides with another molecule or with the sides of the containing vessel. Therefore, a gas has no particular shape or volume but expands until it fills any vessel into which it is introduced.

MECHANICS

Mechanics is the physical science concerned with the state of rest or motion of bodies under the action of forces. This subject has played a major role in the development of engineering throughout history and up to the present day. Modern research and development in the fields of vibration analysis, structures, machines, spacecraft, automatic control, engine performance, fluid flow, electrical apparatus and subatomic, atomic and molecular behaviour are all reliant on the basic principles of mechanics.

The subject of mechanics is conveniently divided into two major areas: *statics*, which is concerned with the equilibrium of rigid bodies under the action of forces, and *dynamics*, which is concerned with the motion of bodies under the action of forces. Dynamics may be further subdivided into the motion of rigid bodies and the motion of fluids.

Statics

Vector representation of forces

You have already met the concept of *force* when we looked at some important fundamentals. You will remember that the effect of a force was dependent on its magnitude, direction and point of application (Figure 3.1), and that a force may be represented on paper as a *vector* quantity (Figure 3.2).

We will now study the vector representation of a force or combination of forces in more detail, noting that all vector quantities throughout this book will be identified using emboldened text.

In addition to possessing the properties of magnitude and direction from a given reference (Figure 3.2), vectors must obey the *parallelogram law* of combination. This law requires that two vectors $\mathbf{v_1}$ and $\mathbf{v_2}$ may be replaced by their equivalent vector $\mathbf{v_T}$ which is the diagonal of the parallelogram formed by $\mathbf{v_1}$ and $\mathbf{v_2}$ as shown in Figure 3.3(a). This vector sum is represented by the vector equation:

$$\mathbf{v_T} = \mathbf{v_1} + \mathbf{v_2}$$

Note that the plus sign in this equation refers to the addition of two vectors, and should not be confused with ordinary scalar addition, which is simply the sum of the magnitudes of these two vectors and is written as $v_T = v_1 + v_2$, in the normal way without emboldening.

Vectors may also be added head-to-tail using the *triangle law* as shown in Figure 3.3(b). It can also be seen from Figure 3.3(c) that the order in which vectors are added does not affect their sum.

The vector difference $\mathbf{v_1} - \mathbf{v_2}$ is obtained by *adding* $-\mathbf{v_2}$ to $\mathbf{v_1}$. The effect of the minus sign is to reverse the direction of the vector $\mathbf{v_2}$ (Figure 3.3(d)). The vectors $\mathbf{v_1}$ and $\mathbf{v_2}$ are known as the components of the vector $\mathbf{v_T}$.

Key point

Two vectors may be added using the parallelogram rule or triangle rule.

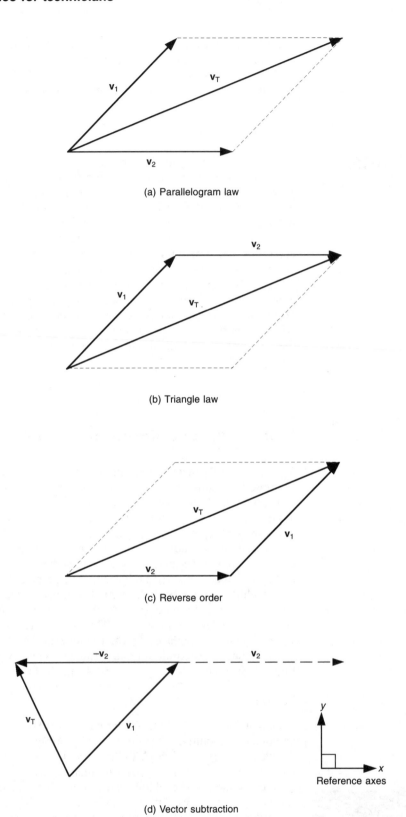

(a) Parallelogram law

(b) Triangle law

(c) Reverse order

(d) Vector subtraction

Figure 3.3 *Vector addition and subtraction*

Example 3.5

Two forces act at a point as shown in Figure 3.4. Find by vector addition their *resultant* (their single equivalent force).

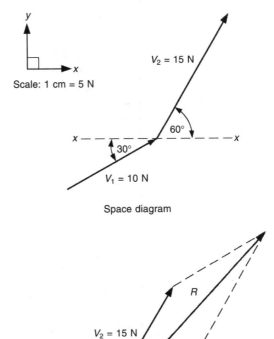

Figure 3.4 *Vector addition using the parallelogram law*

From the vector diagram the resultant vector **R** is 4.8 cm in magnitude which (from the scale) is equivalent to 24 N. So the resultant vector **R** has a magnitude of 24 N at an angle of 48°.

Note that a *space diagram* is first drawn to indicate the orientation of the forces with respect to the reference axes; these axes should always be shown. Also note that the *line of action* of vector v_1 passing through the point O, is shown in the space diagram and may lie anywhere on this line, as indicated on the vector diagram.

Example 3.6

Find the resultant of the system of forces shown in Figure 3.5, using vector addition.

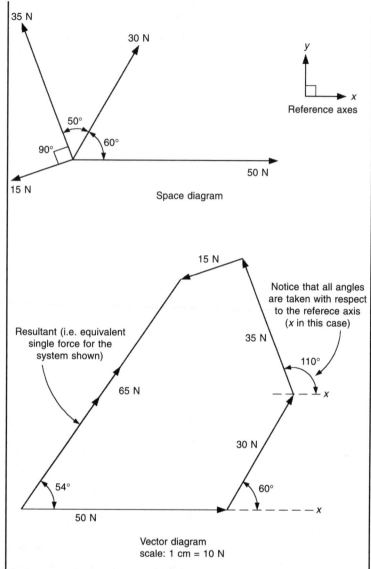

Figure 3.5 *Vector addition using polygon of forces*

From the diagram the resultant = 6.5 cm = 6.5 × 10 N = 65 N. Acting at an angle of 54° from the *x*-reference axis. This result may be written mathematically as resultant = 65 N ∠54°.

Note that for the force system in Example 3.6 vector addition has produced a polygon. Any number of forces may be added vectorially in any order, providing the *head-to-tail rule* is observed. In this example, if we were to add the vectors in reverse order, the same result would be achieved.

If a force, or system of forces, is acting on a body and is balanced by some other force, or system of forces, then the body is said to be in *equilibrium*, so, for example, a stationary body is in equilibrium.

The *equilibrant* of a system of forces is that force which, when added to a system, produces equilibrium. It has been shown in Examples 3.5 and 3.6 that the resultant is the single force which will replace an existing system of forces and produce the same effect. It therefore follows that if the equilibrant is to produce equilibrium it must be equal in magnitude and direction, but opposite in sense to the resultant, Figure 3.6 illustrates this point.

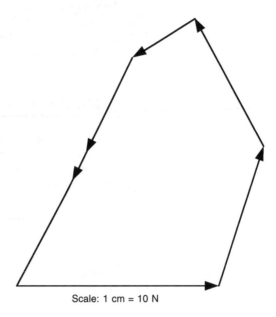

Scale: 1 cm = 10 N

Figure 3.6   *Equilibrant for Example 3.6*

Bow's notation is a convenient system of labelling the forces for ease of reference, when there are three or more forces to be considered. Capital letters are placed in the space between forces in a clockwise direction, as shown in Figure 3.7.

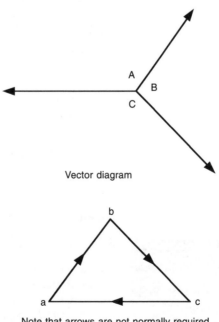

Vector diagram

Note that arrows are not normally required but are shown here for clarity

Figure 3.7   *Bow's notation*

Any force is then referred to by the letters that lie in the adjacent spaces either side of the vector arrow representing that force. The vectors representing the forces are then given the corresponding *lower case* letters. Thus the forces AB, BC and CA are represented by the vectors ab, bc and ca, respectively. This method of labelling applies to any number of forces and their corresponding vectors. Arrowheads need not be used when this notation is adopted, but are shown in Figure 3.7 for clarity.

Resolution of forces

Graphical solutions to problems involving forces are sufficiently accurate for many engineering problems and are invaluable for estimating approximate solutions to more complicated force problems. However, it is sometimes necessary to provide more accurate results, in which case a mathematical method will be required. One such mathematical method is known as the *resolution of forces*.

Consider a force F acting on a bolt (Figure 3.8). The force F may be replaced by two forces P and Q, acting at right angles to each other, which together have the same effect on the bolt.

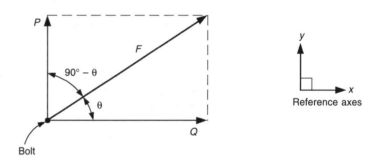

Figure 3.8 *Resolving force F into its components*

From our knowledge of the trigonometric ratios (Unit 4) we know that:

$$\frac{Q}{F} = \cos \theta \text{ and so } Q = F \cos \theta$$

Also, $\dfrac{P}{F} = \cos(90 - \theta)$ and we know that $\cos(90 - \theta) = \sin \theta$

therefore $P = F \sin \theta$

So from Figure 3.8, $P = F \sin \theta$ and $Q = F \cos \theta$

So the single force F has been resolved or split into two equivalent forces of magnitude $F \cos \theta$ and $F \sin \theta$, which act at right angles (they are said to be *orthogonal* to each other). $F \cos \theta$ is known as the *horizontal component of F* and $F \sin \theta$ is known as the *vertical component of F*.

Determination of the resultant or equilibrant using the resolution method is best illustrated by example.

Key point

The resultant of two or more forces is that force which, acting alone, would produce the same effect as the other forces acting together.

Example 3.7

Three *coplanar forces* (forces that act within the same plane), *A, B* and *C*, are all applied to a pin joint (Figure 3.9(a)). Determine the magnitude and direction of the equilibrant for the system.

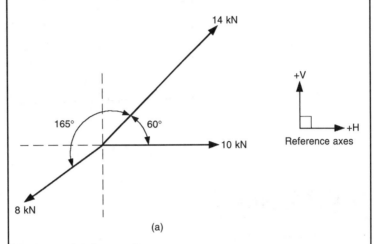

Figure 3.9(a) *Space diagram*

Each force needs to be resolved into its two orthogonal components, which act along the vertical and horizontal axes respectively. Using the normal algebraic sign convention with our axes then, above the origin, *V* is positive and below it is negative. Similarly, *H* is positive to the right of the origin and negative to the left. Using this convention we need only consider acute angles for the sine and cosine functions; these are tabulated below.

Magnitude of force	Horizontal component	Vertical component
10 kN	+10 kN (→)	0
14 kN	+14 cos 60 kN (→)	+14 sin 60 kN (↑)
8 kN	− 8 cos 45 kN (←)	−8 sin 45 kN (↓)

Then total horizontal component = 10 + 7 − 5.66 kN = 11.34 kN (→)

and total vertical component   = 0 + 12.22 − 5.66 kN = 6.46 kN (↑)

Since both the horizontal and vertical components are positive the resultant force will act upwards to the right of the origin. The three original forces have now been reduced to two which act orthogonally. The magnitude of the resultant *R*, or the equilibrant, may now be obtained using Pythagoras' theorem on the right angle triangle obtained from the orthogonal vectors, as shown in Figure 3.9(b).

From Pythagoras we get $R^2 = 6.46^2 + 11.34^2 = 170.33$ and so resultant *R* = 13.05 kN, so the magnitude of the *equilibrant* also = 13.05 kN.

From the right angled triangle shown in Figure 3.9(b), the angle θ that the resultant *R* makes with the given axes may be calculated using the trigonometric ratios.

Then $\tan \theta = \dfrac{6.46}{11.34} = 0.5697$ and θ = 29.67°

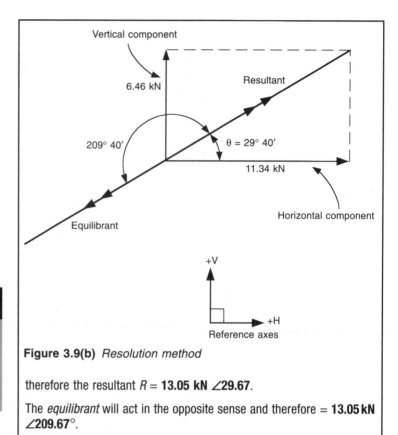

Figure 3.9(b) *Resolution method*

therefore the resultant R = **13.05 kN ∠29.67**.

The *equilibrant* will act in the opposite sense and therefore = **13.05 kN ∠209.67°**.

Key point

The equilibrant is that force which acting alone against the other forces acting on a body in the system places the body in equilibrium.

To complete our initial study on the resolution of forces, we consider one final example concerned with *equilibrium on a smooth plane*. Smooth in this case implies that the effects of friction may be ignored. When we study dynamics latter on in this unit, friction and its effects will be covered in some detail.

A body is kept in equilibrium on a plane by the action of three forces as shown in Figure 3.10, these are:

1. the *weight W* of the body acting vertically down
2. *reaction R* of the plane to the weight of the body. *R* is known as the *normal reaction*, normal in this sense means at right angles to, the plane in this case
3. *force P* acting in some suitable direction to prevent the body sliding down the plane.

Forces *P* and *R* are dependent on:

*   the angle of inclination of the plane
*   the magnitude of *W*
*   the inclination of the force *P* to the plane.

It is therefore possible to express the magnitude of both *P* and *R* in terms of *W* and the trigonometric ratios connecting the angle θ.

In the example that follows we consider the case when the body remains in equilibrium as a result of the force *P* being applied parallel to the plane.

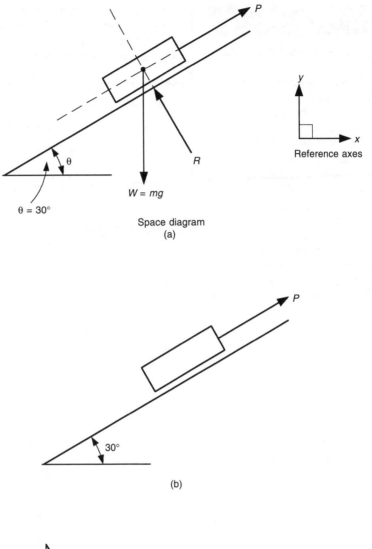

Space diagram
(a)

(b)

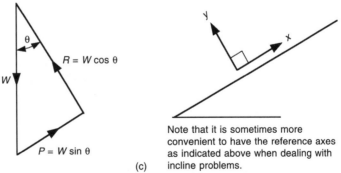

Note that it is sometimes more
convenient to have the reference axes
as indicated above when dealing with
(c)   incline problems.

Figure 3.10 *Equilibrium on a smooth plane*

Example 3.8

A crate of mass 80 kg is held in equilibrium by a force *P* acting
parallel to the plane as indicated in Figure 3.10(a). Determine, using
the resolution method, the magnitude of the force *P* and the normal
reaction *R*, ignoring the effects of friction.

Figure 3.10(b) shows the space diagram for the problem clearly indicating the nature of the forces acting on the body. *W* may therefore be resolved into the two forces *P* and *R*. Since the force component at right angles to the plane = *W* cos θ and the force component parallel to the plane = *W* sin θ (Figure 3.10(c)).

Equating forces gives, $W \cos \theta = R$ and $W \sin \theta = P$

So, remembering the mass/weight relationship, we have:

$W = mg = (80)(9.81) = 784.8$ N then,

$R = 784.8 \cos 30° = \textbf{679.7 N}$ and

$P = 784.8 \sin 30° = \textbf{392.4 N}$

Moments and couples

A moment is a turning force producing a turning effect. The magnitude of this turning force depends on the size of the *force* applied and the *perpendicular distance* from the pivot or axis to the line of action of the force (Figure 3.11(a)).

Examples of a turning force are numerous: opening a door, using a spanner, turning the steering wheel of a motor vehicle, an

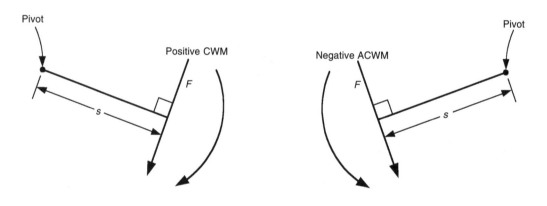

(a) Definition of a moment

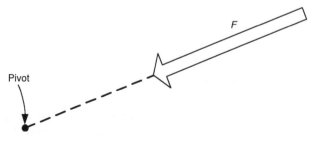

(b) Line of action passing through pivot point

Figure 3.11 *Moment of a force*

aircraft tailplane creating a nose-up and nose-down moment are just four examples.

The moment of a force *M* is defined as:

the product of the magnitude of force F and its perpendicular distance s from the pivot or axis to the line of action of the force.

This may be written mathematically as:

$M = Fs$

The SI unit for a moment is the *newton-metre* (N m).

Moments are always concerned with perpendicular distances.

From Figure 3.11(a), you should note that moments can be clockwise *CWM* or anticlockwise *ACWM*. Conventionally we consider clockwise moments to be positive and anticlockwise moments to be negative.

If the line of action of the force passes through the turning point it has no turning effect and so no moment, Figure 3.11(b) illustrates this point.

Example 3.9

Figure 3.12 shows a spanner being used to loosen a nut. Determine the turning effect on the nut.

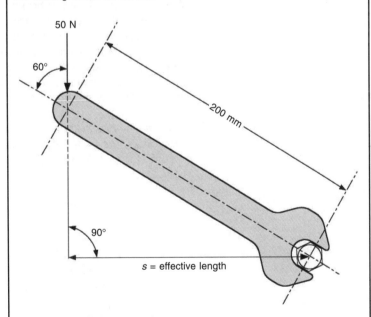

Figure 3.12 *Spanner and nut*

The turning effect on the nut is equal to the moment of the 50 N force about the nut, i.e.

$M = Fs$

Remembering that moments are always concerned with perpendicular distances, the distance *s* is the perpendicular distance or effective length of the spanner. This length is found using trigonometric ratios:

$s = 200 \sin 60°$, therefore $s = (200)(0.866) = 173.2$ mm

Then ACWM $= (50)(173.2) = 8660$ Nmm or 8.66 Nm

So the *turning effect* of the 50 N force acting on a 200 mm spanner at 60° to the centre line of the spanner = **8.66 N m**.

In engineering problems concerning moments you will meet terminology that is frequently used. You are already familiar with the terms CWM and ACWM. Set out below are three more frequently used terms that you are likely to encounter.

Fulcrum:
the *fulcrum* is the point or axis about which rotation takes place. In Example 3.9 above, the geometrical centre of the nut is considered to be the fulcrum.

Moment arm:
the perpendicular distance from the line of action of the force to the fulcrum is known as the *moment arm*.

Resulting moment:
the *resulting moment* is the difference in magnitude between the total clockwise moment and the total anticlockwise moment. Note that if the body is in *static equilibrium* this *resultant will be zero*.

> **Key point**
>
> For static equilibrium the algebraic sum of the moments is zero.

When a body is in equilibrium there can be no resultant *force* acting on it. However, reference to Figure 3.13 shows that a body is not necessarily in equilibrium even when there is no resultant force acting on it. The resultant force on the body is zero but two forces would cause the body to rotate, as indicated. A second condition must be stated to ensure that a body is in equilibrium. This is known as the *principle of moments*, which states:

When a body is in static equilibrium under the action of a number of forces, the total CWM about any point is equal to the total ACWM about the same point.

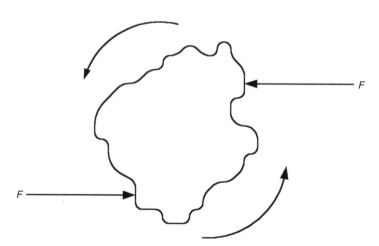

Figure 3.13 *Non-equilibrium condition for equal and opposite forces acting on a body*

This means that for static equilibrium the *algebraic sum of the moments must be zero*.

One other important fact needs to be remembered about bodies in static equilibrium, consider the uniform beam (uniform here means an equal cross-section along its total length) shown in Figure 3.14. We already know from the principle of moments that the sum of the CWM must equal the sum of ACWM. It is also true that the beam would sink into the ground or rise if the upward forces did not equal the downward forces. So a further necessary condition for static equilibrium is that:

upward forces = downward forces

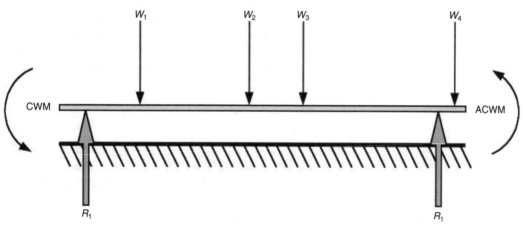

Figure 3.14 *Conditions for equilibrium*

We now have sufficient information to readily solve further problems concerning moments.

Example 3.10

A uniform horizontal beam is supported on a fulcrum (Figure 3.15). Calculate the force F necessary to ensure the beam remains in equilibrium.

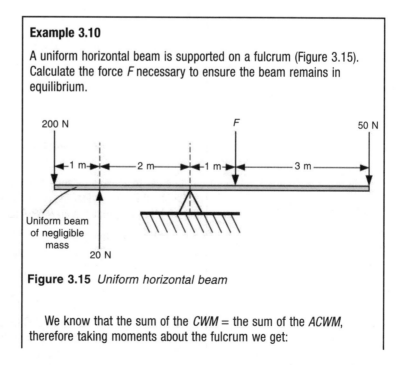

Figure 3.15 *Uniform horizontal beam*

We know that the sum of the *CWM* = the sum of the *ACWM*, therefore taking moments about the fulcrum we get:

$(F \times 1) + (50 \times 4) + (20 \times 2) = (200 \times 3)$ Nm

then, $(F \times 1) + 200 + 40 = 600$ Nm or,

$(F \times 1) = 600 - 200 - 40$ Nm so,

$$F = \frac{360 \text{ Nm}}{1 \text{ m}} = \textbf{360 N}$$

Notes: (a) The 20 N force acting at a distance of 2 m from the fulcrum, tends to turn the beam *clockwise* so is *added* to the sum of the CWM.
(b) The units of F are as required, i.e. they are in newtons, because the RHS is in Nm and is divided by 1 m.
(c) In this example the weight of the beam has been ignored. If the beam is of uniform cross-section, then its mass is deemed to act at its geometrical centre.

Example 3.11

Figure 3.16 shows a motion control system crank lever ABC pivoted at B. AB is 20 cm and BC is 30 cm. Calculate the magnitude of the vertical rod force at C required to balance the horizontal control rod force of magnitude 10 kN applied at A.

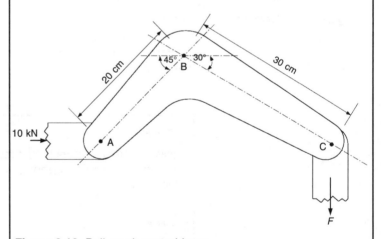

Figure 3.16 *Bell crank control lever*

In order to achieve balance of the forces acting on the lever the CWM about B must equal the ACWM about B. It can also be seen that the 10 kN force produces an ACWM about the fulcrum B. Therefore:

moment of 10 kN force about B = $(10 \times 0.2 \sin 45°)$ kN m (note the manipulation of units)

$= (10)(0.2)(0.7071)$ kN m

$= \textbf{1.414 kN m}$

If we now let the vertical force at C be of magnitude F then F produces a clockwise moment about fulcrum B. Therefore:

moment of force of magnitude F about B $= F \times (0.3 \cos 30°) = 0.26F$

Applying the principle of moments for equilibrium, we get:

$$1.414 = 0.26F$$

therefore $\quad F = \dfrac{1.414 \text{ kN}}{0.26 \text{ m}} = \textbf{5.44 kN}$

Our final example on moments introduces the idea of the *uniformly distributed load* (UDL). In addition to being subject to point loads, beams can be subjected to loads that are distributed for all, or part, of the beam's length. For UDLs the whole mass of the load is assumed to act as a point load through the centre of the distribution.

Example 3.12

For the beam system shown in Figure 3.17, determine the reactions at the supports R_A and R_B, taking into consideration the weight of the beam.

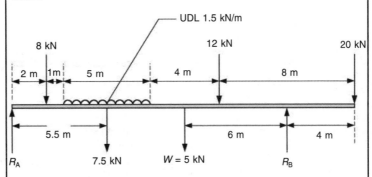

Figure 3.17 *Beam system taking account of weight of beam*

So from what has been said, the UDL acts as a point load of magnitude (1.5 kN $\times$ 5 = 7.5 kN) at the centre of the distribution, which is 5.5 m from R_A.

In problems involved with reaction it is essential to eliminate one reaction from the calculations because only one equation is formed and only one unknown can be solved at any one time. This is achieved by taking moments about one of the reactions and then, since the distance from that reaction is zero, its moment is zero and it is eliminated from the calculations.

So taking moments about A (thus eliminating A from the calculations), we get:

$$(2 \times 8) + (5.5 \times 7.5) + (10 \times 5) + (12 \times 12) + (20 \times 20) = 16R_B$$

or $\hspace{8cm} 651.25 = 16R_B$

so the reaction at $B = \textbf{40.7 kN}$

We could now take moments about B in order to find the reaction at A. However, at this stage, it is easier to use the fact that for static equilibrium:

upward forces = downward forces

so $R_A + R_B = 8 + 7.5 + 5 + 12 + 20$

$R_A + 40.7 = 52.5$

and so the reaction at A = **11.8 kN**

Couples

So far we have restricted our problems on moments to the turning effect of forces taken one at a time. A *couple occurs when two equal forces acting in opposite directions have their lines of action parallel.*

Example 3.13

Figure 3.18 shows the turning effect of a couple on a beam of regular cross-section.

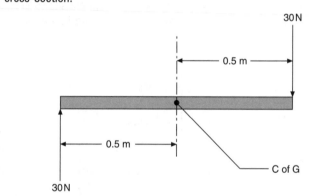

Figure 3.18 *Turning effect of a couple with regular cross section*

Taking moments about the *centre of gravity* (C of G) (the point at which all the weight of the beam is deemed to act), then we get:

$(30 \times 0.5) + (30 \times 0.5)$ = moment

So moment of couple = **30 Nm**

Example 3.14

Figure 3.19 shows the turning effect of a couple on a beam of irregular

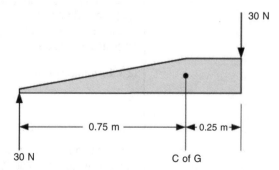

Figure 3.19 *Turning effect of couple, with irregular cross-section beam*

Stress and strain

cross-section, which we will again try to revolve about its centre of gravity.

Again taking moments about the C of G gives:

$(30 \times 0.75) = (30 \times 0.25) = $ moment

So the moment of couple = **30 Nm**

It can be seen from the above two examples that the moment is the same in both cases and is independent of the position of the fulcrum. Therefore if the fulcrum is assumed to be located at the point of application of one of the forces the *moment of a couple* is equal to one of the forces multiplied by the perpendicular distance between them. Thus in both cases shown in Examples 3.13 and 3.14 the *moment of the couple = (30 N × 1 m) = 30 Nm*, as before.

Another important application of the couple is its *turning moment* or *torque*. The definition of torque is as follows:

Torque is the turning moment of a couple and is measured in newton-metres (Nm): torque T = force F × radius r.

The *turning moment* of the couple given above in Example 3.14 is $= F \times r = (30 N \times 0.5 m) = 15 Nm$.

Example 3.15

A nut is to be torque loaded to a maximum of 100 Nm. What is the maximum force that may be applied, perpendicular to the end of the spanner, if the spanner is of length 30 cm.

Since $T = F \times r$ then $F = T/r = 100/30$ therefore $F = $ **333.3 N**

Stress

If a solid, such as a metal bar, is subjected to an external force (or load), a resisting force is set up within the bar and the material is said to be in a state of stress. There are three basic types of stress:

- *tensile stress* – which is set up by forces tending to pull the material apart
- *compressive stress* – produced by forces tending to crush the material
- *shear stress* – resulting from forces tending to cut through the material, i.e. tending to make one part of the material slide over the other.

Figure 3.20 illustrates these three types of stress.

Definition of stress

Stress is defined as force per unit area, i.e.

$$\text{stress, } \sigma = \frac{\text{force, } F}{\text{area, } A}$$

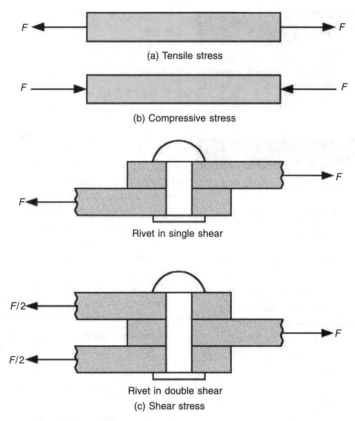

Figure 3.20 *Basic types of stress (a), (b) and (c)*

Note that the Greek letter σ is pronounced sigma.

The basic SI unit of stress is the N m^{-2}; other commonly used units include MN m^{-2}, N mm^{-2} and the pascal (Pa).

In engineering structures components that are designed to carry tensile loads are known as *ties*, while components design to carry compressive loads are known as *struts*.

Strain

A material that is altered in shape due to the action of a force acting on it is said to be *strained*. This may also mean that a body is strained internally even though there may be little measurable difference in its dimensions, just a stretching of the bonds at the atomic level. Figure 3.21 illustrates three common types of strain resulting from the application of external forces (loads).

Definition of strain

Direct strain may be defined as: *the ratio of change in dimension (deformation) over the original dimension*, i.e.

$$\text{direct strain, } \varepsilon = \frac{\text{deformation, } x}{\text{original length, } l}$$

(both *x* and *l* are in metres).

The symbol ε is the Greek lower case letter *epsilon*. Note also that the deformation for tensile strain will be an extension and for compressive strain it will be a reduction.

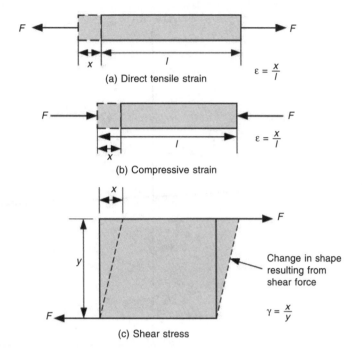

Figure 3.21 *Types of strain (a), (b) and (c)*

Hooke's law

Hooke's law states that: *within the elastic limit of a material the change in shape is directly proportional to the applied force producing it.*

A good example of the application of Hooke's law is the *spring*. A spring balance is used for measuring weight force, where an increase in weight will cause a corresponding extension (see Figure 3.22).

The stiffness (k) of a spring is the force required to cause a certain (unit deflection).

$$\text{stiffness } (k) = \frac{\text{force}}{\text{deflection}}$$

SI units are N m^{-1}.

The concept of elasticity will be looked at in a moment, in the mean time here is a question to consider. What does the slope of the graph in Figure 3.22 indicate?

Modulus

Modulus of elasticity

By considering Hooke's law, it follows that stress is directly proportional to strain, while the material remains *elastic*. That is, while the external forces acting on the material are only sufficient to stretch the atomic bonds, without fracture, so that the material may return to its original shape after the external forces have been removed.

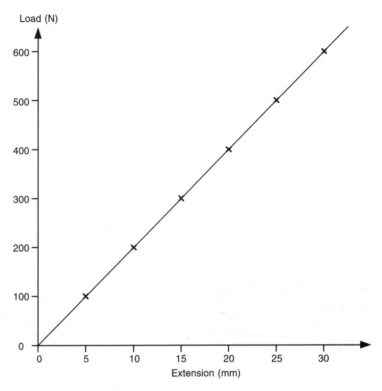

Figure 3.22 *Load–extension graph*

Then from Hooke's law and our definition of stress and strain, we know that *stress is directly proportional to strain in the elastic range*, i.e.

stress ∝ strain   or   stress = (strain × a constant)

so: $\dfrac{\text{stress}}{\text{strain}}$ = a constant, then $\mathbf{\mathit{E} = \dfrac{stress}{strain}}$ (where E is the constant)

This constant of proportionality will depend on the material and is given the symbol E. It is known as the *modulus of elasticity* and because strain has no units it has the same units as stress. Modulus values tend to be very high, for this reason GN m^{-2} or GPa are the preferred SI units.

> ### Key point
>
> The *elastic modulus* of a material may be taken as a measure of the *stiffness* of that material.

Modulus of rigidity

The relationship between the shear stress (τ) and shear strain (γ) is known as the modulus of rigidity (G), i.e.

$$\textbf{modulus of rigidity } (\mathbf{\mathit{G}}) = \frac{\textbf{shear stress } (\boldsymbol{\tau})}{\textbf{shear strain } (\boldsymbol{\gamma})} \text{ (units GPa or GN m}^{-2}\text{)}$$

Note that the symbol τ is the lower case Greek letter tau and the symbol γ is the lower case Greek letter gamma.

Example 3.16

A rectangular steel bar 10 mm × 16 mm × 200 mm long extends by 0.12 mm under a tensile force of 20 kN. Find:

(a) the stress

(b) the strain

(c) the elastic modulus of the bar material.

(a) Now, tensile stress $= \dfrac{\text{tensile force}}{\text{cross-sectional area}}$

Also tensile force $= 20$ kN $= 20 \times 10^3$ N and cross-sectional area $= 10 \times 16 = 160$ mm². Remember tensile loads act against the cross-sectional area of the material.

Then substituting in the above formula we have, tensile stress

$$(\sigma) = \frac{20\,000 \text{ N}}{160 \text{ mm}^2}$$

$$\sigma = \mathbf{125 \text{ N mm}^{-2}}$$

(b) Now, strain $\varepsilon = \dfrac{\text{deformation (extension)}}{\text{original length}}$

Also extension $= 0.12$ mm and the original length $= 200$ mm.

Then substituting gives $\varepsilon = \dfrac{0.12 \text{ mm}}{200 \text{ mm}} = 0.0006$

(c) $E = \dfrac{\text{stress}}{\text{strain}} = \dfrac{125 \text{ N mm}^{-2}}{0.0006} = 208\,000$ N mm⁻² or **208 GN m⁻²**

Example 3.17

A 10 mm diameter rivet holds three sheets of metal together and is loaded as shown in Figure 3.23. Find the shear stress in the bar.

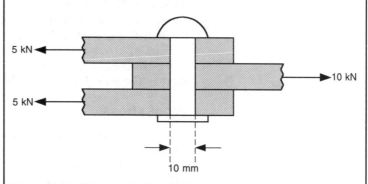

Figure 3.23 *Rivet in double-shear*

We know that each rivet is in double shear.

So the area resisting shear $= 2 \times$ the cross-sectional area

$$= 2\pi r^2 = 2\pi 5^2 = 157 \text{ mm}^2$$

So shear stress $(\tau) = \dfrac{10\,000}{157} = 63.7$ N mm⁻² = **63.7 MN m⁻²**

Note that when a rivet is in double shear, the area under shear is multiplied by 2. With respect to the load we know from Newton's laws that to every action there is an equal and opposite reaction, thus *we only use the action or reaction* of a force in our calculations, *not both*.

Test your knowledge 3.7

1. Define:
   (a) tensile stress;
   (b) shear stress;
   (c) compressive stress.
2. State Hooke's law and explain its relationship to the elastic modulus.
3. Define spring stiffness and quote its SI unit.
4. Define in detail the terms:
   (a) elastic modulus;
   (b) shear modulus;
   (c) bulk modulus.
5. Convert the following into N/m²:
   (a) 240 kN m⁻²;
   (b) 0.228 gap;
   (c) 600 N mm⁻²;
   (d) 0.0033 N mm⁻²;
   (e) 10 kN m⁻².
6. Explain the use of:
   (a) a strut;
   (b) a tie.

Some definitions of mechanical properties

The mechanical properties of a material are concerned with its behaviour under the action of external forces. This is of particular importance to us when considering materials for engineering applications. Here, we will concentrate on a few *simple* definitions of the more important mechanical properties of materials that are needed for our study of statics.

These properties include strength, stiffness, specific strength and stiffness, ductility, toughness, malleability and elasticity, in addition to others given below. We have already considered *stiffness*, which is measured by the *elastic modulus*. Indirectly, we have also defined strength when we considered the various forms of stress that result from the loads applied on a material. However, a more formal definition of strength follows.

Strength

Strength may be defined simply as the applied force a material can withstand prior to fracture. In fact strength is measured by the *yield stress* σ_y or *proof stress* σ_p (see below) of a material. This stress is measured at a known percentage yield for the material under test. Yielding occurs when the material is subject to loads that cause it to extend by a known fraction of its original length. For metals the measure of strength is often taken at the 0.2% yield or 0.2% proof stress.

Working stress

Following on from the argument given above, we now need to define one or two additional types of stress, since these measure the strength characteristics of materials, under varying circumstances.

Working stress is the stress imposed on the material as a result of the worst possible loads that the material is likely to sustain in service. These loads must be within the elastic range of the material.

Proof stress

Proof stress may be formally defined as: *the tensile stress which when applied for a period of 15 seconds and removed, produces a permanent set of a specified amount, usually 0.2 per cent, that is 0.002 or one-five hundredth of the original dimension.*

Ultimate tensile stress (UTS)

The *UTS* of a material is given by the relationship, *maximum load/ original cross-sectional area*. Note that the UTS is a measure of the ultimate tensile strength of the material. The point U on the load–extension graph (Figure 3.24) shows maximum load, this must be divided by the original cross-sectional area (csa), not that directly under the point U where the extension may have altered the original csa.

Specific strength

In many engineering applications such as aircraft and high performance motor vehicles, materials need to be as light and strong

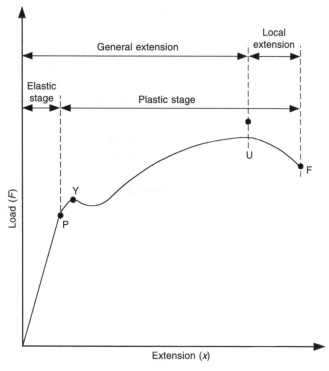

Figure 3.24 *Load–extension curve for a mild steel test piece*

as possible, in order to maximize their performance, while at the same time meeting the stringent safety requirements laid down for their load bearing structures. Thus to be *structurally efficient* aircraft and racing cars need to be made of low density materials, which have the greatest strength. The ratio of the strength of a material (measured by its yield stress) to that of its density is known as *specific strength*, i.e.

$$\text{specific strength} = \frac{\text{yield strength } (\sigma_y)}{\text{density } (\rho)}$$

SI units are joules per kilogram (J kg^{-1}).

Specific stiffness

In a similar manner to the argument given above, the *specific stiffness* of a material is the ratio of its stiffness (measured by its elastic modulus) to that of its density, i.e.

$$\text{specific modulus} = \frac{\text{elastic modulus } (E)}{\text{density } (\rho)}$$

SI units are again joules per kilogram (J kg^{-1}).

Ductility

Ductility is the ability to be drawn out into threads or wire. Wrought iron, aluminum, titanium and low carbon steels are examples of ductile materials.

Brittleness

Brittleness is the tendency to break easily or suddenly with little or no prior extension. Cast iron, high carbon steels and glass are examples of brittle materials.

Toughness

Toughness is the ability to withstand suddenly applied shock loads. Certain alloy steels, some plastics and rubber are examples of tough materials.

Malleability

Malleability is the ability to be rolled into sheets or shaped under pressure. Examples of malleable materials include, gold, copper and lead.

Elasticity

Elasticity is the ability of a material to return to its original shape once external forces have been removed. Internal atomic binding forces are stretched but not broken and act like minute springs to return the material to normal, once force has been removed. Rubber, mild and medium carbon steels are good examples of elastic materials.

Safety factors

The *safety factor* is used in the design of materials subject to service loads, to give a margin of safety and take account of a *certain factor of ignorance*. Factors of safety vary in engineering design, dependent on the structural sensitivity of the member under consideration. They are often around 1.5, but can be considerably higher for joints, fittings, castings and primary load bearing structures in general.

Load–extension graphs

These show the results of mechanical tests used to determine certain properties of a material. For instance as a check to see if heat treatment or processing has been successful, a sample from a batch would be used for such tests.

Load–extension graphs show certain phases. When a material is tested to destruction these include: elastic range, limit of proportionality, yield point, plastic stage and final fracture.

Figure 3.24 shows a typical load–extension curve for a specimen of mild steel which is a ductile material.

The point P at the end of the straight line OP is called the *limit of proportionality*. Between the origin O and P the extension x is directly proportional to the applied force and in this range the material obeys Hooke's law. The *elastic limit* is at or very near the limit of proportionality. When this limit has been passed the extension ceases to be proportional to the load, and at the *yield point* Y the extension suddenly increases and the material enters its *plastic phase*. At point U (the ultimate tensile strength) the load is greatest. The extension of the test piece has been general up to point U, after which waisting or necking occurs and the subsequent extension is local (Figure 3.25).

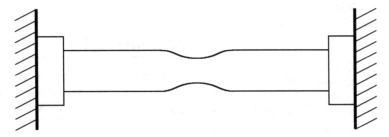

Figure 3.25 *Necking or waisting with local extension*

Since the area at the waist is considerably reduced then from *stress = force/area*, the stress will increase, resulting in a reduced load for a given stress and so fracture occurs at point F, that is at a lower load value than at U.

Remember the elastic limit is at the end of the phase that obeys Hooke's law, after this Hooke's relationship is no longer valid, and full recovery of the material is not possible after removal of the load.

Figure 3.26 shows some typical load–extension curves for some common metals. Where:

HDB = hard drawn 70/30 brass
  CI = cast iron
HDC = hard drawn copper
  AA = aluminium alloy
  AC = annealed copper.

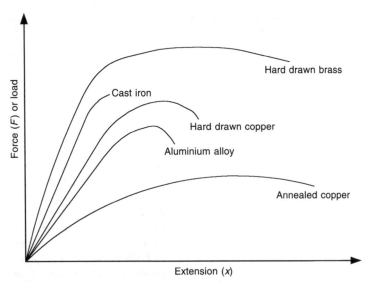

Figure 3.26 *Some typical load–extension graphs*

The above curves show that annealed copper is very ductile, while hard drawn copper is stronger but less ductile. Hard drawn 70/30 brass is both strong and ductile. Cast iron can clearly be seen as brittle and it is for this reason that cast iron is rarely used under tensile load. Aluminium alloy can be seen to be fairly strong yet ductile, it has excellent *structural efficiency* and it is for this reason that it is still used as one of the premier materials for aircraft construction.

Problems 3.2

1. For the force system shown below (Figure 3.27) determine *graphically* the magnitude and direction of the equilibrant. Then use a *mathematical method* to check the accuracy of your result.

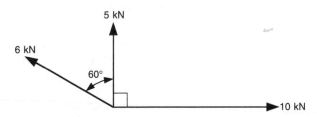

Figure 3.27 *System of forces space diagram*

2. Determine the reactions at the supports for the beam system shown below (Figure 3.28). Assume the beam has negligible mass.

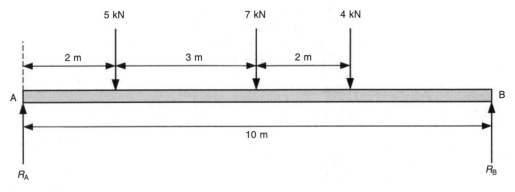

Figure 3.28 *Beam system*

3. A uniform beam of length 5 m and weight 10 kN has to support a uniformly distributed load (UDL) of 1.5 kN m^{-1}. Find the reactions at the supported.
4. An engineering structure contains a steel tie rod that carries a load of 100 kN. If the allowable tensile stress is 75 MN m^{-2}, find the minimum diameter of the tie rod.
5. A hollow copper tube 30 mm outside diameter and 18 mm inside diameter supports an axial compressive load of 60 kN. The axial compression of the cylinder is 0.2 mm and the value of the elastic modulus E for copper is 110 GN m^{-2}. Determine:

   (a) the stress in the copper
   (b) the original length of the tube.

Dynamics

We start our study of dynamics by considering Newton's laws of motion. In order to do this we need to briefly introduce the concepts of speed, velocity, acceleration and momentum, which are fundamental to a proper understanding of these laws.

Speed, velocity and acceleration

Speed may be defined as *distance per unit time*. Speed takes no account of direction and is therefore a *scalar* quantity.

The common SI units of speed are:

kilometres per hour   (kph)
metres per second   (m s^{-1})

Example 3.18

Convert:

(a)   150 kph into m s^{-1}
(b)   120 m s^{-1} into kph.

We will derive the conversion factors for the above speeds by considering the *basic* units.

(a)   We know that there are 10^3 metres in a kilometre so 150 kph = 150 $\times 10^3$ metres per hour. Also there are 3600 seconds in an hour (see Table 3.5). So our required conversion factor is $10^3/3600 = 1/3.6$. So:

$$150 \text{ kph} = (150)(1/3.6) = \textbf{41.67 m s}^{-1}$$

Note that we did not change our conversion factor (1/3.6) into a decimal fraction, because this cannot be done exactly as $1/3.6 = 0.277777$ recurring.

(b)   I hope you can see that all we need do is use the inverse of the conversion factor we have found above to convert m s^{-1} into kph, we are simply converting in reverse. Then the conversion factor becomes $(3.6/1) = 3.6$. So:

$$120 \text{ m s}^{-1} = (120)(3.6) = \textbf{432.0 kph}$$

It will aid your understanding of unit conversion if you attempt to derive your own *conversion factors* from basic units.

Key point

Speed is a scalar quantity whereas velocity is a vector quantity.

Velocity is defined as: *distance per unit time in a specified direction*. Therefore, velocity is a *vector quantity* and the SI units for the magnitude of velocity are the SI units for speed, i.e. m s^{-1}.

The direction of a velocity is not always quoted but it should be understood that the velocity is in some defined direction, even though this direction is unstated.

Acceleration is defined as: *change in velocity per unit time or rate of change of velocity*. Acceleration is also a *vector quantity* and the SI unit of acceleration is m s^{-1}/s or m s^{-2}.

Equilibrium, momentum and inertia

A body is said to be in *equilibrium* when *its acceleration continues to be zero*, that is, when *it remains at rest or when it continues to move in a straight line with constant velocity*.

Momentum may be described as the quantity of motion of a body. *Momentum is the product of the mass of a body and its velocity.* Any change in momentum requires a change in velocity, that is an acceleration. It may be said that for a fixed quantity of matter to be in equilibrium, it must have constant momentum. A more rigorous definition of momentum is given next, when we consider Newton's second law.

All matter resists change. The force resisting change in momentum (that is, acceleration) is called *inertia*. The inertia of a body depends on its mass, the greater the mass, the greater the inertia. The inertia of a body is an innate force that only becomes effective when acceleration occurs. An applied force acts against inertia so as to accelerate (or tend to accelerate) a body.

Newton's laws of motion

Before we consider Newton's laws we need to revisit the concept of force. We already know that force cannot exist without opposition, i.e. action and reaction. If we apply a 100 N pulling force to a rope, this force cannot exist without opposition.

Force is that which changes, or tends to change, the state of rest or uniform motion of a body. Forces that act on a body may be external (applied from outside the body) such as weight, or internal (such as the internal resistance of a material subject to a compression).

The difference between the forces tending to cause motion and those opposing motion is called the *resultant* or *out-of-balance force*. A body that has no out-of-balance external force acting on it is in equilibrium and will not accelerate. A body that has such an out-of-balance force will accelerate at a rate dependent on the mass of the body and the magnitude of the out-of-balance force. The necessary opposition that permits the existence of the out-of-balance force is provided by the force of inertia (Figure 3.29(a) and (b)).

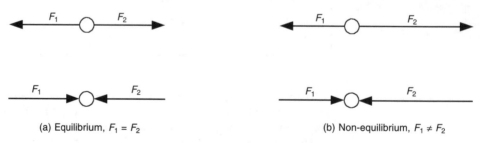

(a) Equilibrium, $F_1 = F_2$ (b) Non-equilibrium, $F_1 \neq F_2$

Figure 3.29 *(a) Equilibrium forces; (b) non-equilibrium forces*

Newton's first law of motion states that: *a body remains in a state of rest, or of uniform motion in a straight line, unless it is acted upon by some external resultant force.*

Newton's second law of motion states that: *the rate of change of momentum of a body is directly proportional to the force producing the change, and takes place in the direction in which the force acts.*

We defined force earlier as force = mass × acceleration. We also know that acceleration may be defined as change in velocity per unit time or rate of change in velocity. If we assume that a body has an *initial velocity* u and a *final velocity* v, then the change in velocity is given by $(v - u)$ and so the rate of change of velocity or acceleration may be written as $\dfrac{(v - u)}{t}$ where t is unit time.

So since $F = ma$ then this may be written as:

$$F = \frac{m(v - u)}{t}$$

and multiplying out the brackets gives:

$$F = \frac{mv - mu}{t}$$

Now we also know that momentum was defined earlier as mass × velocity. So the product mu gives the initial momentum of the body, prior to the application of the force and mv gives the final momentum of the body. Thus the expression $(mv - mu)$ is the change in momentum and so $\dfrac{mv - mu}{t}$ is the rate of change of momentum and so Newton's second law may be expressed as:

$$F = \frac{mv - mu}{t}$$

or

$$F = ma$$

Newton's third law states that: *to every action there is an equal and opposite reaction.*

So, for example, the compressive forces that result from the weight of a building, the *action*, are held in equilibrium by the *reaction* forces that occur inside the materials of the building's foundation. Another example is that of propulsion. An aircraft jet engine produces a stream of high velocity gases at its exhaust, the *action*, these act on the airframe of the aircraft causing a *reaction*, which enables the aircraft to accelerate and increase speed for flight.

Linear equations of motion

You have already been introduced to the concept of force, velocity and acceleration – and now, Newton's laws. All these concepts are further exploited through the use of the *equations of motion*. Look back now and remind yourself of the relationship between mass, force, acceleration and Newton's laws.

The linear equations of motion rely for their derivation on the one very important fact that the *acceleration is assumed to be constant*. We will now consider the derivation of the four standard equations of motion using a graphical method.

Key point

$F = ma$ is a consequence of Newton's second law of motion.

Test your knowledge 3.9

1. Use Table 3.5 to convert the following units:

   (a) 600 km/hr into m/s^1
   (b) 140 km/hr into kph.

2. Define 'inertia' and quote its units.
3. What may we write as the equivalent to the *rate of change of momentum* in Newton's second law.

Key point

For the linear equations of motion to be valid, the acceleration is assumed to be constant.

Velocity/time graphs

Even simple linear motion, motion along a straight line, can be difficult to deal with mathematically. However, in the case where acceleration is constant it is possible to solve problems of motion by use of a *velocity/time graph*, without recourse to the calculus. The equations of motion use standard symbols to represent the variables, these are shown below:

s = distance in metres   (m)
u = initial velocity   (m s^{-1})
v = final velocity   (m s^{-1})
a = acceleration in metres/second^2 (m s^{-2})
t = time in seconds   (s)

The *velocity is plotted on the vertical axis and time on the horizontal axis.* Constant velocity is represented by a *horizontal straight line* and acceleration by a *sloping straight line*. Deceleration or *retardation* is also represented by a sloping straight line but with a *negative slope*.

By considering the velocity/time graph shown in Figure 3.30, we can establish the equation for distance.

The distance travelled in a given time is equal to the velocity $m\,s^{-1}$ multiplied by the time s, this is found from the graph by the *area under the sloping line*. In Figure 3.30, a body is accelerating from a velocity u to a velocity v in time t seconds.

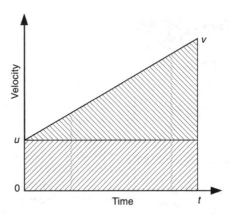

Figure 3.30 *Velocity/time graph for uniform acceleration*

Now the distance travelled s = area under graph

$$s = ut + \frac{(v - u)}{2} \times t$$

$$s = ut + \frac{vt}{2} - \frac{ut}{2}$$

$$s = \frac{(2u + v - u)t}{2}$$

and so: $s = \frac{(u + v)t}{2}$

In a similar manner to the above, one of the velocity equations can also be obtained from the velocity/time graph. Since the acceleration is the rate of change of velocity with respect to time, the value of the acceleration will be equal to the gradient of a velocity/time graph. Therefore, from Figure 3.30, we have:

$$\textbf{Gradient} = \frac{\textbf{velocity}}{\textbf{time taken}} = \textbf{acceleration}$$

therefore acceleration is given by

$$a = \frac{(v - u)}{t}$$

or $\quad v = u + at$

The remaining equations of motion may be derived from the two equations found above. As an exercise in manipulating formulae try to obtain:

(a) the equation $\quad t = \dfrac{(v - u)}{a}$

(b) $s = ut + \dfrac{1}{2}at^2$

using the above equations.

Example 3.19

A body starts from rest and accelerates with constant acceleration of 2.0 m s^{-2} up to a speed of 9 m s^{-1}. It then travels at 9 m s^{-1} for 15 seconds after which time it is retarded to a speed of 1 m s^{-1}. If the complete motion takes 24.5 seconds, find:

(a) the time taken to reach 9 m s^{-1}.
(b) the retardation
(c) the total distance travelled.

The solution is made easier if we sketch a graph of the motion, as shown in Figure 3.31.

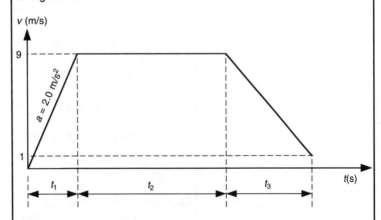

Figure 3.31 *Velocity/time graph of the motion*

(a) We first tabulate the known values:

$u = 0$ m s^{-1} (we start from rest)
$v = 9$ m s^{-1}
$a = 2.0$ m s^{-2}
$t_1 = ?$

All we need now do is select an equation which contains all the variables listed above, i.e.

$v = u + at$

and on transposing for t and substituting the variables we get

$$t = \frac{9 - 0}{2}$$

so

$t_1 = \textbf{4.5 s}$

(b) The retardation is found in a similar manner:

$u = 9$ m s^{-1}
$v = 2$ m s^{-1}
$t_3 = 5$ s
$a = ?$

We again select an equation which contains the variables, i.e.

$v = u + at$

and on transposing for a and substituting the variables we get

$$a = \frac{1 - 9}{5}$$

so

$a = -\textbf{1.6 m s}^{-2}$ (the –ve sign indicates a retardation)

(c) The total distance travelled requires us to sum the component distances travelled for the times t_1, t_2 and t_3. Again we tabulate the variable for each stage:

$u_1 = 0$ m s^{-1}	$u_2 = 9$ m s^{-1}	$u_3 = 9$ m s^{-1}
$v_1 = 9$ m s^{-1}	$v_2 = 9$ m s^{-1}	$v_3 = 1$ m s^{-1}
$t_1 = 4.5$ s	$t_2 = 15$ s	$t_3 = 5$ s
$s_1 = ?$	$s_2 = ?$	$s_3 = ?$

The appropriate equation is:

$$s = \frac{(u + v)t}{2}$$

and in each case we get

$$s_1 = \frac{(0 + 9)4.5}{2}, \quad s_2 = \frac{(9 + 9)15}{2}, \quad s_3 = \frac{(9 + 1)5}{2}$$

$s_1 = 20.25$ $s_2 = 135$ $s_3 = 25$

Then total distance $S_T = 20.5 + 135 + 25 = \textbf{180.25 m}$

Using Newton's laws

You saw earlier that *Newton's second law* may be defined as:

$F = ma$

Test your knowledge 3.10

With reference to the velocity time graphs shown in Figure 3.32, answer questions 1 to 8.
   Fill in the gaps for questions 1 to 8.

1. The slope of the velocity/time graph measures
   _____.

2. The area under a velocity/time graph determines
   _____.

3. Average velocity may be determined by dividing the _____ by _____.

4. Graph (a) is a graph of constant velocity therefore acceleration is given by
   _____ and the distance travelled is equal to
   _____.

5. Graph (b) shows uniformly accelerated motion therefore the distance travelled is equal to
   _____.

6. Graph (c) shows
   _____
   _____.

7. Graph (d) represents uniformly accelerated motion having initial velocity *u*, final velocity *v* and acceleration (a). So distanced travelled is equal to
   _____.

8. Graph (e) represents
   _____
   acceleration.

or that

$$F = \frac{mv - mu}{t}$$

In words, we may say that force is equal to the rate of change of momentum of a body. Look back again and make sure you understand the relationship between force, mass and the momentum of a body. Remembering that *momentum* may be defined as *the mass of a body multiplied by its velocity*. Also that *the inertia force is such as to be equal and opposite to the accelerating force that produced it*; this essentially is Newton's third law.

Example 3.20

A racing car of mass 1965 kg accelerates from 160 km/hr to 240 km/hr in 3.5 seconds. If the air resistance is 2000 N per tonne, find the:

(a) average acceleration
(b) force required to produce the acceleration
(c) inertia force on the car.

(a) We first need to convert the velocities to standard units.

$$u = 160 \text{ kph} = \frac{160 \times 1000}{60 \times 60} = 44.4 \text{ m s}^{-1}$$

$$v = 240 \text{ kph} = \frac{240 \times 1000}{60 \times 60} = 66.6 \text{ m s}^{-1}$$

also *t* = 3.5 s, and we are required to find the acceleration *a*. Then using the equation *v = u + at* and transposing for *a* we get:

$$a = \frac{v - u}{t} \text{ and substituting values } a = \frac{66.6 - 44.4}{3.5}$$

$$a = \textbf{6.34 m s}^{-2}$$

(b) The accelerating force is readily found using Newton's second law, where:

$$F = ma = 1965 \text{ kg} \times 6.34 \text{ m s}^{-2}$$

$$= \textbf{12.46 kN}$$

(c) From what has already been said you will be aware that the inertia force = the accelerating force, therefore the **inertia force = 12.46 kN**.

Angular motion

You previously met the equations for linear motion. A similar set of equations exists to solve engineering problems that involve angular motion as experienced, for example, in the rotation of a drive shaft. The linear equations of motion may be transformed to represent angular motion using a set of equations that we will refer to as the *transformation equations*. These are given below, followed by the equations of angular motion, which are compared with their linear equivalents.

9. Define the terms:
   (a) inertia force;
   (b) momentum.
10. What is the essential difference between speed and velocity?
11. If a rocket is sent to the moon its mass remains constant but its weight changes, explain this statement.
12. Explain how the expression $F = ma$ is related to the rate of change of momentum with respect to Newton's second law.

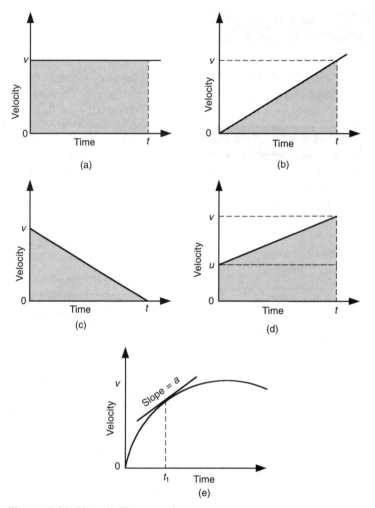

Figure 3.32 *Velocity/time graphs*

Transformation equations

$s = \theta r$

$v = \omega r$

$a = \alpha r$

where r = radius of body from centre of rotation and θ, ω and α are the angular distance, angular velocity and angular acceleration, respectively.

Angular equation of motion	Linear equation of motion
$\theta = \dfrac{(\omega_1 + \omega_2)t}{2}$	$s = \dfrac{(u + v)t}{2}$
$\theta = \omega_1 t + \frac{1}{2}\alpha t^2$	$s = ut + \frac{1}{2}at^2$
$\omega_2^2 = \omega_1^2 + 2\alpha\theta$	$v^2 = u^2 + 2as$
$\alpha = \dfrac{(\omega_2 - \omega_1)}{t}$	$a = \dfrac{(v - u)}{t}$

Angular velocity

Angular velocity refers to a body moving in a circular path and may be defined as:

$$\text{angular velocity} = \frac{\text{angular distance moved (radians)}}{\text{time taken (seconds)}}$$

or in symbols $\omega = \theta/s$ (radians per second).

Angular distance is measured in radians, you should refer to 404 if you cannot remember the definition of the radian, or how to convert radians to degrees and vice versa.

We are often given rotational velocity in the non-SI units of *revolutions per minute* (rev/min). It is therefore useful to be able to converts revs/min into radians per second and vice versa.

So, for example, to convert 350 rev/min into radians per second we multiply by $2\pi/360$, i.e.

$$350 \text{ rev/min} = 350 \times 2\pi/60 = 36.65 \text{ rad} \cdot \text{s}^{-1}$$

Example 3.21

A 540 mm diameter wheel is rotating at $1500/\pi$ rev/min. Determine the angular velocity of the wheel in rad.s^{-1} and the linear velocity of a point on the rim of the wheel.

All we need do to find the angular velocity is convert from rev/min to rad/sec, i.e.

angular velocity (rad $\cdot$ s^{-1}) = $1500/\pi \times 2\pi/60$ = **50 rad.s^{-1}**

Now from the transformation equations, linear velocity v
= angular velocity, $\omega \times$ radius, $r = 50$ rad/sec $\times 0.270$ m

$v = \textbf{13.5 m s}^{-1}$

Angular acceleration

Angular acceleration is defined as the rate of change of angular velocity with respect to time, i.e.

$$\text{angular acceleration, } \alpha = \frac{\text{change in angular velocity (rad.s}^{-1})}{\text{time (s)}}$$

So units for angular acceleration are $\alpha = \theta \text{ s}^{-2}$.

Example 3.22

The pinion shown in Figure 3.33 is required to move with an initial angular velocity of 300 revs/min and final angular velocity of 600 rev/min. If the increase takes place over 15 seconds, determine the linear acceleration of the rack. Assume a pinion radius of 180 mm.

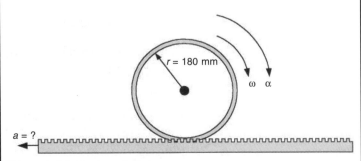

$r = 180$ mm

ω α

$a = ?$

Figure 3.33 *Rack and pinion*

In order to solve this problem we first need to convert the velocities into radians per second.

300 rev/min = 300 × 2π/60 = 31.4 rads⁻¹
600 rev/min = 600 × 2π/60 = 62.8 rads⁻¹

We can use the equation

$$\alpha = \frac{\omega_1 - \omega_2}{t}$$

to find the angular acceleration.
So

$$\alpha = \frac{62.8 - 31.4}{15} = \textbf{2.09 rads}^{-2}$$

Now we can use the transformation equation $a = \alpha r$ to find the linear acceleration, i.e.

$$\alpha = (2.09 \text{ rad.s}^{-1})(0.18 \text{ m}) = \textbf{0.377 m s}^{-2}$$

Torque and angular acceleration

We can apply Newton's third law of motion to angular motion, if it is realized that the distribution of mass relative to the axis of rotation has some bearing on the calculation. For this reason it is not possible to deal directly with a rotating wheel, but rather with a small element of mass whose radius of rotation can be more easily defined.

Figure 3.34 shows a small element of mass δm rotating at a radius r from the centre O, with uniform angular velocity w (rad.s). We know from the transformation equations that the linear velocity at any instant is given by:

$$v = \omega r$$

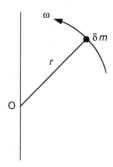

Figure 3.34 *A point mass subject to rotational velocity*

and from Newton's third law, to accelerate this mass would require a force such that:

$$F = ma$$

In this case the force would be applied at the radius r and thus would constitute a moment or more correctly a torque T about the centre of rotation thus:

$$T = Fr \quad \text{or} \quad T = mar$$

Since the linear acceleration, $a = \alpha r$, then:

$T = m(\alpha r)r$

or

$T = m\alpha r^2$

The quantity mr^2 is a concentrated mass multiplied by its radius of rotation squared and is known as the *moment of inertia I*. The quantity I is an important property of a rotating body, in the SI system it has units = kg m^2. Therefore substituting I for mr^2 in our above equation $T = m\alpha r^2$, gives:

$T = I\alpha$

The last relationship may be compared with $F = ma$ for linear motion.

Key point

Think of the moment of inertia of a rotating body as being equivalent to the mass of a body subject to linear motion.

Example 3.23

A flywheel has a moment of inertia of 130 kg m^2. Its angular velocity drops from 12 000 rpm to 9000 rpm in 6 seconds, determine (a) the retardation; (b) the braking torque.
Then,

$\omega_1 = 12\,000 \times 2\pi/60 = 1256.6$ rads^{-1}
$\omega_2 = 9000 \times 2\pi/60 = 942.5$ rads^{-1}

and from
$$\alpha = \frac{\omega_2 - \omega_1}{t}$$

$$\alpha = \frac{942.5 - 1256.6}{6}$$

$$\alpha = -52.35 \quad \text{or retardation} = \textbf{52.35 rads}^{-2}$$

Now torque
$$T = I\alpha$$

$$T = (130)(52.35)$$

so braking torque
$$T = \textbf{6805.5 N m}$$

Centripetal acceleration and force

If we consider Figure 3.35, again we can see that the direction of the mass must be continually changing to produce the circular motion, therefore it is being subject to an acceleration, which is acting towards the centre; this acceleration is known as the *centripetal acceleration* and is equal to $\omega^2 r$. When acting on a mass this acceleration produces a force known as centripetal force, thus:

centripetal force (F_r) = mass × centripetal acceleration

$F_r = m\omega^2 r$

and since $v = \omega r$

$$F_r = \frac{mv^2}{r}$$

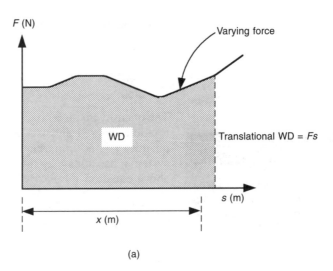

(a)

Test your knowledge 3.11

1. Define the following, stating their SI units:
   (a) angular velocity;
   (b) angular acceleration.
2. A body acting at a radius of 175 mm has a tangential (linear) velocity of 25 m s^{-1}, find its angular velocity.
3. Convert the following angular velocities into standard SI units:
   (a) 250 rev/min;
   (b) 12 500 rev/hr;
   (c) 175 rev/sec.
4. Define:
   (a) torque;
   (b) moment of inertia.
5. Explain why the moment of inertia is used instead of the total mass of the body, when considering objects subject to angular motion?
6. Define the terms:
   (a) centripetal acceleration;
   (b) centrifugal force.
7. If a racing car is in a steady turn on a banked track, explain the nature of the forces acting on the car during the turn. Which one of these forces holds the racing car in the turn?
8. Define the terms:
   (a) momentum;
   (b) inertia.

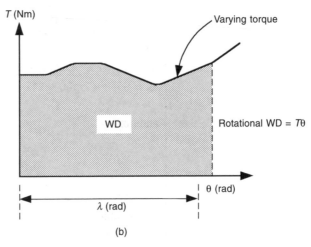

(b)

Figure 3.35 *Work done*

From Newton's third law there must be an *equal and opposite force* opposing the centripetal force; this is known as the *centrifugal force* and *acts outwards* from the centre of rotation.

Example 3.24

An aircraft with a mass of 80 000 kg is in a steady turn of radius 300 m, flying at 800 km/hr. Determine the centripetal force required to hold the aircraft in the turn.

Then the linear velocity of the aircraft = 800 × 1000/3600 m s^{-1}
= 222.2 m s^{-1}

and from $F_r = \dfrac{mv^2}{r}$ we get $F_r = \dfrac{(80\,000)(222.2)^2}{300}$ = **13.166 MN**.

Problems 3.3

1. A body weighs 562 N on the surface of the earth:

   (a) What force is required to give an acceleration of 8 m s^{-2}?
   (b) What will be the inertia force of the body when given this acceleration?

2. A car starts from rest and accelerates uniformly at 2 m s^{-2} for 10 seconds. Calculate the distance travelled.

3. A vehicle accelerates at a constant rate from rest to 20 m s^{-1} in a time of 4 seconds. Draw to scale a velocity/time graph and determine the acceleration and distance travelled.

4. Convert 350 rev/min into rad s^{-1}.

5. A Cessna 172 aircraft and a Boeing 747 aircraft are each given an acceleration of 5 m s^{-2}. To achieve this the thrust force produced by the Cessna's engines is 15 kN and the thrust force required by the Boeing 747 is 800 kN. Find the mass of each aircraft.

6. A wheel has a diameter of 0.54 m and is rotating at $\frac{1500}{\pi}$ rev/min. Calculate the angular velocity of the wheel in rad s^{-1} and the linear velocity of a point on the rim of the wheel.

7. A flywheel rotating at 20 rad s^{-1} increases its speed uniformly to 40 rad s^{-1} in 1 min. Sketch the angular velocity/time graph and determine:

   (a) the angular acceleration of the flywheel
   (b) the angle turned through by the flywheel in 1 minute and so calculate the number of revolutions made by the flywheel in this time.

8. The flywheel of a cutting machine has a moment of inertia of 130 kg m^2. Its speed drops from 120 rev/min to 90 rev/min in 2 seconds. Determine:

   (a) the deceleration of the flywheel
   (b) the braking torque.

Energy

Introduction

In this section we start our study on energy by considering the interrelationship between mechanical work, energy and power and apply this knowledge to typical engineering problems. We will then look at heat energy considering, in particular: change of state, heat and temperature, expansion of gases and the gas laws. As mentioned previously, electrical energy will be covered separately when we study the outcome on electrical principles.

Energy may exist in many different forms, for example mechanical, electrical, nuclear, chemical, heat, light and sound.

The principle of the conservation of energy states that: energy may neither be created nor destroyed only changed from one form to another.

There are many engineering examples of devices that transform energy, these include the:

- loudspeaker which transforms electrical to sound energy
- petrol engine which transforms heat to mechanical energy
- microphone which transforms sound to electrical energy
- dynamo which transforms mechanical to electrical energy
- battery which transforms chemical to electrical energy
- filament bulb which transforms electrical to light energy.

In our study of energy we start by looking at the various forms of mechanical energy and its conservation. Provided no mechanical energy is transferred to or from a body, the total amount of mechanical energy possessed by a body remains constant, unless mechanical work is done; this concept is looked at next.

Mechanical work, energy and power

Work done

The *energy* possessed by a body is its capacity to do work. Mechanical work is done when a force overcomes a resistance and it moves through a distance.

Mechanical work may be defined as:

mechanical work done (J) = force required to overcome the resistance (N) × distance moved against the resistance (m)

The SI unit of work is the newton-metre (N m) or joule where 1 joule = 1 N m.

Note:

(a) No work is done unless there is both resistance and movement.
(b) The resistance and the force needed to overcome it are equal.
(c) The distance moved must be measured in exactly the opposite direction to that of the resistance being overcome.

The more common resistances to be overcome include: *friction*, *gravity* (the weight of the body itself) and *inertia* (the resistance to acceleration of the body) where:

the work done (WD) against friction = friction force × distance moved
WD against gravity = weight × gain in height
WD against inertia = inertia force × distance moved

Note:

(a) Inertia force is the out-of-balance force multiplied by the distance moved or: the *inertia force = mass × acceleration × distance moved*.
(b) Work done in overcoming friction will be discussed in more detail later.

In any problem involving calculation of work done, the first task should be to identify the type of resistance to overcome. If, and only if, there is motion between surfaces in contact, is work done against friction. Similarly, only where there is a gain in height is

> **Key point**
>
> Mechanical energy may be defined as the capacity to do work.

the work done against gravity and only if a body is accelerated is work done against inertia (look back at our definition of inertia).

Example 3.25

A body of mass 30 kg is raised from the ground at constant velocity through a vertical distance of 15 m. Calculate the work done (WD).
   If we ignore air resistance, then the only WD is against gravity.

WD against gravity = weight × gain in height or *WD = mgh* (and assuming $g = 9.81$ m s^{-2})

then

WD = (30)(9.81)(15)

WD = 4414.5 J or **4.414 kJ**

Work done may be represented graphically and, for linear motion, this is shown in Figure 3.35(a). Where the force needed to overcome the resistance is plotted against the distance moved. The WD is then given by the area under the graph.

Figure 3.35(b) shows the situation for angular motion, where a varying torque T in Nm is plotted against the angle turned through in radians. Again the WD is given by the area under the graph, where the units are Nm × radian. Then noting that the radian has no dimensions, the unit for work done remains as Nm or joules.

Mechanical energy

Mechanical energy may be subdivided into three different forms of energy: *potential energy*, *strain energy* and *kinetic energy*.

Potential energy

Potential energy (PE) *is energy possessed by a body by virtue of its position*, relative to some datum. The change in PE is equal to its weight multiplied by the change in height. Since the weight of a body = mg, then the change in PE may be written as:

change in PE = mgh

which of course is identical to the work done in overcoming gravity. So the work done in raising a mass to a height is equal to the PE it possesses at that height, assuming no external losses.

Strain energy

Strain energy is a particular form of PE possessed by an elastic body that is deformed within its elastic range, for example a stretched or compressed spring possesses strain energy.

Consider the spring arrangement shown in Figure 3.36. The force required to compress or extend the spring is $F = kx$, where k is the spring constant.

Figure 3.36(a) shows a helical coil spring in the unstrained, compressed and extended positions. The force required to move the spring varies in direct proportion to the distance moved (Figure 3.36(b)). Therefore: *strain energy of spring when compressed or extended = area under graph (force × distance moved)*

> **Key point**
>
> Strain energy is a particular form of potential energy.

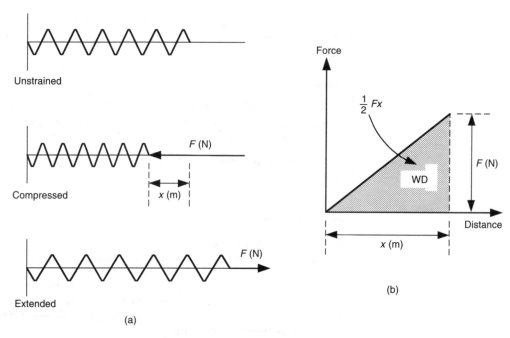

Figure 3.36 *Spring system demonstrating strain energy*

$$= \tfrac{1}{2} Fx \text{ joules}$$

and since $F = kx$, then substituting for F gives,

strain energy of spring in tension or compression $= \tfrac{1}{2} kx^2$ joules

A similar argument can be given for a spring which is subject to twisting or torsion about its centre (or polar axis). It can be shown that:

strain energy of a spring when twisted $= \tfrac{1}{2} k_{tor} \theta^2$ J (where $\theta =$ the angle of twist)

Kinetic energy

Kinetic energy (KE) is energy possessed by a body by virtue of its motion. Translational KE, that is the KE of a body travelling in a linear direction (straight line), is:

$$\text{translational kinetic energy (J)} = \frac{\text{mass (kg)} \times (\text{velocity})^2 \, (\text{m s}^{-1})}{2}$$

translation KE $= \tfrac{1}{2} \mathbf{m v}^2$

Flywheels are heavy wheel-shaped masses fitted to shafts in order to minimize sudden variations in the rotational speed of the shaft, due to sudden changes in load. A flywheel is therefore a store of rotational KE.

Rotational kinetic energy can be defined in a similar manner to translational KE, i.e.

rotational KE of mass $= \tfrac{1}{2} I \omega^2$

where $I =$ mass moment of inertia (which was introduced in your earlier work on torque).

Note:
The moment of inertia of a rotating mass I can be defined in

general terms by the expression $I = Mk^2$ where $M =$ the *total mass* of the rotating body and $k =$ the *radius of gyration*, that is the radius from the centre of rotation where all of the mass is deemed to act. When we studied torsion earlier we defined I for *concentrated or point masses*, where $I = mr^2$. You should remember that I has different values for different rotating shapes. We will only be considering circular cross-sections, where I is defined as above. One final point, try not to mix up k for the radius of gyration with k for the spring constant!

Example 3.26

Determine the total kinetic energy of a four wheel drive car which has a mass of 800 kg and is travelling at 50 kph. Each wheel of the car has a mass of 15 kg, a diameter of 0.6 m and a radius of gyration of 0.25 m.

The total KE $\quad$ = translational (linear) KE + angular KE

and linear KE $\quad = \frac{1}{2}mv^2$ where $v =$ 50 kph = 13.89 m s^{-1}

$\qquad\qquad = (\frac{1}{2})(800)(13.89)^2 =$ **77.16 kJ**

and angular KE $\quad = \frac{1}{2}I\omega^2$ where $I = Mk^2$

$\qquad\qquad = (15)(0.25)^2 = 0.9375$ kg m^2 (for each wheel!)

and from $v \qquad = \omega r$ then $\omega = v / r = 13.89/0.3 = 46.3$ rad.s

$\qquad\qquad = (\frac{1}{2})(4 \times 0.9375)(46.3)^2$

$\qquad\qquad =$ **4.019 kJ**

Therefore total KE of the car = 77.16 + 4.019 = **81.18 kJ**.

Conservation of mechanical energy

From the definition of the conservation of energy we can deduce that the total amount of energy within certain defined boundaries will remain the same. When dealing with mechanical systems, the potential energy possessed by a body is frequently converted into kinetic energy and vice versa. If we ignore air frictional losses, then:

potential energy + kinetic energy = a constant

Thus, if a mass m falls freely from a height h above some datum, then at any height above that datum:

total energy = potential energy + kinetic energy

This important relationship is illustrated in Figure 3.37, where at the highest level above the datum the potential energy is a maximum and is gradually converted into kinetic energy, as the mass falls towards the datum, immediately before impact when height $h = 0$, the potential energy is zero and the kinetic energy is equal to the initial potential energy.

Since the total energy is constant then:

$$mgh_1 = mgh_2 + \tfrac{1}{2}mv_2^2 = mgh_3 + \tfrac{1}{2}mv_3^2 = \tfrac{1}{2}mv_4^2$$

Immediately after impact with the datum surface, the mechanical kinetic energy is converted into other forms such as heat strain and sound.

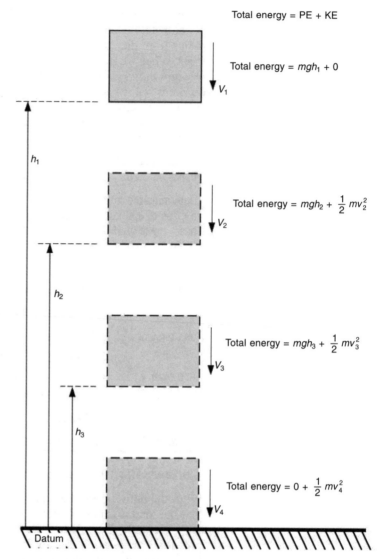

Total energy = PE + KE

Total energy = $mgh_1 + 0$

Total energy = $mgh_2 + \frac{1}{2}mv_2^2$

Total energy = $mgh_3 + \frac{1}{2}mv_3^2$

Total energy = $0 + \frac{1}{2}mv_4^2$

Figure 3.37 *PE + KE = a constant*

If friction is present then work is done overcoming the resistance due to friction and this is dissipated as heat. Then:

initial energy = final energy + work done in overcoming frictional resistance

Note:
Kinetic energy is not always conserved in collisions. Where kinetic energy is conserved in a collision we refer to the collision as *elastic*, when kinetic energy is not conserved we refer to the collision as *inelastic*.

Example 3.27

Cargo weighing 2500 kg breaks free from the top of the cargo ramp (Figure 3.38). Ignoring friction, determine the velocity of the cargo the instant it reaches the bottom of the ramp.

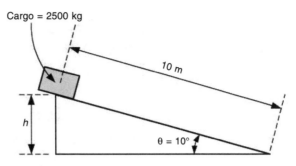

Figure 3.38 *Cargo ramp*

The vertical height h is found using the sine ratio, i.e.

$10 \sin 10 = h$ so $h = 1.736$ m

so increase in potential energy $= mgh$

$$= (2500)(9.81)(1.736) \text{ J}$$

$$= 42\,587.2 \text{ J}$$

Now using the relationship PE + KE = total energy. Then immediately prior to the cargo breaking away KE = 0 and so PE = total energy also; immediately prior to the cargo striking the base of slope PE = 0 and KE = total energy (all other energy losses being ignored).

So at base of slope: $42\,587.2$ J = KE

and $42\,587.2 = \frac{1}{2}mv^2$

so $\dfrac{(20)(42\,587.2)}{2500} = v^2$

and so velocity at bottom of ramp = **18.46 m s⁻¹** (check this working for yourself!).

Power

Power is a measure of the rate at which work is done or the rate of change of energy. Power is therefore defined as: *the rate of doing work*. The SI unit of power is the watt (W), i.e.

$$\textbf{power (W)} = \frac{\textbf{work done (J)}}{\textbf{time taken (s)}} = \frac{\textbf{energy change (J)}}{\textbf{time taken (s)}}$$

or, if the body moves with constant velocity,

power (W) = force used (N) × velocity (m s⁻¹)

Note units are N m s⁻¹ = J s⁻¹ = watt (W).

Example 3.28

A packing crate weighing 1000 N is loaded onto the back of a lorry by being dragged up an incline of 1 in 5 at a steady speed of 2 m s⁻¹. The frictional resistance to motion is 240 N. Calculate:

(a) the power needed to overcome friction
(b) the power needed to overcome gravity
(c) the total power needed.

(a) Power = friction force × velocity along surface

$$= 240 \times 2$$

$$= \textbf{480 W}$$

(b) Power = weight × vertical component of velocity

$$= 1000 \times 2 \times 1/5$$

$$= \textbf{400 W}$$

(c) Since there is no acceleration and therefore no work done against inertia,

Total power = power for friction + power for gravity

$$= 480 + 400$$

$$= \textbf{880 W}$$

Let us now consider power transmitted by a torque. You have already met the concept of torque. Figure 3.39 shows a force F (N) applied at radius r (m) from the centre of a shaft that rotates at n rev/min.

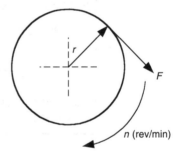

Figure 3.39 *Power transmitted by a torque*

Since the work done is equal to the force multiplied by the distance, then the WD in one revolution is given by:

WD in one revolution = $F \times 2\pi r$ J

but Fr is the torque T applied to the shaft, therefore the work done in one revolution is,

WD in one revolution = $2\pi T$ J

In 1 minute the work done = work done per revolution × number of rev/min (n)

$$= 2\pi nT$$

and WD in 1 second $= 2\pi nT/60$ and since WD per second is equal to power (1 J s^{-1} = 1 W)

then,

Power (W) transmitted by a torque = $2\pi nT/60$

Friction

We have already met *friction*, in terms of the *frictional force that tends to oppose relative motion*, but up till now we have not fully defined the nature of friction.

When a surface is moved over another surface with which it is in contact, a resistance is set up opposing this motion. The value of the resistance will depend on the materials involved, the condition of the two surfaces, and the force holding the surfaces in contact; but the opposition to motion will always be present. This resistance to movement is said to be the result of *friction* between the surfaces.

We require a slightly greater force to start moving the surfaces (*static friction*) than we do to keep them moving (*sliding friction*). As a result of numerous experiments involving different surfaces in contact under different forces, a set of rules or laws has been established which, for all general purposes, materials in contact under the action of forces seem to obey. These rules are detailed below, together with one or two limitations for their use.

Laws of friction

1. The frictional force always opposes the direction of motion, or the direction in which a body is tending to move.
2. The sliding friction force F opposing motion, once motion has started, is proportional to the normal force N that is pressing the two surfaces together, i.e. $F \propto N$.
3. The sliding frictional force is independent of the area of the surfaces in contact. Thus two pairs of surfaces in contact made of the same materials and in the same condition, with the same forces between them, but having different areas, will experience the same frictional forces opposing motion.
4. The frictional resistance is independent of the relative speed of the surfaces. This is not true for very low speeds nor in some cases for fairly high speeds.
5. The frictional resistance at the start of sliding (*static friction*) is slightly greater than that encountered as motion continues (*sliding friction*).
6. The frictional resistance is dependent on the nature of the surfaces in contact. For example, the type of material, surface geometry, surface chemistry, etc.

Key point

Friction always opposes the motion that produces it.

Solving problems involving friction

From the above laws we have established that the sliding frictional force F is proportional to the normal force N pressing the two surfaces together, that is $F \propto N$. You will remember from your mathematical study of proportion that in order to equate these forces we need to insert a constant, the constant of proportionality, i.e. $F = \mu N$. This constant μ is known as the coefficient of friction and in theory it has a maximum value of one. Figure 3.40 shows the space diagram for the arrangement of forces on two horizontal surfaces in contact.

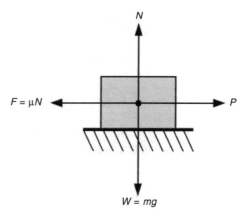

Figure 3.40 *Forces on two horizontal surfaces in contact*

You may find the solution of problems involving friction rather difficult. This is because it is often difficult to visualize the nature and direction of all the forces that act on two bodies in contact, as well as resolving these forces into their component parts. Problems involving friction may be solved by calculation or by drawing. The following generalized example involving the simple case of a block in contact with a horizontal surface should help you understand both methods of solution.

Example 3.29

(a) *Solution by calculation*

Consider again the arrangement of forces shown in Figure 3.40. If the block is in equilibrium, i.e. just on the point of moving, or moving with constant velocity then we can equate the horizontal and vertical forces as follows:

resolving horizontally gives $\qquad\qquad P = F \qquad\qquad\qquad$ (1)

resolving vertically $\qquad\qquad\qquad N = mg \qquad\qquad\quad$ (2)

but from the laws of dry friction $\qquad F = \mu N \qquad\qquad\quad$ (3)

substituting (2) into (3) gives $\qquad F = \mu mg \qquad\qquad\;$ (4)

substituting (4) into (1) gives $\qquad P = \mu mg$

(b) *Solution by vector drawing*

You know from your previous work on resolution of coplanar forces (page 188) that two forces can be replaced by a single resultant force in a vector diagram. The space diagram for our horizontal block is shown in Figure 3.41(a), where F and N can be replaced by a resultant R at an angle ϕ to the normal force N.

From Figure 3.41 it can be seen that:

$$\frac{F}{R} = \sin \phi$$

$$F = R \sin \phi$$

and $\qquad\qquad\qquad\qquad \dfrac{N}{R} = \cos \phi$

$$N = R \cos \phi$$

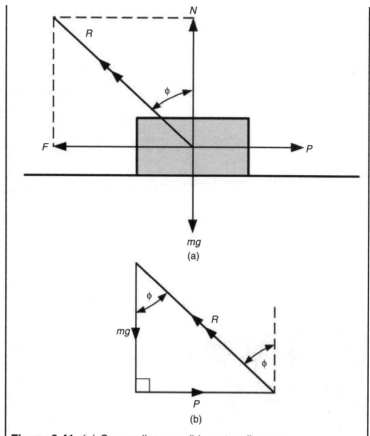

Figure 3.41 *(a) Space diagram, (b) vector diagram*

$$\frac{F}{N} = \frac{R \sin \phi}{R \cos \phi} = \tan \phi$$

however $$\frac{F}{N} = \mu$$

therefore

$$\mu = \tan \phi$$

ϕ is known as the *angle of friction.*

Once F and N have been replaced by R the problem becomes one of three coplanar forces mg, P and R and can therefore be solved using the triangle of forces you met earlier.

Then choosing a suitable scale the vector diagram is constructed as shown in Figure 3.41(b).

Example 3.30

For the situation illustrated in Figure 3.42(a), find the value of the force P to maintain equilibrium.

We can solve this problem by calculation resolving the forces into their horizontal and vertical components or we can solve by drawing both methods of solution as detailed below.

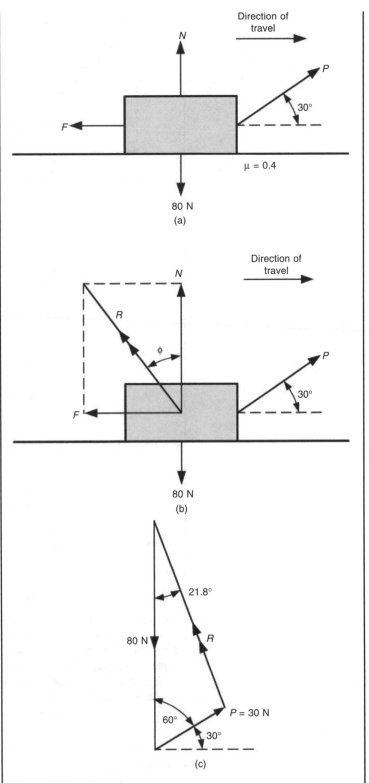

Figure 3.42 *a, b and c*

Test your knowledge 3.13

1. On what variables does the value of frictional resistance depend?
2. 'The frictional resistance is independent of the relative speed of the surfaces, under all circumstances.' Is this statement true or false? You should give reasons for your decision.
3. Define (a) the angle of friction; (b) the coefficient of friction. Explain how they are related.
4. Sketch a space diagram that shows all the forces that act on a body moving with uniform velocity along a horizontal surface.

(a) *Solution by calculation*

resolving forces horizontally $F = P \cos 30$

resolving forces vertically $\qquad\qquad N + P\sin 30 = 80$

but $\qquad\qquad\qquad\qquad\qquad\qquad F = mN$

and substituting for N from above gives $F = m(80 - P\sin 30)$

We are told that $m = 0.4$ and replacing F in the above equation by $P\cos 30$, in a similar manner to the general example, gives:

$P\cos 30 = 0.4(80 - P\sin 30)$

and by multiplying out the brackets and rearrangement we get

$P\cos 30 + 0.4\,P\sin 30 = 0.4 \times 80$

so $P(\cos 30 + 0.4\sin 30) = 32$

and $P = \mathbf{30.02\ N}$

Make sure you can follow the above trigonometric and algebraic argument.

(b) *Solution by drawing*
The magnitude and direction of all known forces for our block is shown in Figure 3.42(b).

Remembering that $\mu = \tan\phi$ then

$\tan\phi = \mu = 0.4$ so $\phi = \tan^{-1} 0.4$ (the angle whose tangent is) and $\phi = 21.8°$

From the resulting vector diagram Figure 3.42(c), we find that $P = \mathbf{30\ N}$.

Problems 3.4

1. Find the kinetic energy of a mass of 2000 kg moving with a velocity of 40 kph.
2. The scale of a spring balance which indicates weights up to 20 N extends over a length of 10 cm. Calculate the work done in pulling the balance out until it indicates 12 N.
3. A crane raises a load of 1640 N to a height of 10 m in 8 seconds. Calculate the average power developed.
4. A wooden packing case weighing 560 N rests on a concrete floor. The coefficient of friction is 0.3. Calculate the least horizontal force required to move the case.
5. A drilling machine of mass 300 kg is pulled along a horizontal floor by means of a rope inclined at 20° to the horizontal. If a tension in the rope of 600 N is just sufficient to move the machine, determine the coefficient of friction between the machine and the floor.
6. A motor vehicle starting from rest freewheels down a slope whose gradient is 1 in 8. Neglecting all resistances to motion, find its velocity after travelling a distance of 200 m down the slope.
7. A train having a mass of 15 tonnes is brought to rest when striking the buffers of a terminus. The buffers consist of two springs in parallel, each having a spring constant of 120 kN m^{-1} and able to be compressed to a maximum of 0.75 m. Find:

    (a) the strain energy gained by the buffers
    (b) the velocity of the train at the instant it strikes the buffers.

HEAT ENERGY

Heat

The study of *heat energy* is a necessary foundation for an understanding of the area of science known as *thermodynamics*. Engineering thermodynamics is concerned with the relationship between heat, work and the properties of systems. As engineers we are concerned with the machines (engines) that convert heat energy from fuels into useful mechanical work. It is therefore appropriate to start our study of heat energy by considering the concept of *heat* itself.

Energy is the most important and fundamental physical property of the universe. We have already defined energy as *the capacity to do work*. A more accurate definition is *the capacity to produce an effect*. These effects are apparent during the process of energy transfer.

A modern idea of heat is that it is energy in transition and cannot be stored by matter. *Heat (Q)* may be defined as: *transient energy brought about by the interaction of bodies by virtue of their temperature difference when they communicate.* Matter possesses stored energy but not transient (moving) energy such as heat or work. Heat energy can only travel or *transfer* from a hot body to a cold body, it cannot travel up hill. Figure 3.43 illustrates this fact.

> **Key point**
>
> Heat and work is energy in transit and cannot be stored by matter.

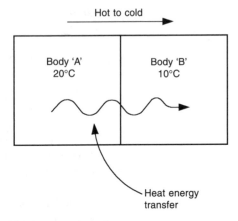

Figure 3.43 *Heat energy transfer*

Within matter the amount of molecular vibration determines the amount of *kinetic energy* a substance possesses. For incompressible fluids (liquids) the amount of molecular vibration is relatively small and can be neglected. For compressible fluids and gases the degree of vibration is so large that it has to be accounted for in thermodynamics. This kinetic energy is classified as *internal energy (U)* and is a form of stored energy.

The gas laws

In the study of gases we have to consider the interactions between temperature, pressure and volume (remembering that density is mass per unit volume). A change in one of these characteristics always produces a corresponding change in at least one of the other two.

Unlike liquids and solids, gases have the characteristics of being easily compressible and of expanding or contracting readily in response to changes in temperature. Although the characteristics themselves vary in degree for different gases, certain basic laws can be applied to what we call a perfect gas. A *perfect* or *ideal gas* is simply one which has been shown, through experiment, to follow or adhere very closely to these gas laws. In these experiments one factor, for example volume, is kept constant while the relationship between the other two is investigated. In this way it can be shown that:

1. *The pressure of a fixed mass of gas is directly proportional to its absolute temperature, providing the volume of the gas is kept constant.*

In symbols:

$\dfrac{P}{T}$ = **constant   (providing V remains constant)**

The above relationship is known as the *pressure law*.

Gas molecules are in a state of perpetual motion, constantly bombarding the sides of the gas containing vessel. Each molecule produces a minute force as it strikes the walls of the container, since many billion molecules hit the container every second, this produces a steady outward pressure.

Figure 3.44 shows how the pressure of the gas varies with temperature.

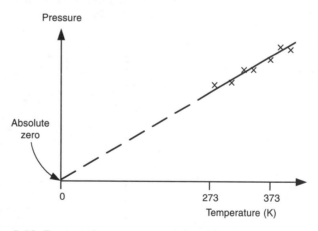

Figure 3.44 *Pressure/temperature relationship of gas*

Key point

When dealing with the gas equations or any thermodynamic relationship we always use absolute temperature (*T*) in degrees kelvin.

If the graph is 'extrapolated' downwards, in theory we will reach a temperature where the pressure is zero. This temperature is known as *absolute zero* and is approximately equal to *–273 kelvin*. Each one degree kelvin *K* is equivalent to one degree celsius *C*. The relationship between the kelvin scale and the celsius scale is shown in Figure 3.45.

Returning to the gas laws, it can also be shown experimentally that:

2. *The volume of a fixed mass of gas is directly proportional to its absolute temperature providing the pressure of the gas remains constant.*

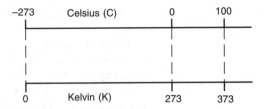

Figure 3.45 *Kelvin/Celsius scales*

So for a fixed mass of gas:

$$\frac{V}{T} = \text{constant} \quad \text{(providing } M \text{ is fixed and } P \text{ remains constant)}$$

This relationship is known as *Charles' law*.

A further relationship exists when we keep the temperature of the gas constant, this states that:

> *The volume of a fixed mass of gas is inversely proportional to its pressure providing the temperature of the gas is kept constant.*

In symbols:

$$p \propto \frac{1}{V}$$

or, for a fixed mass of gas:

pV = constant

This relationship is better known as *Boyle's law*, it is illustrated in Figure 3.46.

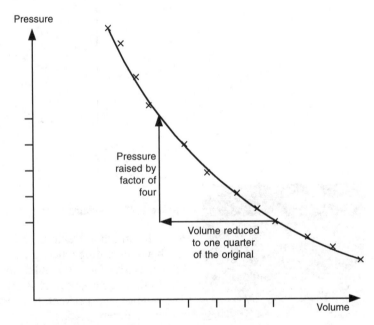

Figure 3.46 *Boyle's pressure/volume relationship*

In dealing with problems associated with the gas laws, remember that we assume that all gases are *ideal*; in reality no gas is ideal but at low and medium pressures and temperatures, most gases behave in an ideal way.

The pressure law, Charles' law and Boyle's law can all be expressed in terms of one single equation known as the *combined gas equation*, this is for a fixed mass of gas:

$$\frac{pV}{T} = \text{constant}$$

If we consider a fixed mass of gas before and after changes have taken place, then from the combined gas equation, it follows that:

$$\frac{p_1 V_1}{T_1} = \frac{p_2 V_2}{T_2}$$

where subscript 1 is used for the initial state and subscript 2 for the final state of the gas.

The above relationship is very useful when solving problems concerned with the gas laws.

Key point

A perfect gas is one that is assumed to obey the ideal gas laws.

Example 3.31

A quantity of gas occupies a volume of 0.5 m³. The pressure of the gas is 300 kPa, when its temperature is 30°C. What will be the pressure of the gas if it is compressed to half its volume and heated to a temperature of 140°C?

When solving problems involving several variables, always tabulate the information given, in appropriate units:

$p_1 = 300$ kPa $p_2 = ?$
$V_1 = 0.5$ m² $V_2 = 0.25$ m²
$T_1 = 303$ K $T_2 = 413$ K

Remember to convert temperature to kelvin by adding 273°C.
Using the combined gas equation and after rearrangement:

$$p_2 = \frac{p_1 V_1 T_2}{T_1 V_1} = \frac{(300)(0.5)(413)}{(303)(0.25)} = \textbf{817 kPa}$$

Temperature and its measurement

We have already met the idea of temperature when we considered the pressure law, but as yet, we have not fully defined it. A more formal definition of temperature is as follows:

Temperature is a measure of the quantity of energy possessed by a body or substance. It measures the vibration of the molecules which form the substance.

These molecular vibrations only cease when the temperature of the substance reaches *absolute zero*, that is −273.15°C.

You have already met the celsius temperature scale and the way in which we convert degrees centigrade into kelvin and vice versa.

Key point

Temperature measures the energy possessed by the vibration of the molecules that go to make up a substance.

Example 3.32

Convert 60°C into kelvin
You already know that 1°C = 1 K and that to convert degrees celsius into kelvin, we simple add 273. Therefore 60°C + 273 = **333 K**.
Note that to be strictly accurate we should add 273.15, but for all practical purposes the approximate value of 273 is adequate.

The method used to measure temperature depends on the degree of hotness of the body or substance being measured. Measurement apparatus includes liquid-in-glass thermometers, resistance thermometers, thermistor thermometers and thermocouples.

All *thermometers* are based on some property of a material that changes when the material becomes colder or hotter. Liquid-in-glass thermometers use the fact that most liquids expand slightly when they are heated. Two common types of liquid-in-glass thermometer are the mercury thermometer and alcohol thermometer, both have relative advantages and disadvantages.

Alcohol thermometers are suitable for measuring temperatures down to −115°C and have a higher expansion rate than mercury, so a larger containing tube may be used. They have the disadvantage of requiring the addition of a colouring in order to be seen easily. Also, the alcohol tends to cling to the side of the glass tube and may separate.

Mercury thermometers conduct heat well and respond quickly to temperature change. They do not wet the sides of the tube and so flow well in addition to being easily seen. Mercury has the disadvantage of freezing at −39°C and so is not suitable for measuring low temperatures. Mercury is also poisonous and special procedures must be followed in the event of spillage.

Resistance thermometers are based on the principle that current flow becomes increasingly more difficult with increase in temperature. They are used where a large temperature range is being measured, approximately −200°C to 1200°C. *Thermistor thermometers* work along similar lines, except in this case they offer less and less resistance to the flow of electric current as temperature increases.

Thermocouple thermometers are based on the principle that when two different metal wires are joined at two junctions and each junction is subjected to a different temperature, a small current will flow. This current is amplified and used to power an analogue or digital temperature display. Thermocouple temperature sensors are often used to measure the temperatures inside engines, they can operate over a temperature range from about −200°C to 1600°C.

Thermal expansion

We have mentioned in our discussion on thermometers that certain liquids expand with increase in temperature, this is also the case with *solids*. Thermal expansion is dependent on the nature of the material and the magnitude of the temperature increase. We normally measure the linear expansion of solids, such as the increase in length of a bar of the material; with gases (as you have already seen) we measure volumetric or cubic expansion.

Every solid has a *linear expansivity value, that is the amount the material will expand in metres per kelvin or per degree celsius.* This expansivity value is often referred to as the *coefficient of linear expansion* (α), some typical values of α are given below.

Given the length of a material (l), its linear expansion coefficient (α) and the temperature rise (Δt), the *increase in its length* can be calculated using:

increase in length $= \alpha l(t_2 - t_1)$

Material	Linear expansion coefficient $\alpha/°C$
Invar	1.5×10^{-6}
Glass	9×10^{-6}
Cast iron	10×10^{-6}
Concrete	11×10^{-6}
Steel	12×10^{-6}
Copper	17×10^{-6}
Brass	19×10^{-6}
Aluminium	24×10^{-6}

Note that we are using lower case t to indicate temperature because when we find a *temperature difference* (Δt) we do not need to convert to kelvin.

For solids an estimate of the cubic or volumetric expansion may be found using:

change in volume $= 3\alpha V(t_2 - t_1)$

where V is the original volume.

A similar relationship exists for surface expansion, where a body experiences a change in area. In this case the linear expansion coefficient is multiplied by 2, therefore:

change in area $= 2\alpha A(t_2 - t_1)$ where A is the original area.

Example 3.33

A steel bar has a length of 4.0 m at 10°C. What will be the length of the bar when it is heated to 350°C? If a sphere of diameter 15 cm is made from the same material what will be the percentage increase in surface area, if the sphere is subject to the same initial and final temperatures?

Using $\alpha = 12 \times 10^{-6}$ from the above table, then increase in length of the bar is given by:

$x = \alpha l(t_2 - t_1) = (12 \times 10^{-6})(4.0)(350 - 10) = 0.0163$ m

This can now be added to the original length, so

final length $= 4.0 + 0.0163 = $ **4.0163 m**

Increase in surface area of the sphere $= 2\alpha A (t_2 - t_1)$.
We first need to find the original surface area which is given by:

$A = 4\pi r^2 = 4\pi \times (0.075)^2 = 0.0707$ m^2

and from above the increase in surface area
$= (2)(12 \times 10^{-6})(0.0707)(340) = 5.769 \times 10^{-4}$ m.

Therefore, percentage increase in area $= \left(\dfrac{\text{increase in area}}{\text{original area}} \right) \times 100$

$= \dfrac{(300)(0.5)(413)}{0.0707} = $ **0.82%**.

Heat energy transfer

Literature on heat transfer generally recognizes three distinct modes of heat transmission, the names of which will be familiar to you, i.e. *conduction, convection* and *radiation.* Technically only conduction and radiation are true heat transfer processes, because both of these depend totally and utterly on a temperature difference being present. Convection also depends on the transportation of a mechanical mass. Nevertheless, since convection also accomplishes transmission of energy from high to low temperature regions, it is conventionally regarded as a heat transfer mechanism.

Thermal conduction in solids and liquids seems to involve two processes; the first is concerned with atoms and molecules (Figure 3.47), the second with free electrons.

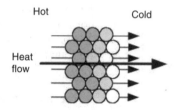

Solid molecules/atoms– direct conduction

(a) Conduction by molecular transfer in solids and gases

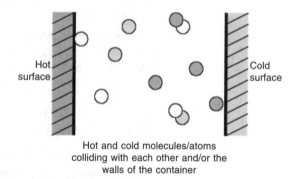

Hot and cold molecules/atoms colliding with each other and/or the walls of the container

(b) Conduction in gases

Figure 3.47 *Conduction by molecular transfer in solids and gases*

Atoms at high temperatures vibrate more vigorously about their equilibrium positions than their cooler neighbours. Since atoms and molecules are bonded to one another, they pass on some of their vibrational energy. This energy transfer occurs from atoms of high vibrational energy to those of low vibrational energy, without appreciable displacement. This energy transfer has a knock-on effect, since high vibrational energy atoms increase the energy in adjacent low vibrational energy atoms, which in turn causes them to vibrate more energetically, causing thermal conduction to occur. In solids (Figure 3.47(a)) the energy transfer is by direct contact between one molecule and another. In gases the conduction process occurs as a result of collisions between hot and cold molecules and the surface of the containing vessel.

The second process involves material with a ready supply of free electrons. Since electrons are considerably lighter than atoms, then any gain in energy by electrons results in an increase in the electron's velocity and it is able to pass this energy on quickly to cooler parts of the material. This phenomenon is one of the reasons why electrical conductors that have many free electrons are also good thermal conductors, Do remember that metals are not the only good thermal conductors; the first mechanism described above which does not rely on free electrons is a very effective method of thermal conduction, especially at low temperatures.

Heat transfer by convection consists of two mechanisms. In addition to energy transfer by random molecular motion (diffusion), there is also energy being transferred by the bulk motion of the fluid.

So in the presence of a temperature difference large numbers of molecules are moving together in bulk (Figure 3.48), at the same time as the individual motion of the molecules takes place. The cumulative effect of both of these energy transfer methods is referred to as heat transfer by convection.

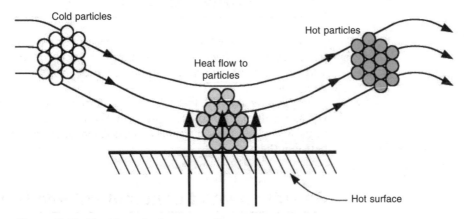

Figure 3.48 *Heat transfer by convection*

Radiation may be defined as the transfer of energy *not requiring a medium* through which the energy must pass, thus radiation can be transferred through empty space. Thermal radiation is attributed to the electron energy changes within atoms or molecules. As electron energy levels change energy is released which travels in the form of electromagnetic waves of varying wavelength. You will meet electromagnetic waves again when you study light. When striking a body the emitted radiation is either absorbed by, reflected by, or transmitted through the body.

Specific heat

From what has been said about heat transfer above, it will be apparent that different materials have different capacities for absorbing and transferring thermal energy. The thermal energy needed to produce a temperature rise depends on the mass of the material, the type of material and the temperature rise to which the material is subjected.

Thus the inherent ability of a material to absorb heat for a given mass and temperature rise is dependent on the material itself. This property of the material is known as its *specific heat capacity*. In the SI system, *the specific heat capacity of a material is the same as the thermal energy required to produce a 1 K rise in temperature in a mass of 1 kg*. Therefore knowing the mass of a substance and its specific heat capacity, it is possible to calculate the thermal energy required to produce any given temperature rise, from:

thermal energy, $Q = mc\Delta t$

where c = specific heat capacity of the material (J/kg K) and ΔT is the temperature change.

Example 3.34

How much thermal energy is required to raise the temperature of 5 kg of aluminium from 20°C to 40°C? Take the specific heat capacity for aluminium as 900 J/kg K.

   All that is required is to substitute the appropriate values directly into the equation:

$Q = mc\Delta t = (5)(900)(40 - 20) = 90\,000$ J = **90 kJ**

Another way of defining the specific heat capacity of any substance is: *the amount of heat energy required to raise the temperature of unit mass of the substance through one degree, under specific conditions*.

In *thermodynamics*, two specified conditions are used, those of constant volume and constant pressure. With *gases* the two specific heats do not have the same value and it is essential that we distinguish between them.

Specific heat at constant volume (c_v)

If 1 kg of a gas is supplied with an amount of heat energy sufficient to raise the temperature by 1 degree centigrade or kelvin while the volume of the gas remains constant, then the amount of heat energy supplied is known as the *specific heat capacity at constant volume and is denoted by c_v*.

Note that under these circumstances (Figure 3.49(a)) no work is done, but the gas has received an increase in internal energy (U). The specific heat at constant volume for air is c_v *air = 718 J/kg K*, this value is well worth memorizing!

Specific heat at constant pressure (c_p)

If 1 kg of a gas is supplied with a quantity of heat energy sufficient to raise the temperature of the gas by 1 degree centigrade or kelvin while the pressure is held constant, then the amount of heat energy supplied is known as the *specific heat capacity at constant pressure and is denoted by c_p*.

This implies that when the gas has been heated it will expand a distance h (Figure 3.49(b)), so work has been done. Thus for the same amount of heat energy there has been an increase in internal energy (U), plus work. The value of c_p is, therefore, greater than the corresponding value of c_v.

Key point

Specific heat capacity at constant pressure for air is 1005 J/kg K.

Key point

Specific heat at constant pressure will be greater than specific heat at constant volume, since work is done.

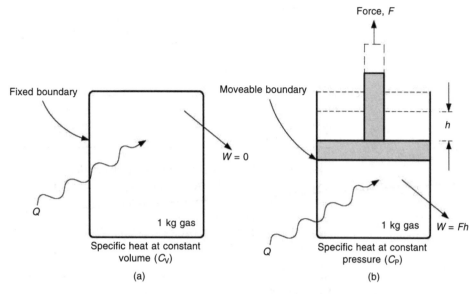

Figure 3.49 *Comparison of constant pressure and constant volume specific heats*

The specific heat capacity at constant pressure for air is c_p *air* = *1005 J/kg K*, again this is a value worth remembering.

The characteristic gas equation

The combined gas law, which you met earlier, stated that for a perfect gas with unit mass:

$$\frac{pV}{T} = \text{a constant}$$

This relationship is of course true for any fixed mass of gas and so we can write that

$$\frac{pV}{T} = \text{mass} \times \text{a constant}$$

Now for any perfect gas which obeys the ideal gas laws this constant R is specific to that particular gas, i.e. R *is the characteristic gas constant* or specific gas constant for the individual gas concerned. Therefore, the characteristic gas equation may be written as:

$$\frac{pV}{T} = mR$$

or

$$pV = mRT$$

The unit for the characteristic gas constant is the joule per kilogram kelvin (J/kg K).

Note, that when the above equation is used both *absolute pressure and absolute temperature must be used*.

The characteristic gas constant for a number of gases is given in the table below.

Gas	Characteristic gas constant (J/kg K)
Hydrogen	4124
Helium	2077
Nitrogen	297
Air	287
Oxygen	260
Argon	208
Carbon dioxide	189

The characteristic gas constant for air, from the above table, is $R = 287$ J/kg K. This is related to the specific heat capacities for air in the following way, i.e. $R = c_p - c_v$, you should check this relationship by noting the above values of R, c_p and c_v for air. This relationship ($R = c_p - c_v$) is not only valid for air, *it is also valid for any perfect gas that follows the gas laws.*

Example 3.35

0.22 kg of gas at a temperature of 20°C and pressure of 103 kN/m^2 occupies a volume of 0.18 m^3. If the c_v for the gas = 720 J/kg K, find:

(a) the characteristic gas constant
(b) the specific heat capacity at constant pressure.

(a) Using $pV = mRT$
then on rearrangement,

$$R = pV/mT = \frac{(103 \times 10^3)(0.18)}{(0.22)(293)} = \textbf{288J/kg K}$$

(b) from $R = c_p - c_v$ then $c_p = R + c_v = 288 + 720 = \textbf{1008 J/kg K}$

Latent heat

When a substance changes state, that is when heat is applied to a solid and it turns into a liquid and with further heating the liquid turns into a gas, we say the substance has undergone a *change in state*. The three states of matter are *solid*, *liquid* and *gas*. Therefore, the heat energy added to a substance does not necessarily give rise to a measurable change in temperature, it may be used to change the state of a substance; under these circumstances we refer to the heat energy as *latent* or *hidden heat*.

We refer to the thermal energy required to change a solid material into a liquid as the latent heat of fusion. For water, 334 kJ of thermal energy are required to change 1 kg of ice at 0°C into water at the same temperature. Thus *the specific latent heat of fusion for water is 334 kJ*. In the case of latent heat, *specific* refers to unit mass of the material, i.e. per kilogram. So we define the specific latent heat of fusion of a substance as: *the thermal energy required to turn 1 kg of a substance from a liquid into a solid without change in temperature.*

If we wish to find the thermal energy required to change any amount of a substance from a solid into a liquid, then we use the relationship:

$$Q = mL$$

where L is the *specific latent heat* of the substance.

In a similar manner to the above argument: *the thermal energy required to change 1 kg of a substance from a liquid into a gas without change in temperature, is known as the specific latent heat of vaporization.* Again, if we wish to find the thermal energy required to change any amount of a substance from a liquid into a gas we use the relationship $Q = mL$, but in this case L = the specific latent heat of vaporization.

The specific latent heat of vaporization for water = 2.26 MJ/kg K.

Example 3.36

(a) How much heat energy is required to change 3 kg of ice at 0°C into water at 30°C?
(b) What thermal energy is required to condense 0.2 kg of steam into water at 100°C?

(a) The thermal energy required to convert ice at 0°C into water at 0°C is calculated using the equation:

$Q = mL$ and substituting values we get

$Q = (3)(334 \times 10^3) = 1.002$ MJ

The 3 kg of water formed has to be heated from 0°C to 30°C. The thermal energy required for this is calculated using the equation

$Q = mc\Delta t$, you have already met this equation when we studied specific heat earlier.

So in this case $Q = (3)(4200)(30) = 378\,000$ J

Then total thermal energy required = 1.002 + 0.378 = **1.38 MJ**

(b) In this case we simply use $Q = mL$, since we are converting steam to water at 100°C which is the vaporization temperature for water into steam.

Then $Q = (0.2)(2.226 \times 10^6) = $ **445.2 kJ**

Note the large amounts of thermal energy required to change the state of a substance.

A liquid does not have to boil in order for it to change state; the nearer the temperature is to the boiling point of the liquid, the quicker the liquid will turn into a gas. At much lower temperatures the change may take place by a process of *evaporation*. The steam rising from a puddle, when the sun comes out after a rainstorm, is an example of evaporation. Where water vapour forms as steam, well below the boiling point of the water.

There are several ways that a liquid can be made to evaporate more readily. These include, an *increase in temperature* that increases the molecular energy of the liquid sufficient for the more energetic molecules to escape from the liquid. *Reducing the pressure above the liquid* in order to allow less energetic molecules to escape as a gas. *Increasing the surface area*, thus providing more opportunity for the more energetic molecules to escape or *by passing a gas over the surface* of the liquid to assist molecular escape.

Problems 3.5

1. A quantity of gas occupies a volume of 4 m^3. The pressure of the gas is 350 kPa when its temperature is 30°C. What will be its pressure if it is compressed into half its volume and heated to 130°C?

2. A metal bar is heated from 20°C to 120°C and as a result its length increases from 1500 mm to 1503 mm. Determine the linear expansion coefficient of the metal.

3. (a) Write down the formula for the thermal energy input into a solid and explain the meaning of each term.
   (b) If 3 kg of aluminium requires 54 kJ of energy to raise its temperature from 10°C to 30°C, find the specific heat capacity for aluminium.

4. 0.5 kg of a gas at a temperature of 20°C and at standard atmospheric pressure occupies a volume of 0.4 m^3. If the c_p for the gas = 1000 J/kg K find:

   (a) the characteristic gas constant
   (b) the specific heat capacity at constant volume.

5. How much heat energy is required to change 2 kg of ice at 0°C into water at 40°C?

ELECTRICAL PRINCIPLES

Introduction

Throughout our study of electrical principles the concept of *electrical energy* will be apparent. Electrical energy, like heat energy, requires a difference in potential for transfer. For heat energy transfer, this difference in potential was generated by a difference in temperature between two bodies. When electrical energy is transferred this difference in potential can be compared to that of a waterfall, where the water at the top of a waterfall is at a higher potential than that at the bottom. Exactly how this difference in electrical potential is created will be discussed shortly.

In order to discover the nature of this potential difference (pd), we first investigate the fundamental ideas of charge, potential, resistance, current and power. To do this, we need to be familiar with the ideas you first met when we made a brief study of matter at the beginning of this unit.

Therefore, we start our study of electricity by looking at the fundamental properties identified above. This section will include a brief discussion on the ways in which electricity can be produced by chemical means using batteries. We then look at the production of electrical energy in simple circuits where voltage, current, resistance, electrical work and power within these circuits is considered. In the final topic of the outcome the concept of magnetism is studied. The principle of electromagnetic induction is then applied to electrical machines that produce and use electrical energy for many engineering applications.

Electrical fundamentals

Atomic structure and conduction

As you already know, all matter is made up of atoms or groups of atoms (molecules) bonded together in a particular way. In order to understand something about the nature of electrical charge we need to consider a simple model of the atom. This model, known as the Bohr model, shows a single atom (Figure 3.50) consisting of a central nucleus with orbiting electrons.

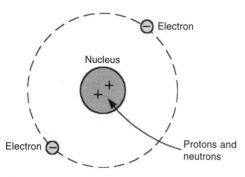

Figure 3.50 *Bohr's model of the atom*

Within the nucleus there are *protons* which are *positively charged* and *neutrons* which, as their name implies, are electrically neutral and *have no charge*. Orbiting the nucleus are *electrons that have a negative charge, equal in magnitude (size) to the charge on the proton.* These electrons are approximately two thousands times lighter than the protons and neutrons in the nucleus.

In a stable atom the number of protons and electrons is equal, so that overall, the atom is neutral and has no charge. However, if we rub two particular materials together, electrons may be transferred from one to another. This alters the stability of the atom, leaving it with a net positive or negative charge. When an atom within a material *loses electrons* it becomes positively charged and is known as a positive ion, when an atom *gains an electron* it has a surplus negative charge and so is known as a *negative ion*. These differences in charge can cause *electrostatic* effects. For example, combing your hair with a nylon comb may result in a difference in charge between your hair and the rest of your body, resulting in your hair standing on end when your hand or some other differently charged body is brought close to it.

The number of electrons occupying a given orbit within an atom is predictable and is based on the position of the element within the periodic table. The electrons in all atoms sit in a particular position (shell) dependent on their energy level. Each of these shells within the atom is filled with electrons from the nucleus outwards (Figure 3.51). The first, inner most, of these shells can have up to two electrons, the second shell can have up to eight and the third up to 18.

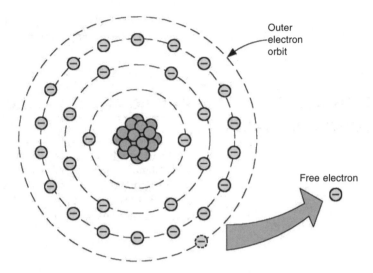

Figure 3.51 *The copper electron with a loosely bound outer electron*

Figure 3.51 shows the copper atom with one outer electron that can become easily detached from the parent atom. It requires a small amount of external energy to overcome the attraction of the nucleus. Sources of such energy may include heat, light or electrostatic fields. The electron once detached from the atom is able to move freely around the structure of the material and is called a *free electron*. It is these free electrons that become the *charge carriers* within a material. Materials that have large numbers of free electrons make good *conductors* of electrical energy and heat.

In a material containing free electrons their direction of motion is random, as shown in Figure 3.52(a), but if an external force is applied that causes the free electrons to move in a uniform manner (Figure 3.52(b)) an electric *current* is said to flow.

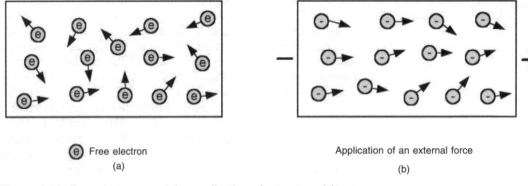

Figure 3.52 *Free electrons and the application of an external force*

Metals are the best conductors, since they have a very large number of free electrons available to act as charge carriers. Materials which do not conduct charge are called *insulators*; their electrons are tightly bound to the nuclei of their atoms.

The effects of electric current flow can be detected by the presence of one or more of the following effects: light, heat, magnetism, chemical, pressure and friction. Thus, for example, if a *piezo-*

electric crystal is subject to an electrical current it can change its shape and exert pressure. Heat is another, more obvious, effect, from electric heating elements. Therefore, as free electrons are the mechanism by which electric current flows, the electrical properties of a material are determined by the number of free electrons present.

Electrical terminology

1. Charge

All electrons and protons have an electrostatic *charge*; its value is so small that a more convenient unit of charge is needed for practical use, which we call the *coulomb*. One coulomb C is the total charge Q of 6.21×10^{18} electrons. Thus a single electron has a charge of 1.61×10^{-19} C.

2. Current

You have already met a definition for the *unit of current* at the beginning of this unit when we considered the SI system of units, it stated that: *the ampere* is that *constant current* which, if maintained between two straight parallel conductors of infinite length, of negligible circular cross-section, and placed 1 metre apart in a vacuum, would produce between these conductors a force equal to 2×10^{-7} newton per metre of length. Figure 3.53 illustrates the physical set-up for this definition.

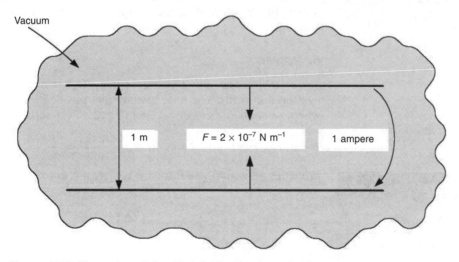

Vacuum

1 m

$F = 2 \times 10^{-7}$ N m^{-1}

1 ampere

Figure 3.53 *Illustration of the SI definition for the unit of current*

Current I itself is defined as *the rate of flow of charge* and its unit is the ampere (A). One ampere is equal to one coulomb C per second, or:

one ampere of current $I = \dfrac{Q}{t}$ where t = time in seconds

So, for example: if a steady current of 3 A flows for two minutes, then the amount of charge transferred will be:

$Q = I \times t = 3\text{ A} \times 120\text{ s} = 360$ coulombs

Key point

Electromotive force (emf) is that electrical force which is initially available for moving electrons round a circuit.

If the same 3 A of current was flowing in the conductors then the force set-up between them would be 6×10^{-7} newton per unit length.

3. Voltage

The *force which creates the flow of current* or rate of flow of charge carriers in a circuit is known as the *electromotive force* (emf) and it is measured in volts. The *potential difference* (pd) is the voltage difference, or voltage drop between two points.

Key point

The difference in *electrical pressure* between any two points in a circuit is called the *potential difference* (pd) between those points.

> *One volt is the potential difference between two points if one joule of energy is required to move one coulomb of charge between them.*

$V = \dfrac{J}{Q}$ where J = energy and Q = charge, as before. Energy is defined later.

4. Resistance

All materials at normal temperatures oppose the movement of electric charge through them. This opposition to the flow of the charge carriers is known as the *resistance* (R) of the material. This resistance is due to collisions between the charge carriers (electrons) and the atoms of the material. The unit of resistance is the *ohm*, with symbol Ω.

Note that 1 volt is equal to the amount of force required to move 6.21×10^{18} electrons (1 coulomb) through a resistance of 1 Ω in 1 second.

Key point

Resistance (R) is opposition to current flow.

5. Power

Power (P), is the rate at which energy is transferred and it is measured in *watts*. You have already met this definition of power when we studied mechanical energy and power earlier. So:

1 watt = 1 joule per second or, $P = \dfrac{\text{energy}}{\text{time}} = \dfrac{J}{s}$

From this definition, and those for voltage and current given above, then:

Key point

1 watt of power = 1 joule per second.

$$\text{power} = \text{voltage}\left(\dfrac{J}{Q} \right) \times \text{current}\left(\dfrac{Q}{t} \right)$$

$$= \text{voltage}(v) \times \text{current}(I) \text{ with units of } \left(\dfrac{J}{s} \right) \text{ or watts}$$

6. Electrical energy

Electrical energy is like all other forms of energy you have met before. Energy is the capacity to do work and is always in transit. This means that it can only be transferred when a difference in energy levels exists. As before the unit of energy is the joule. Then, from our definition of power, where 1 watt = 1 joule/second we know that the energy in *joules* = watts × *seconds* or *watt-seconds*.

If the power were to be measured in kilowatts and the time in hours, then the unit of electrical energy would be the *kilowatt-hour* (kWh), often known as the *unit of electricity*. The electricity meter in your home records the number of kilowatt-hours and is therefore an *energy* meter.

Example 3.37

An electric fire, when switched to its highest setting, consumes 3 kilowatts of power. If the fire is left on this setting for 2 hours, how much energy is transferred in joules?

We know that our electric fire will consume 3 kilowatts × 2 hour or 6 kWh of electrical energy.

1 kWh = 1000 watt-hours = 1000 × 3600 watt-seconds or joules. Therefore, the number of joules of energy transferred is 6 kWh = 6 × 1000 × 3600 = 2.16×10^7 joules or 21.6 MJ. Quite a lot of electrical energy!

---

7. Conductors, semiconductors and insulators

A material which has many free electrons available to act as charge carriers and thus allows current to flow freely is known as a *conductor*. Examples of good conductors include silver, copper, aluminium and iron.

Conversely, a material with few free electrons that cannot pass a significant current is known as an *insulator*. Examples of insulators include ceramics, glass, plastics and rubber.

In a *semiconductor* there may be a number of electrons, sufficient to allow a small current to flow. It is possible to add foreign atoms to the semiconductor material which modify the properties of the semiconductor. Varying combinations of these additional atoms are used to produce various electrical devices such as diodes and transistors. Typical semiconductors are silicon and germanium.

8. Conductance

The reciprocal of resistance R is called conductance, with symbol G. Conductance is a measure of the ease with which electricity flows. The greater the conductance of a material the easier it conducts electric current. The unit of conductance is the *siemen* (S). When considering resistors in parallel, it is sometimes more convenient to work in terms of conductance rather than resistance where $G = 1/R$ *siemen*.

9. Variation in conduction properties with temperature

All materials, as stated earlier, offer some kind of resistance to current flow. In *conductors* the free electrons, rather than passing unobstructed through the material, collide with the relatively large and solid nuclei of the atoms. As the temperature increases, the nuclei vibrate more energetically further obstructing the path of the free electrons, causing more frequent collisions. The result is that the *resistance of conductors increases with temperature*.

Due to the nature of the bonding in *insulators*, there are no free electrons, except that when thermal energy increases as a result of a temperature increase, a few outer electrons manage to break free from their fixed positions and act as charge carriers. The result is that the *resistance of insulators decreases as temperature increases*.

Semiconductors behave in a similar manner to insulators. Where at absolute zero (−273°C) the insulator and semiconductor insulate in a near perfect manner. However, unlike the insulator, as temperature increases in a semiconductor *large numbers* of electrons break free to act as charge carriers. Therefore *in a semiconductor as temperature increases the resistance decreases rapidly*.

By producing special alloys, such as Eureka and Manganin which combine the effects of insulators and conductors, it is possible to produce a material where the resistance remains constant with increase in temperature. Figure 3.54 illustrates the resistance behaviour of insulators, semiconductors, conductors and special alloys, with change in temperature.

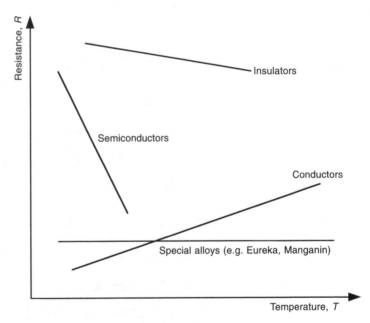

Figure 3.54 *Variation in the resistance of materials with change in temperature*

10. Capacitance

When a conductor is given *more charge*, its positive or negative *potential rises* in value. This happens because the increased charge repels any incoming charge more strongly than before, so large quantities of work have to be done to increase the charge on the conductors still further. Thus the larger the charge on a conductor the larger the potential difference required to get it there. The amount of charge deposited on a conductor is known as its *capacitance*. The capacitance of a conductor is defined as follows:

$$\text{capacitance} = \frac{\text{charge on conductor}}{\text{potential on conductor}} \quad \text{or in symbols } C = \frac{Q}{V}$$

If charge is measured in coulombs C and the potential in volts V, capacitance is measured in coulombs/volt or *farads* (F).

For example, if a potential of 350 000 V is required to deposit a charge of 5×10^{-6} C on a conductor then the capacitance in farads is:

$$\text{capacitance} = \frac{\text{charge}}{\text{potential}} = \frac{5 \times 10^{-6}\,\text{C}}{350\ 000} = 1.42 \times 10^{-11}\,\text{F}$$

We will use some of the definitions given above when we consider electric circuits in the next section.

Load

A load is simply an electrical component which converts electrical energy into either mechanical movement, light or heat. For example, motors, lamps and heaters are all loads. In all cases, the load will offer resistance to current flow and without resistance there would be no work done.

The factors that govern the value of the load resistance are: the material from which they are made, their cross-sectional area, their length and their temperature. So, for example, copper has a far lower resistance to current flow than, say, glass. A thin wire has greater resistance than a thick wire, a long wire has greater resistance than a short wire and a hot wire has greater resistance than a cold wire. The inherent resistance of the material is known as its resistivity and it is given the symbol ρ. The units of resistivity are Ω m or $\mu\Omega$ mm. Ignoring the change in resistance with temperature, the above relationships may be combined to give a definition of resistance as:

$$R = \rho \frac{l}{A}$$

where ρ = resistivity of the material (Ω m), l = length of material in metres and A = area of cross-section of the material (m^2).

Values of resistivity for some materials are given in the table below.

Material	Resistivity Ω m
Conductors	
Silver	1.6×10^{-8}
Copper	1.7×10^{-8}
Aluminium	2.8×10^{-8}
Iron	10.7×10^{-8}
Manganin	46×10^{-8}
Eureka	49×10^{-8}
Nichrome (Ni, Cr)	109×10^{-8}
Carbon	4000×10^{-8}
Semiconductors	
Germanium	6.0×10^{-1}
Silicon	2300
Insulators	
Glass	10^{10}–10^{14}
Polystyrene	10^{15}

Example 3.38

The element of an immersion heater is to be made from nichrome wire of diameter 0.25 mm. Calculate the length of wire required in order that its resistance is 90 Ω.

Then using the relationship:

$$R = \rho\, \frac{l}{A}$$

we have

$$R = \rho\frac{l}{A}, \quad \text{then } l = \frac{RA}{\rho} \quad \text{where } R = 90 \text{ and } A = \frac{\pi(0.25)^2}{4} \text{ mm}^2$$

So:

$$l = \frac{90\pi(0.00025)^2}{(109 \times 10^{-8})(4)} \text{ metres}$$

and the required length $l = $ **4.053 m**.

Key point

The resistivity of a material is an inherent property of that material and is measured in ohm-metres.

Key point

Connected cells form a battery.

Key point

Primary cells use up the chemicals from which they are made.

Production of electrical energy by chemical means

A *cell* is a device that pushes out charge when a chemical reaction takes place, it is thus a source of electromotive force (emf). When several of these cells are connected together they form a *battery*. We will now look briefly at the nature of cells and batteries. Cells are divided into two main groups known as primary and secondary cells.

Primary cells produce electrical energy at the expense of the chemicals from which they are made and once these chemicals are used up, no more electricity can be obtained from the cell. All cells consist of two *electrodes* which are dissimilar metals, or carbon and a metal, which are placed into an *electrolyte*.

One of the simplest examples of a primary cell is the *voltaic* type. This cell (Figure 3.55) consists of a plate of zinc forming the negative electrode, a plate of copper forming the positive electrode and dilute sulphuric acid as the electrolyte. The negative electrode is known as the *cathode* and the positive electrode is known as the *anode*.

When the electrodes are connected outside the cell, a current flows from the copper electrode, through the external circuit to the zinc and from the zinc to the copper, through the electrolyte in the cell.

One of the problems with the voltaic cell is that it only works for a short time before a layer of hydrogen bubbles builds up on the positive copper electrode, drastically reducing the emf of the cell and increasing its internal resistance. This effect is called *polarization*. The removal of this hydrogen layer from the copper electrode may be achieved by mechanical brushing or adding a depolarizer such as potassium dichromate to the acid solution. The removal of this hydrogen layer is known as *depolarization*.

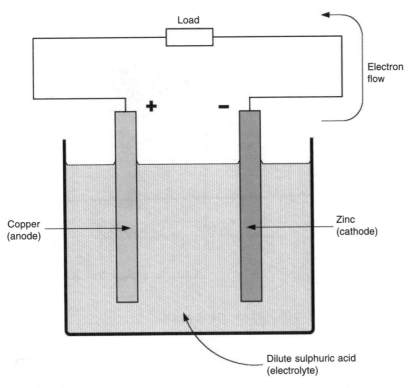

Figure 3.55 *The voltaic cell*

Test your knowledge 3.15

1. Does a negative ion have a surplus or shortage of electrons, when compared with the protons in the nucleus?
2. A conductor will have many free electrons, what is required to make these electrons flow?
3. A material is charged with 186.3×10^{18} electrons. How many coulombs of charge is this?
4. 180 joules of energy is required to move 15 coulombs of charge between two conductors. What is the potential

If the zinc electrode is not 100 per cent pure, which for cost reasons is often the case. Then the impurities react with the zinc and the sulphuric acid to produce miniature cells on the surface of the zinc electrode. This reaction takes place in the voltaic cell, irrespective of whether a current is being taken from the cell or not. This *local action*, as it is known, is obviously wasteful and may be eliminated by coating the zinc plate with mercury, or by using the more expensive pure zinc. The emf of a cell of this type is approximately equal to 1.0 V.

A second type of *primary* cell is the *dry* cell. In this type of cell instead of using a dilute acid electrolyte we use ammonium chloride in thick paste form. In one variant of this cell the positive electrode is a centrally positioned carbon rod (Figure 3.56) while the negative electrode is the zinc outer casing of the cell. Carbon and manganese dioxide act as the depolarizing agent that surrounds the carbon electrode.

This type of cell is often used to power torches, each cell has an emf of approximately 1.5 V.

The lead–acid cell is one of the most common secondary cells. In this type of cell, the electrical energy is initially supplied from an external source and converted and stored in the cell as chemical energy. This conversion of energy is reversible and when required this stored chemical energy can be released as a direct electric current. This process of storage leads to the alternative name for this type of cell, the lead–acid accumulator.

The manufacture of this cell is quite complex. The positive plate consists of a grid of lead and antimony filled with lead peroxide (Figure 3.57). The negative plate uses a similar grid, but its open spaces are filled with spongy lead. Thus the cells are made up of a

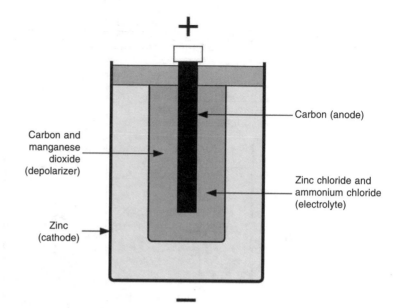

Figure 3.56 *The dry cell*

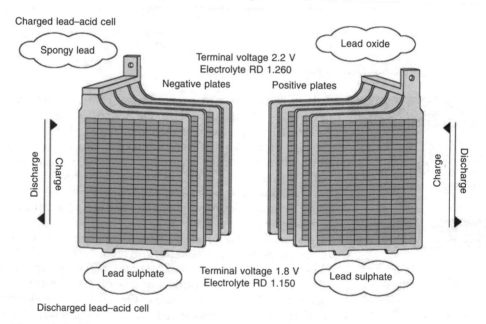

Figure 3.57 *The lead–acid cell*

difference (pd) in volts set up between them?

5. Convert 80 kilowatt-hours into joules.

6. The total conductance of two equal resistors in parallel is 0.125 seimens. What is the resistance in ohms of each resistor?

7. What happens to the resistance of

group of positive plates, joined together and interlaced between a stack of negative plates. Porous separators keep the plates apart and hold a supply of electrolyte in contact with the active materials. The electrolyte consists of a mixture of sulphuric acid and water, which covers the plates and takes an active part in the charging and discharging of the cell.

A fully charged lead–acid cell has an emf of approximately 2.2 V, but when in use this value falls rapidly to about 2.0 V. In the fully charged condition the negative plate is spongy lead and the positive plate is lead peroxide. In the discharged condition, where the emf is about 1.8 V, the chemical action of the cell converts both positive and negative plates into a lead sulphate mix. When

semiconductors with decrease in temperature?
8. Define the resistivity of a material.
9. How do primary cells differ from secondary cells?
10. What is polarization and what effect does it have on the performance of a cell?

discharged, the cell may then be recharged from an external source and made ready for further use. The condition of this type of cell may be checked by measuring the relative density of the electrolyte. In the fully charged condition this will be around 1.26, while in the discharged condition it drops to around 1.15. This type of cell, when joined together as a battery, has many commercial uses, the most familiar of which is as a motor vehicle battery.

Electrical circuits

The basic circuit

A basic electrical circuit transfers electrical energy from a source to a load, in order to do work. Every electrical circuit has some form of device to control the transfer of this energy. As you have already seen the transfer of energy occurs by the flow of electric current, which in turn requires an emf to force it around the circuit. The emf necessary to produce current flow is provided by the source which is usually a battery or generator. If the current flow is unidirectional (flows in one direction only), then it is referred to as direct current (DC), if the current flows alternatively from one direction to the other, it is referred to as alternating current (AC). These two types of current flow may be illustrated graphically, as shown in Figure 3.58. *During this unit only DC circuits will be considered.*

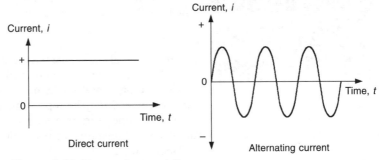

Figure 3.58 *Types of current flow*

Whether current flow is DC or AC it will, by *convention*, always flow from the positive terminal of the source back to the negative terminal of the source (Figure 3.59). You should note that electrons

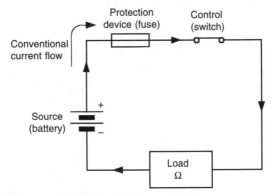

Figure 3.59 *Basic circuit showing current flow*

actually flow in the opposite direction. However, when considering current flow in circuits, *we will* stick to this *convention*.

The *control* for the circuit can simply be a switch, which completes the circuit when in the closed position, to enable current to flow. The components of the circuit are joined by conductors, that is the electrical wiring. Circuit components are represented by symbols which must adhere to British Standard 3939. You met these symbols when you studied the unit on Engineering communications.

Ohm's law

George Simon Ohm, a German physicist, introduced a law which provided a very important relationship between *potential difference* in volts (*V*), *resistance* (*R*) in ohms and *current* (*I*) in amperes.

In words, *Ohm's law* states that: *the potential difference (pd) between the ends of a conductor is directly proportional to the current flowing through it, providing that the temperature remains constant.*

Mathematically Ohm's law may be expressed as $V \propto I$, where *V* is the potential difference and *I* is the current. All that we need do to turn this inequality into an equation is to insert a *constant of proportionality k*, such that

$$V = kI$$

The *constant k is defined as the resistance R* of the circuit. Ohm's law may therefore be represented by the equation

$$V = IR$$

where, as stated above, the potential difference *V* must be expressed in volts, resistance *R* in ohms and current *I* in amperes.

Conductors that obey Ohm's law are known as *ohmic* or *linear conductors* and may be represented by a straight-line graph, since constant temperature is assumed. The resistance of non-ohmic or non-linear conductors varies with the potential difference, in the manner shown in Figure 3.60.

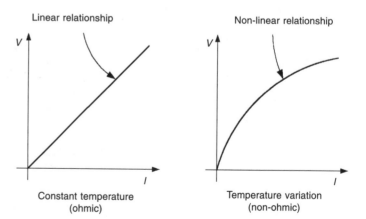

Figure 3.60 *Ohmic and non-ohmic conductors*

Example 3.39

A 12 V car battery provides the potential difference to light a car head lamp that offers a resistance of 8 Ω. What current is drawn from the battery?

The solution to this problem simply requires us to find current *I* using Ohm's law. Therefore from $V = IR$ we get

$$I = \frac{V}{R} \quad \text{and so} \quad I = \frac{12}{8} = 1.5 \text{ amperes}$$

The current drawn from the battery = **1.5 amperes**.

Measurement of circuit resistance

In order to understand the behaviour of resistors in electric circuits, we can take electrical measurements to determine their values. One method of measuring resistance involves the use of an *ammeter* and a *voltmeter*, which may be combined into an instrument known as a *multimeter*. These instruments may be *analogue* (continually variable signal values) or *digital* (discrete single point values).

The *ammeter* is used to measure *current* flowing in a circuit. To do this *the ammeter is connected into the circuit*, so as to measure the current flowing through it (Figure 3.61). The *voltmeter* is used for measuring the *potential difference* between any two points in a circuit. The voltmeter is connected *across* a load, with one terminal either side of the load, as shown in Figure 3.61 and the potential difference is measured between these points.

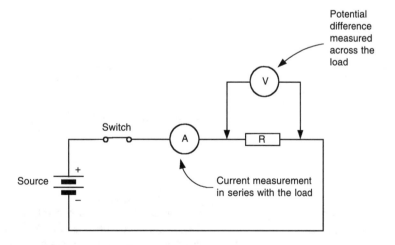

Figure 3.61 *Measurement of current and potential difference in a circuit*

So, to find the resistance of a load all we need do is take measurements of the current flowing through the resistor and the potential difference across the resistor. We then apply Ohm's law, where

$$R = \frac{V}{I}$$

and on substitution of the measured values of potential difference V and current I, we determine the value of the resistive load.

This method of determining resistance values is only suitable when an *approximation* of resistance values is acceptable. This is because the resistance of the measuring instruments themselves alters the readings obtained. Nevertheless, for practical measurements of the resistance of loads such as lamps and heating elements, the above method is sufficiently accurate.

Resistors in series circuits

The various components forming a circuit may be connected in two ways, *series* or *parallel*. The components are said to be in *series* when they are connected end to end providing only one path for the current to flow. The same current must, therefore, pass through all components, including the power supply, as illustrated in Figure 3.62.

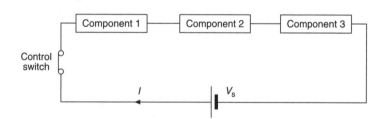

Figure 3.62 *A simple series circuit*

When a current flows through a resistor, or component with resistance, there is a potential difference between its ends (Figure 3.63). This potential difference may be measured by placing a voltmeter *across* the resistor. When more than one resistor is connected in series, the total potential difference produced by the source can be measured, as shown in Figure 3.63, by placing the voltmeter across all the resistors in the circuit. We assume that the current carrying conductors in the circuit offer negligible resistance when compared with the circuit components and, therefore, we see that the sum of the volts drops across each of the individual resistors is equal to the voltage at the source.

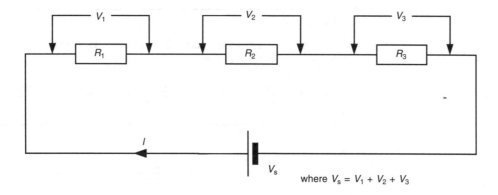

where $V_s = V_1 + V_2 + V_3$

Figure 3.63 *Potential difference across resistors in series*

So, for the above situation we may write:

$V_s = V_1 + V_2 + V_3$ (where V_s = the voltage at source)

Now from Ohm's law we know that $V = IR$. So applying Ohm's law to each of the resistors in turn, we get:

$V_1 = IR_1$, $V_2 = IR_2$ and $V_3 = IR_3$ since the current is common to all resistors.

Therefore:

$V_T = V_1 + V_2 + V_3 = IR_T = IR_1 + IR_2 + IR_3$ (where V_T = total resistance)

or

$IR_T = I(R_1 + R_2 + R_3)$ current I being a common factor

and

$R_T = R_1 + R_2 + R_3$

Key point

For resistors connected in series the total resistance is equal to the sum of the individual resistors.

So, *for resistors in series*:

1. the total resistance is equal to the sum of the individual resistors
2. the current is common to all resistors
3. the sum of the individual volts drops is equal to the applied voltage (voltage at source).

In any series circuit Ohm's law applies to each component and so remembering that $V_1 = IR_1$, $V_2 = IR_2$ ... etc. for any source voltage we can produce any smaller voltage we wish, in the circuit, by inserting resistors of the appropriate value, in series. This arrangement is known as *voltage division*.

Example 3.40

For the circuit shown in Figure 3.64, calculate the potential difference in volts across each resistor.

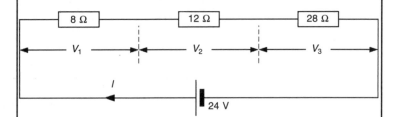

Figure 3.64 *Resistors in series circuit*

Then from Ohm's law $I_T = \dfrac{V_s}{R_T} = \dfrac{24}{48} = 0.5$ amperes

where $R_T = R_1 + R_2 + R_3$.

So $V_1 = IR_1 = (0.5)(8) = \mathbf{4\ V}$, $\qquad V_2 = IR_2 = (0.5)(12) = \mathbf{6\ V}$,

$V_3 = IR_3 = (0.5)(28) = \mathbf{14\ V}$

We may check this result by remembering that $V_s = V_1 + V_2 + V_3$ = 4 + 6 + 14 = 24 V.

Resistors in parallel circuits

Resistances are said to be connected in *parallel* when they provide *separate pathways* for the *current flow*, the total current being divided between the individual resistors in each pathway or branch.

Consider the circuit shown in Figure 3.65, in which three resistors R_1, R_2 and R_3 are connected in parallel.

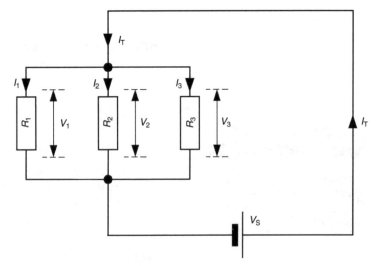

Figure 3.65 *Resistors connected in parallel*

Now the total current I_T is separated as it passes through each resistor into currents I_1, I_2 and I_3. Now:

$$I_T = I_1 + I_2 + I_3$$

Also remembering how we measure voltage across resistors, it can be seen that the voltage drop V across each resistor is the same. Then, applying Ohm's law to the resistors in each individual pathway, we have:

$$I_1 = \frac{V}{R_1}, \quad I_2 = \frac{V}{R_2}, \quad I_3 = \frac{V}{R_3}$$

and on adding the equations,

$$I_T = I_1 + I_2 + I_3 = \frac{V}{R_1} + \frac{V}{R_2} + \frac{V}{R_3} = \frac{V_s}{R_T}$$

Now we know from above that the voltage at source is equal to the volts drop across each of the resistors. So we may divide the right-hand side of the above equation by V to give:

$$\frac{1}{R_T} = \frac{1}{R_1} + \frac{1}{R_2} + \frac{1}{R_3}$$

Then, *for resistors or resistive components in parallel:*

1. the voltage across each resistor is the same
2. the reciprocal of the total equivalent resistance is equal to the sum of the reciprocals of the individual resistors
3. Ohm's law applies to each individual resistor or resistive component.

Effective resistance

Now, remembering that for two or more resistors connected in parallel the voltage drop or potential difference across each is the same. Also, that each additional resistor added will draw current, so the *total current* being draw *increases with the addition of further resistors*. Therefore the effective total resistance R_T must decrease ($V = IR$), with V always being the same. The net effect is that the *supply current* drawn, *increases* with each additional resistor added. So the *effective resistance* of the circuit *decreases*, with the addition of each individual resistor. This *effective resistance* will, therefore, *always be less* than that of any individual resistor paralleled into the circuit. An important point to note is that *the supply current increases, as each additional resistor is added*. If not carefully monitored, this supply current could well reach a value where possible wiring damage might occur!

> **Key point**
>
> For resistors in parallel the total resistance is equal to the sum of the reciprocals of the individual resistors.

> **Key point**
>
> For two or more resistors connected in parallel, the total effective resistance is always smaller than any one individual resistor.

Example 3.41

Three resistors of 10 Ω, 20 Ω and 30 Ω are connected into a circuit in parallel, where the supply voltage is 120 V. Find the circuit total equivalent resistance and the total current drawn from the source.

Then total resistance may be found using the formula

$$\frac{1}{R_T} = \frac{1}{R_1} + \frac{1}{R_2} + \frac{1}{R_3}$$

where

$$\frac{1}{R_T} = \frac{1}{10} + \frac{1}{20} + \frac{1}{30} = \frac{3+2+1}{30} = \frac{6}{30} \text{ or } \frac{1}{5} \text{ so } R_T = 5\,\Omega, \text{ with a}$$

total circuit resistance = **5 Ω**

Now to find total current we use

$$I = \frac{V_s}{R_T} = \frac{120}{5} = \mathbf{24\ A}$$

This is a fairly large current!

We may combine the way in which we connect resistive components, so that in many circuits, some components may be in series and others in parallel with the supply. The techniques we have developed for the individual circuits can be combined to solve problems involving *series/parallel circuits.*

Example 3.42

Three resistors of 4 Ω, 6 Ω and 12 Ω are connected in parallel with each other, then in series with the remaining resistor of 4 Ω across a supply of 24 V. Determine:

(a)   the total current taken from the supply
(b)   the potential difference across the single 4 Ω resistor
(c)   the current flowing in each of the paralleled resistors.

The resistor combination is shown in Figure 3.66.

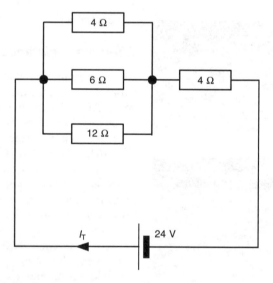

Figure 3.66 *Resistors in parallel circuit*

(a)   To establish the *total circuit current* we need first to find the total equivalent resistance R_T. To do this we find the equivalent resistance of the parallel combination and add this equivalent resistance to the one remaining resistor which is in series with the supply. Then:

$$\frac{1}{R_E} = \frac{1}{4} + \frac{1}{6} + \frac{1}{12} = \frac{3+2+1}{12} = \frac{6}{12} = \frac{1}{2} \quad \text{therefore } R_E = 2\ \Omega$$

and so $R_T = 2\ \Omega + 4\ \Omega = 6\ \Omega$ and again using Ohm's law as before:

$$I_T = \frac{V_s}{R_T} = \frac{24}{6} = \mathbf{4\ A}$$

(b) From our analysis above, we now have the equivalent of two resistors in series; one is 2 Ω, which is the single equivalent resistance of the three in parallel and the other is the remaining 4 Ω resistor in series. We have also established the current passing through both of them as 4 A. Therefore the potential difference across the 4 Ω resistor is found from Ohm's law as:

$$V_4 = IR_4 = (4)(4) = \textbf{16 V}$$

(c) Now to find the current flowing in each of the paralleled resistors, all we need do is establish the potential difference across the combination and then apply Ohm's law to each.

Then the potential difference (volts drop) across the single equivalent resistance of 2 Ω is given by Ohm's law as $V_2 = IR_2 = (4)(2) = 8$ V. This volts drop will be the same across each of the original resistors in parallel. Note also that the sum of the volts drops across these two resistors is equal to the supply voltage; this is no coincidence as you will see later.

All we need do now is to reapply Ohm's law to each of the parallel resistors in order to find the current in each pathway. Then:

$$\text{Current flowing through the 4 Ω resistor} = \frac{V_E}{R_4} = \frac{8}{4} = \textbf{2 A}$$

Similarly the current flowing in the 6 Ω resistor

$$= \frac{V_E}{R_6} = \frac{8}{6} = \frac{\textbf{4}}{\textbf{3}} \textbf{ A}$$

And in the 8 Ω resistor $= \dfrac{V_E}{R_8} = \dfrac{8}{12} = \dfrac{\textbf{2}}{\textbf{3}} \textbf{ A}$

It is also no coincidence that the sum of these individual currents is equal to the total current flowing into the parallel combination, i.e.

$$2A + \frac{4}{3}A + \frac{2}{3}A = 4A$$

notice also, the advantage of leaving these currents in fractions!

Kirchhoff's laws

Although you may not have realized it, you have already meet the consequences of Kirchhoff's laws, when looking at the solution to the problems given in Example 3.42. Kirchhoff identified two important relationships, which have developed into useful laws when solving circuit problems.

Kirchhoff's first law states that: *the total current entering any point (junction) in a circuit must equal the sum of the currents leaving that junction.* Figure 3.67(a) illustrates this law for three pathways leaving the junction. The law is valid no matter what number of pathways are used.

Figure 3.67(b) shows the use of Kirchhoff's law when considering Example 3.42 above. The sum of the individual currents in the parallel pathways equals the input current, as we found.

Kirchhoff's second law states that: *the algebraic sum of the individual voltage drops (potential differences) must equal the applied*

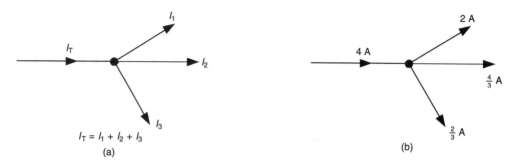

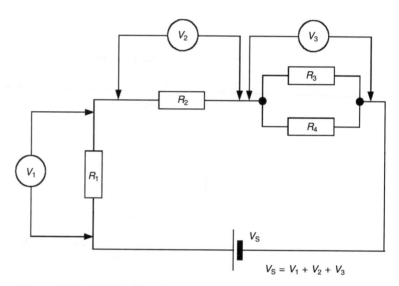

Figure 3.67 *Kirchhoff's current law*

voltage, from the source. Figure 3.68 illustrates this law in general terms.

Figure 3.68 *Kirchhoff's voltage law*

Both of Kirchhoff's laws, apart from simplifying some of the calculations, also provide a useful means of checking the accuracy of your working.

Power in series/parallel circuits

You have already met the definitions of electrical power and electrical energy. Here we are going to apply these definitions.

When current flows through a component offering resistance, electrical energy is converted into heat. You know that the rate at which electrical energy is converted into heat is power. Electrical components such as resistors and bulbs have a *power rating*. In the case of the bulb, this power rating may be sustained for long periods of time, without failure. In the case of a resistor, this power rating is the absolute maximum that the resistor can sustain, therefore we must always ensure that we keep *below this stated value* when considering the use of resistors in circuits.

The surface area and therefore the size of a component determines the rate at which heat is dissipated from the component to its surroundings. So, in general, the larger the component the larger its power rating. Carbon resistors with the same resistance values are commonly available in ratings between 0.25 W and 2 W. For a higher power requirement, different types of resistor may be used.

In order to calculate the power dissipated in circuit components we may use the following formula:

power (in watts) $P = $ (voltage V)(current I)   that is $P = IV$ watts

Now dependent on the variables we are using we may use Ohm's law to express power in terms of resistance and voltage or resistance and current, as well as in terms of current and voltage, shown above. Then using Ohm's law we have:

$V = IR$

so substituting IR for V in $P = IV$ gives

$P = IIR$ or $P = I^2R$

Also from

$I = \dfrac{V}{R}$

then again substituting $\dfrac{V}{R}$ for I in $P = IV$ gives

$P = \dfrac{VV}{R}$ or $\dfrac{V^2}{R}$

These power relationships are summarized below.

$$P = IV \ \text{ or } P = I^2R \ \text{ or } P = \dfrac{V^2}{R}$$

where power P is in watts, current is in amperes and potential difference or electromotive force is in volts.

Example 3.43

Calculate the total power dissipated in the circuit shown in Figure 3.69 *and* given that the power rating for resistor $R_1 = 10$ watts, check the rating of this resistor.

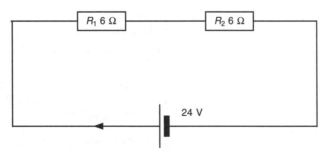

Figure 3.69 *Resistor current rating*

This is a simple series circuit where the total equivalent resistance,

$R_T = R_1 + R_2 = 6 + 6 = 12 \ \Omega$

Also from Ohm's law the total current drawn from the supply is

$$I = \frac{24}{12} = 2 \text{ A}$$

Then we have all three variables for the circuit, so we may use any of our power relationships. Using $P = IV$, then **P = (2)(24) = 48 W**.

Now the power dissipated by R_1 may be found using $P = I^2R = (22)(6) = 24$ watts. We are told that the maximum allowable power dissipation is 10 watts, thus the power rating is exceeded. This will most likely lead to premature resistor failure.

Example 3.44

A DC motor with a series resistor of 10 Ω is connected in series with a 240 V supply. When the motor is running the pd across the series resistor is found to be 60 V:

(a) determine the pd across the motor and the current passing through it
(b) find the power drawn from the supply
(c) find the power dissipated by the resistor as heat.

(a) The potential difference across the motor is easily found by remembering Kirchhoff's voltage law. Where the pd across the supply (240 V) is equal to the sum of the pds, i.e.

240 V = 60 V + X, where X, the pd across the motor, is = **180 V**

The set-up is illustrated in Figure 3.70.

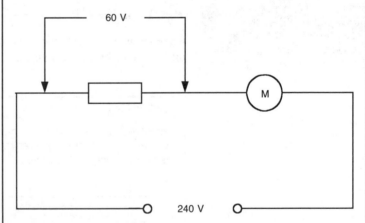

Figure 3.70 *Motor circuit*

Since we have a simple series circuit, the current through the motor is equal to the current through the series resistor and, from Ohm's law:

$$I = \frac{V}{R} = \frac{60}{10} = \textbf{6 A}$$

(b) The power drawn from the mains supply may be found using $P = IV$, then

(6)(240) = 1440 W

(c) The power dissipated by the resistor $P_R = IV_R$, where V_R = volts drop across resistor, then $P_R = $ **(6)(60) = 360 W**.

Cells and batteries

You are already familiar with the nature of chemical cells that produce an electromotive force (emf) and the fact that when cells are connected together they form a battery.

Cells are said to be connected in *series when the positive plate of one cell is connected to the negative plate of the other*; several cells connected in this way form a battery of cells in series (Figure 3.71).

Figure 3.71 *Battery of cells in series*

This arrangement of cells has the following characteristics:

1. The battery emf equals the sum of the emfs of the individual cells.
2. The maximum current from the battery is the same as that from one individual cell.
3. The internal resistance of the battery equals the sum of the individual cell resistances.

When all the positive terminals of a group of cells are connected together and all the negative terminals are connected together, the cells are said to be connected in parallel (Figure 3.72).

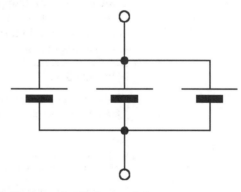

Figure 3.72 *Battery of cells in parallel*

The characteristics of this arrangement are:

1. The emf of several cells in parallel equals the emf of one individual cell.
2. The reciprocal of the total internal resistance equals the sum of the reciprocals of each internal resistance.
3. The total current equals the sum of the individual currents.

The above characteristics need a little explanation. All electrical supplies have internal resistance; in batteries this is mainly due to the resistance of the electrolyte. The circuit in Figure 3.73 shows a single cell battery with emf E, internal resistance r, connected in series with a circuit load resistance R.

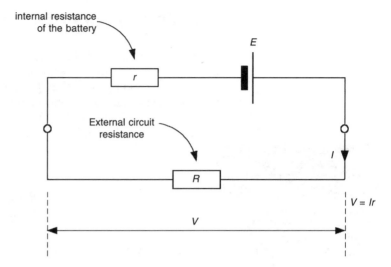

Figure 3.73 *Circuit showing internal resistance of battery*

Now if we measure the voltage V across the battery, when the battery is disconnected from the circuit it will equal the emf in a circuit with *no* resistive load applied. When a resistive load R is connected to the battery, the load current I flows through the internal resistance r of the battery and causes an internal resistance volts drop. So that the *on-load* terminal voltage V, is equal to the emf E, minus the internal resistance volts drop Ir. Or in symbols:

$V = E - Ir$

It can be seen from the above formula that as the current drawn from the battery (supply) increases, the terminal voltage at the battery falls. If we assume that the internal resistance remains constant, the fall in terminal voltage is directly proportional to the resistive load current.

Example 3.45

A cell is connected in series with an external resistance $R = 0.1\ \Omega$. If the internal resistance of the battery $r = 0.025\ \Omega$ and the emf = 2 V, find:

(a)   the current in the circuit
(b)   the terminal voltage of the cell.

(a)   The total circuit resistance is $R + r = 0.1 + 0.025 = 0.125\ \Omega$
      Now from $V = E - Ir$ and from Ohm's law $V = E - Ir = IR$, then
      $E = IR + Ir$, and transposing for I, then:

$$\text{circuit current} = I = \frac{E}{R + r} = \frac{2}{0.125} = \mathbf{16\ A}$$

(b)   The terminal voltage V may be found from $V = E - Ir$, and so $V = 2 - (16)(0.025) = 1.6\ V$.

So terminal voltage = **1.6 V**.

Capacitance and capacitors

We have already defined the unit of capacitance as the *farad*, which we redefine here.

A capacitor has a capacitance of 1 farad when a charging current of one ampere flowing for one second causes a change of one volt across the plates of the capacitor. Or in symbols:

$$C = \frac{Q}{V}$$

where Q is in coulombs (remembering that one ampere is equal to one coulomb per second) and C is the capacitance in farads.

The farad is a very large and impractical unit and in practice capacitors have much smaller values, normally measured in microfarads µF or picofarads pF. Where, from your study of units, you will know that:

$1 \ \mu F = 10^{-6}$ farads   and $1 \ pF = 10^{-12}$ farads

In practice, a *capacitor* is an electronic component that can store electrical energy in the form of an electric field. In its simplest form, a capacitor consists of two conducting plates separated by an insulator called the *dielectric*.

You have already seen how by applying a voltage to the capacitor it is *charged*. A battery may be used as the voltage source to charge the capacitor, as shown in Figure 3.74.

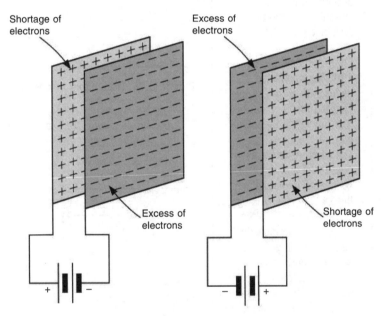

Shortage of electrons

Excess of electrons

Excess of electrons

Shortage of electrons

Figure 3.74 *Charging a capacitor with a battery*

Once charged, the capacitor has a large difference in potential energy, which may be *discharged* by connecting (shorting) the two plates together.

Capacitors in circuits

Capacitors, like resistors, may be connected in parallel or in series with a supply. *When capacitors are connected in parallel the total charge Q drawn from the supply is the sum of the individual charges in the capacitor.* Figure 3.75 shows three capacitors connected in parallel with a supply.

Then, from what has been said, in symbols:

Key point

For capacitors connected in series the charge on each capacitor is the same.

Key point

For capacitors connected in series, the voltage across each capacitor is inversely proportional to its capacitance.

Test your knowledge 3.16

1. How does *conventional* current flow differ from *real* current flow?
2. Under what circumstances is Ohm's law valid?
3. Describe how circuit resistance may be measured.
4. Three resistors of 4, 6 and 10 Ω are connected in series with a 24 V supply. Find the total equivalent resistance and the circuit current.
5. The same resistors as in question 4 are now connected in parallel with the 24 V supply. What is the current drawn from the supply *and* the individual currents flowing through each resistor?
6. A circuit with a 12 V supply has four loads connected in series with it. The potential difference across the first three loads is 2 V, 3 V and 4 V, respectively. If the current drawn from the supply is 6 A, what is the resistance of *each* load in the circuit?
7. An electric motor offers a resistance of 10 Ω when operating at full power from a 240 V supply. What is the power rating of this motor in kW?

$Q_T = Q_1 + Q_2 + Q_3$

and since supply voltage is the same across each paralleled capacitor then:

$C_T V = C_1 V + C_2 V + C_3 V$

and so,

$C_T = C_1 + C_2 + C_3$

Thus the combined capacitance of capacitors connected in parallel is found in the same way as that of resistors connected in series. Note also that the charge on each capacitor is directly proportional to the value of its capacitance.

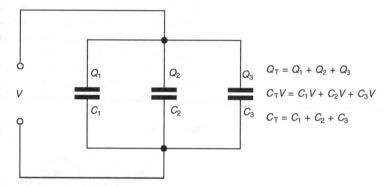

Figure 3.75 *Capacitors in parallel*

Now when *capacitors are connected in series* (Figure 3.76) *the charge on each capacitor is the same as drawn by the supply.*

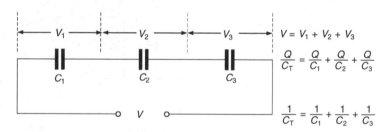

Figure 3.76 *Capacitors in series*

So from Kirchhoff's voltage law, $V = V_1 + V_2 + V_3$ and from our definition of capacitance and what has been said above:

$$\frac{Q}{C_T} = \frac{Q}{C_1} + \frac{Q}{C_2} + \frac{Q}{C_3}$$

and so,

$$\frac{1}{C_T} = \frac{1}{C_1} + \frac{1}{C_2} + \frac{1}{C_3}$$

A consequence of the above argument is that the total capacitance of capacitors connected in series is found in the same way as resistors connected in parallel.

Magnetism

Elementary theory of magnetism

Magnetism is an effect created by moving the elementary atomic particles in certain materials such as iron, nickel and cobalt. Iron has outstanding magnetic properties and materials that behave magnetically, in a similar manner to iron, are known as *ferromagnetic* materials. These materials experience forces which act on them when placed near a magnet.

The atoms within these materials group in such a way that they produce tiny individual magnets with their own *North and South poles*. When subject to the influence of a magnet or when an electric current is passed through a coil surrounding them, these individual tiny magnets line up and the material as a whole exhibits magnetic properties.

Figure 3.77(a) shows a ferromagnetic material that has not been influenced by the forces generated from another magnet. In this case, the miniature magnets are oriented in a random manner. Once the material is subject to the influence of another magnet, then these miniature magnets line up (Figure 3.77(b)) and the material itself becomes magnetic with its own North and South poles.

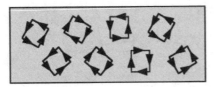

Miniature magnets, randomly oriented, no observable magnetic effect

(a)

Miniature magnets aligned (material magnetized). Production of magnetic lines of force with polarity observed in the material

(b)

Figure 3.77 *The behaviour of ferromagnetic materials*

A magnetic *field of flux* is the region in which the forces created by the magnet have influence. This field surrounds a magnet in all directions, being strongest at the end extremities of the magnet, known as the *poles*. Magnetic fields are mapped by an arrangement of lines that give an indication of strength and direction of the flux as illustrated in Figure 3.78. When freely suspended horizontally a magnet aligns itself North–South parallel with the earth's magnetic field. Now because unlike poles attract, the North of the magnet aligns itself with the South magnetic pole of the earth and the South pole of the magnet aligns itself with the earth's North pole. This is why the extremities of the magnet are known as poles.

Whenever an electric current flows in a conductor a magnetic field is set up around the conductor in the form of concentric circles. The field is present along the whole length of the conductor and is strongest nearest to the conductor. Now like permanent magnets, this field also has direction. The direction of the magnetic field is dependent on the direction of the current passing through the conductor and may be established using the right-hand grip or right-hand screw rule, as shown in Figure 3.79.

If the right-hand thumb is pointing in the direction of current flow in the conductor, then when gripping the conductor in the

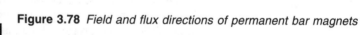

Figure 3.78 *Field and flux directions of permanent bar magnets*

right hand, the fingers indicate the direction of the magnetic field. In a *cross-sectional view* of the conductor *a point or dot (•) indicates the current is flowing towards you*, out of the page, and *a cross (×) indicates the current is flowing away from you*. This convention mirrors arrow flight, where the dot is the tip of the arrow and the cross is the feathers at the tail of the arrow, as the arrow leaves the bow.

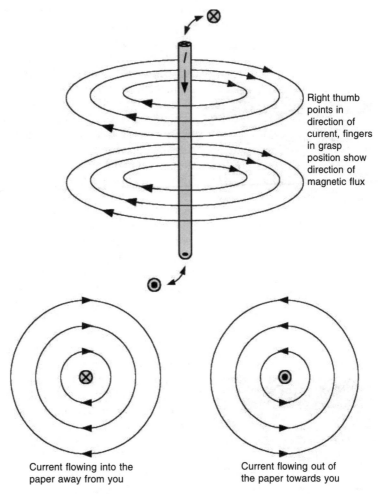

Right thumb
points in
direction of
current, fingers
in grasp
position show
direction of
magnetic flux

Current flowing into the
paper away from you

Current flowing out of
the paper towards you

Figure 3.79 *Right-hand screw rule*

The conductor in a magnetic field

If we place a current carrying conductor in a magnetic field, the
conductor has a force exerted on it.

Consider the set-up shown in Figure 3.80, in which a current

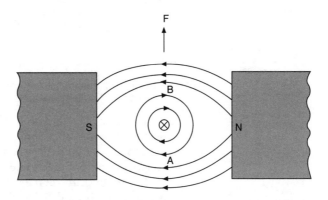

Figure 3.80 *Current carrying conductor in a magnetic field*

Key point

The direction of the magnetic field surrounding a current carrying conductor can be found using the right-hand grip rule.

carrying conductor is placed between the N–S poles of two permanent magnets. The direction of the current passing through it is into the page going away from us. Then by the *right-hand grip rule*, the direction of the magnetic field, created by the current in the conductor, is clockwise, as shown. We also know that the *flux lines* from the permanent magnet *exit at a North pole* and *enter at a South pole*, in other words, they travel from North to South, as indicated by the direction arrows. The net effect of the coming together of these two magnetic force fields is that *at position A*, they both travel in the same direction and *reinforce one another*. While *at position B*, they travel in the opposite direction and tend to *cancel one another*. So with a *stronger force field at position A* and *a weaker force at position B* the conductor is *forced upwards* out of the magnetic field.

If the direction of the current was reversed, i.e. it travelled towards us out of the page, then the direction of the magnetic field in the current carrying conductor would be reversed and therefore so would the direction of motion of the conductor.

A convenient way of establishing the *direction of motion* of the current carrying conductor is to use *Fleming's left-hand (motor) rule*.

This rule is illustrated in Figure 3.81, where the left hand is extended with the thumb, first finger and second finger pointing at right angles to one another. From the figure it can be seen that the *first finger* represents the magnetic *field*, the *second finger* represents the direction of the *current* in the conductor and the *thumb* represents the *motion* of the conductor, due to the forces acting on it. As an *aide-mémoire*, we say:

Key point

The direction of motion of a current carrying conductor in a magnetic field is given by Fleming's *left-hand motor* rule.

First finger = Field

SeCond finger = Current

ThuMb   = Motion

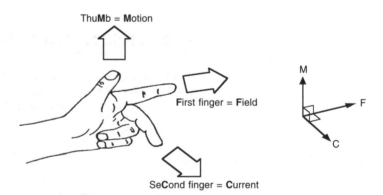

Figure 3.81 *Fleming's left-hand rule*

So much for the direction of a current carrying conductor in a magnetic field. What about the *magnitude of the force* acting on the conductor? Well, the magnitude of the force trying to move the conductor out of the field depends on: the current passing through it, the length of the conductor in the field and the amount of flux per unit area (flux density) that is created by the magnet. Then the force depends upon:

Key point

The force on a conductor in a magnetic field is equal to the product of the current flow in the conductor, the magnetic flux density and the length of the conductor in the field.

Key point

Flux density is the total flux divided by the area of the magnetic field over which it acts.

- the flux density β in tesla
- the current I in amperes
- the length of the conductor in the field L.

Or in symbols

$F = \beta IL$

The term flux density needs just a little more explanation. The total flux in a magnetic field is really a measure of the total magnetic intensity present in the field and it is measured in webers (Wb) and represented by the symbol Φ (phi). Therefore:

$$\text{flux density} = \frac{\text{total flux}}{\text{area of field}} \text{ (tesla)}, \quad \text{or } \beta \text{ (T)} = \frac{\Phi \text{ (Wb)}}{A \text{ (m}^2)}$$

In practice, values of flux density are generally small and millitesla (mT) and microtesla (μT) are often used.

Example 3.46

In Figure 3.82, a straight current carrying conductor lies at right angles to a magnetic field of flux density 1.2 T such that 250 mm of its length lies within the field. If the current passing through the conductor is 15 A, determine the force on the conductor and the direction of its motion.

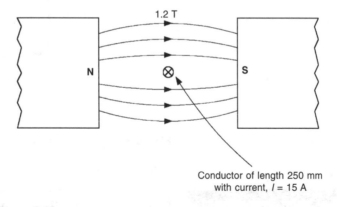

Figure 3.82

In order to find the magnitude of the force we use the relationship $F = \beta IL$ ensuring the correct use of units. Then:

$F = (1.2)(15)(250 \times 10^{-3}) = \textbf{4.5 N}$

Now the direction of motion is easily found using *Fleming's left-hand rule*, where we know that the first finger points in the direction of the magnetic field *N–S*, the second finger points inwards into the page in the direction of the current, which leaves your thumb pointing *downwards* in the direction of *motion*.

The DC motor principle

If we wrap our current carrying conductor around a rectangular former (Figure 3.83), pivoted at its centre, and place the conductor in the magnetic field, at right angles to the flux, then when an electric current is passed through the conductor as shown, we set up two equal and opposite forces which act on the conductor in the directions indicated.

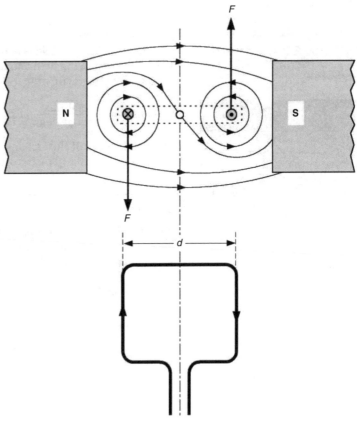

Torque = Force, F × Distance, d

Figure 3.83 *Torque on a coil*

The direction of the forces acting on each arm of the conductor can be established by again using the right-hand grip rule and Fleming's left-hand rule. Now because the conductors are equidistant from their pivot point and the forces acting on them are *equal and opposite*, then they form a couple. You have already met the idea of a couple when you studied *forces* during your study of statics. The important point to remember here is that *the moment of a couple is equal to the magnitude of a single force multiplied by the distance between them* and that this moment is known as *torque*. In symbols:

torque $T = Fd$

We already know that the magnitude of the force F is given by $F = \beta IL$, therefore the torque produced by the current carrying conductor is:

torque $T = \beta ILd$

Now one of the most useful applications for this torque (turning moment) is in the electric motor. In an electric motor the input is electrical energy and the output is motion. A simple direct current motor consists of a rectangular coil of wire mounted on a former and free to rotate about a spindle in a magnetic field, in a similar manner to that illustrated in Figure 3.84.

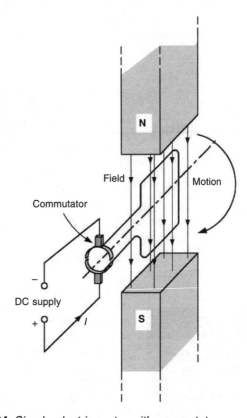

Figure 3.84 *Simple electric motor with commutator*

In real motors, this rotating coil is know as the *armature* and consists of many hundreds of turns of conducting wire. This arrangement is needed in order to maximize the force imposed on the conductor by introducing the *longest possible conductor* into the magnetic field. Also from the relationship $F = \beta IL$ it can be seen that the force used to provide the torque in a motor is directly proportional to the size of the *magnetic flux* β. Instead of using a permanent magnet to produced this flux, in a real motor, an *electromagnet* is used. An electromagnetic field can be set up by utilizing the *solenoid* principle (Figure 3.85). A long length of conductor is wound into a coil consisting of many turns and a current passed through it. Now each of the turns in the conductor assists each other to produce a strong magnetic field, as shown in Figure 3.85.

This field may be intensified by inserting a ferromagnetic core inside the coil. Once the current is applied to the conducting coil, the core is magnetized and all the time the current is on it acts in combination with the coil to produce a permanent magnet, having its own N–S poles.

Now returning to the simple motor illustrated in Figure 3.84. We know that when current is supplied to the *armature* (rotor) a *torque* is produced. In order to produce continuous rotary motion, this torque (turning moment) must always act in the same direction.

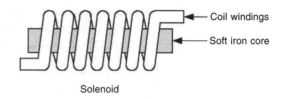

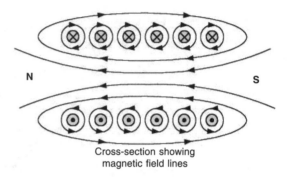

Figure 3.85 *Magnetic field produced by a solenoid*

Therefore, the current in each of the armature conductors must be reversed as the conductor passes between the North and South magnetic field poles. The *commutator* acts like a rotating switch, reversing the current in each armature conductor at the appropriate time to achieve this continuous rotary motion. Without the presence of a commutator in a DC motor, only a half turn of movement is possible!

In Figure 3.86(a) the rotation of the armature conductor is given by Fleming's left-hand rule. When the coil reaches a position mid-

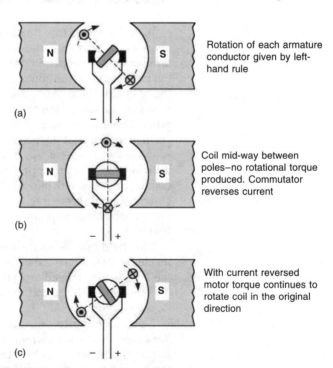

Figure 3.86 *Action of the commutator*

way between the poles (Figure 3.86(b)), no rotational torque is produced in the coil. At this stage the commutator reverses the current in the coil. Finally (Figure 3.86(c)) with the current reversed, the motor torque now continues to rotate the coil in its original direction.

Example 3.47

The rectangular armature shown in Figure 3.87 is wound with 500 turns of wire. When situated in a uniform magnetic field of flux density 300 mT, the current in the coil is 20 mA.

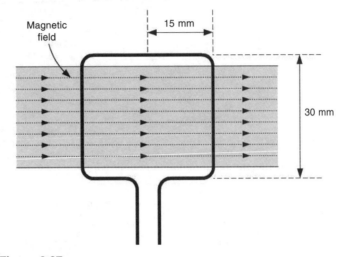

Figure 3.87

Calculate the force acting on the side of the coil *and* the maximum torque acting on the armature.

With this arrangement the ends of the conductor are not within the influence of the magnetic field and therefore have no force exerted on them. Therefore the force acting on one length of conductor is given from $F = \beta IL$ as:

$$F = (300 \times 10^{-3})(20 \times 10^{-3})(30 \times 10^{-3}) = \mathbf{1.8 \times 10^{-4}} \textbf{ N}$$

Then the force on one side of the coil

$$= (500)(1.8 \times 10^{-4}) = 0.09 \text{ N} = 9 \times 10^{-2} \text{ N}$$

Now from our definition of torque $T = Fd$, the torque acting on the armature windings is,

$$T = (9 \times 10^{-2})(30 \times 10^{-3}) = \mathbf{2.7 \times 10^{-3}} \textbf{ Nm}$$

This is a relatively small amount of torque. Practical motors can be made to produce output torques with very small values as demonstrated here, up to several hundred newton-metres!

One other application of the motor principle may be used in simple *analogue* measuring instruments.

Some meters, including multimeters used to measure current, voltage and resistance, operate on the principle of a coil rotating in a magnetic field. The basic construction is shown in Figure 3.88, where the current *I* passes through a pivoted coil and the resultant *motor force, the deflecting force, is directly proportional to the current flowing in the coil windings* which of course is the current being measured. The magnetic flux is concentrated within the coil by a solid cylindrical ferromagnetic core, in exactly the same manner as the flux is concentrated within a solenoid.

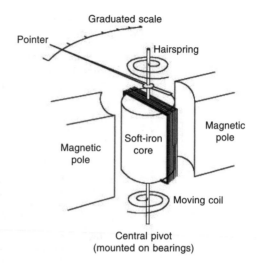

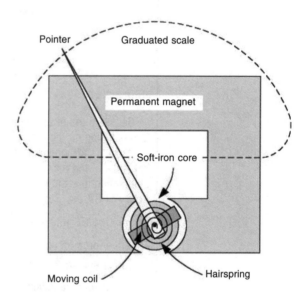

Figure 3.88 *The moving-coil meter*

Electromagnetic induction and electrical generation

The way in which electricity is generated in a conductor may be viewed as being the exact opposite to that which produces the motor force. In order to generate electricity we require *movement in to get electricity out*. In fact we need the same components to generate electricity as those needed for the electric motor, namely a closed conductor, a magnetic field and movement.

Whenever relative motion occurs between a magnetic field and a conductor acting at right angles to the field, an *emf is induced, or generated* in the conductor. The manner in which this emf is generated is based on the *principle of electromagnetic induction*.

Consider Figure 3.89, which shows relative movement between a magnet and a closed coil of wire.

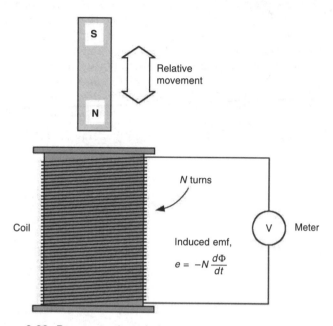

Figure 3.89 *Demonstration of electromagnetic induction*

An emf will be induced in the coil whenever the magnet is moved in or out of the coil (or the magnet is held stationary and the coil moved). The *magnitude* of the induced emf depends on the number of turns N and the rate at which the flux changes in the coil $\dfrac{d\Phi}{dt}$. Do not be concerned about this expression, it is simply a mathematical way of expressing *the rate of change* of flux with respect to time. You will meet similar expressions when you study the differential calculus in your mathematics!

In general this relationship is written as

$$E = -N\frac{d\Phi}{dt} \text{ volts}$$

The minus sign is a consequence of the way the induced emf acts, which we will discuss later.

Put another way, the *magnitude of the induced emf depends on*:

- the velocity of the conductor v in m s^{-2}
- the strength of the magnetic field β in tesla
- the length of the conductor L in metres.

This is so if we realize that the number of turns N *is directly related to the length of the conductor L*, moving through the magnetic field, with flux density β. Also *the velocity with which the conductor moves through the field* determines the *rate at which the flux changes in the coil* as it cuts the flux field. Thus the magnitude of the induced (generated) emf E is proportional to the flux density, length of conductor and relative velocity between the field and the conductor, or in symbols:

$$E \propto \beta L v \text{ volts}$$

Now you are probably wondering why the above relationship has the *proportionality* sign. In order to generator an emf the conductor must cut the lines of magnetic flux.

If they cut the lines of flux at a right angle (Figure 3.90(a)) then the *maximum* emf is generated; cutting them at any other angle θ (Figure 3.90(b)), reduces this value until θ = 0, at which point the *lines of flux are not being cut at all and no emf is induced* or generated in the conductor. So the magnitude of the induced emf is also dependent on sin θ. So we may write:

induced (generated) emf $E = \beta L v \sin \theta$

Key point

The magnitude of the induced emf in a coil is equal to the product of the flux density, length of the conductor, velocity of the conductor and the angle at which the conductor cuts the flux lines.

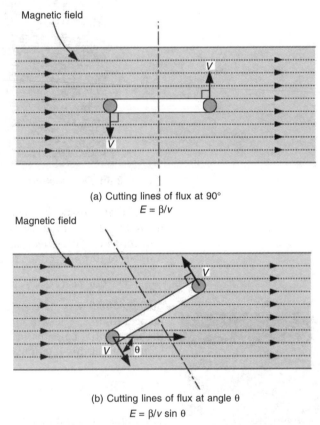

(a) Cutting lines of flux at 90°
$$E = \beta / v$$

(b) Cutting lines of flux at angle θ
$$E = \beta / v \sin \theta$$

Figure 3.90 *Cutting lines of flux and the generated emf*

Key point

The direction of the induced emf in a closed conductor can be found using Fleming's *right-hand generator* rule.

So much for the magnitude of the generated emf, what about its direction in the conductor? Since the conductor offers some resistance, the generated emf will initiate current flow as a result of the potential difference and the *direction of this current* can be found using *Fleming's right-hand rule*.

Note that for *generators* we use the *right*-hand rule (Figure 3.91), for motors we used the left-hand rule.

The first finger, second finger and thumb represent the field, current and motion, respectively, as they did when we looked at the motor rule.

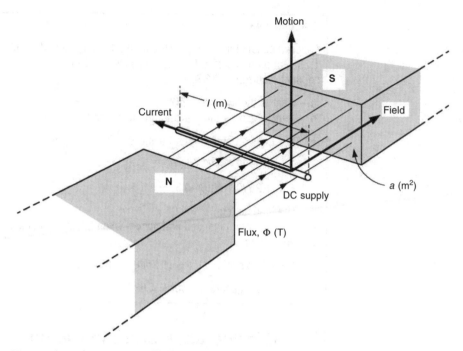

Figure 3.91 *Fleming's right-hand generator rule*

To finish our theory on induced emf, we need to consider two important laws which describe the nature of *electromagnetic induction*.

Faraday's law:

> When a magnetic flux through a coil is made to vary, an emf is induced. The magnitude of this emf is proportional to the rate of change of flux.

What this law is saying in effect is that *relative movement* between the magnetic flux and the conductor is essential to generate an emf. The voltmeter shown in Figure 3.89 reads the induced (generated) emf and *if the direction of motion changes the polarity of the induced emf in the conductor changes*. Faraday's law also tells us that the magnitude of the induced emf is dependent on the *relative velocity* with which the conductor cuts the lines of magnetic flux.

Lenz's law:

> The induced current always acts in such a direction so as to oppose the change in flux producing the current.

This is the reason for the minus sign in the formula

$$E = -\frac{d\Phi}{dt}$$

because E tends to oppose any change in the current I, in other words it is a *back* emf.

Example 3.48

A closed conductor of length 15 cm cuts the magnetic flux field of 1.25 T, at an angle of 60°, with a velocity of 25 m s^{-1}. If the conductor current is 13 A:

(a) what is the induced emf?
(b) what would be the maximum induced emf?

(a) The induced emf is found using $E = \beta L v \sin \theta$ so

$E = (1.25)(0.15)(25)(\sin 60)$

$E = (1.25)(0.15)(25)\ (0.866)$

The induced emf $E = 4.06$ V

(b) The maximum induced emf occurs when the lines of flux are cut at 90°. Then:

$E = \beta L v \sin 90 = \beta L v = (1.25)(0.15)(25)$ and

maximum generated emf = **4.6875 V**

Mutual inductance and the transformer

We know that the relative movement between a closed coil of wire and a magnetic field induces an emf in the coil. The same effect occurs if *two coils* are in close proximity and a current in one of the coils is varied, as shown in Figure 3.92.

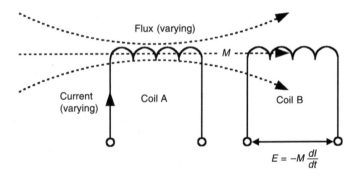

Figure 3.92 *Mutual inductance*

The polarity of E depends upon whether I is increasing or decreasing. It is called a *back emf* because by Lenz's law, E tends to oppose any change in I. In practice this means that if the coil B circuit is closed, the current due to E flows in such a direction as to produce a flux which opposes any change in the coil A flux.

The magnitude of E depends partly upon factors such as the angle and distance between the coils and the *reluctance* (magnetic resistance) of the flux path but, for a given pair of coils, E is directly proportional to the rate of change of I, which we represent as $\dfrac{dI}{dt}$.

The ratio between E and $\dfrac{dI}{dt}$ is the mutual inductance of the two coils, which has the symbol M. Then

$$E = -M\frac{dI}{dt}\text{ volts}$$

The negative sign has the same meaning as with the induced emf generated by a coil and a magnetic field. The negative sign indicating the back emf, in accordance with Lenz's law. The unit of mutual inductance M is the *henry H. A circuit has a mutual inductance M of one henry if one volt is induced by a rate of change of current* $\frac{dI}{dt}$ *of one ampere per second.*

Notice that mutual inductance is proportional to the rate of change of current in one of the coils. Therefore, no induction will take place mutually unless there is a continually changing current. This can be alternating polarity DC current activated by continuous switching or, more usually, alternating current (AC) that continually changes polarity at its frequency of generation.

Transformers

The main application of mutual inductance is in voltage transformers for *alternating current* (AC) power supply circuits. The basic construction of such a transformer is illustrated in Figure 3.93.

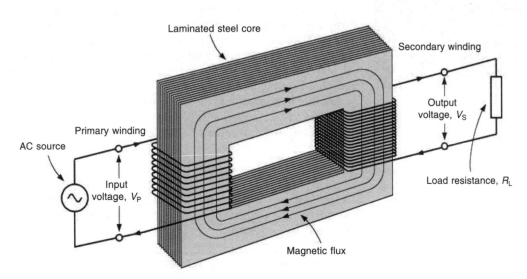

Figure 3.93 *The voltage transformer*

The *input* and *output windings* are called the *primary* and *secondary*, respectively. The ferromagnetic core provides a low reluctance (think of reluctance at this stage as AC resistance) flux path. Our mutual inductance formula for transformers may be written as:

$$V_s = -M\frac{dI_p}{dt}$$

where V_s = voltage induced in the secondary coil in volts, M = the mutual inductance in henries H, I_p = current in the primary coil in amperes A and t = time in seconds, s.

Example 3.49

Two coils of a transformer have a mutual inductance of 0.3 H. If the current in the primary coil is varied uniformly from 5 A to 2 A in 0.4 s, calculate the emf induced into the secondary coil.

The solution to this problem simply requires us to substitute the given values into our equation, ensuring that we use the correct value of $\dfrac{dI_p}{dt}$ which is 3 A/0.4 s or 7.5 A/s, then:

$$V_s = -M\frac{dI_p}{dt} = -(0.3)\frac{3}{0.4} = -(0.3)(7.5) = -2.25\,\text{V}$$

remembering that the minus sign indicates a back emf.
So emf in the secondary coil = **2.25 V**.

Transformers are easily the most efficient machines built by man, 98% or better! So it is reasonable to say that the *power input = power output*. Then, remembering our power formulae, we may write:

$$I_p V_p = I_s V_s$$

where V_p = rms voltage across the primary, I_p = rms current through the primary, V_s = rms voltage across the secondary and I_s = rms current through the secondary, all with the normal units.

Note that the *rms (root mean squared) values* of current and voltage *give their direct current equivalent* for the purpose of power calculations. For example, the rms equivalent voltage for our domestic household supply is *240 V (rms)*.

Another important transformer formula recognizes that the same flux through each coil, that is the voltage across each coil, must be proportional to the number of turns, or:

$$\frac{V_p}{V_s} = \frac{N_p}{N_s}$$

where V_p = rms voltage across primary, V_s = rms voltage across secondary, N_p = number of turns on primary, N_s = number of turns on secondary.

> **Key point**
>
> For mutual inductance to occur between two coils one coil must be subject to a continually changing current.

Example 3.50

240 V is applied to the primary windings of a transformer that has 1280 windings and draws a current of 13 A. Given that the required output voltage from the transformer is to be 12 V, determine:

(a) the number of windings on the secondary
(b) the output power from the transformer, given that the transformer is 98% efficient.

(a) This is simply found using the volts-to-turns ratio formula, where:

$$N_s = \frac{N_p V_s}{V_p} = \frac{(1280)(12)}{240} = \textbf{64}$$

(b) To find the output power, that is the power available from the secondary windings, we need first to find the input power because we are given the current flowing in the primary windings of the transformer.

Then from our power formula where $P = IV$ watts, the power at the primary windings, $I_p V_p = (13)(240) = 3120$ watts. We are told that the transformer is 98% efficient, so the power at the secondary will be 0.98 times the power at the primary or,

power at the secondary (output power) = **(0.98)(3120) = 3057.6 W**

Self-inductance

Up till now, we have concentrated on mutual inductance involving two coils; however, a back emf E is also produced by a flux change in a *single* coil. The back emf is proportional to the rate of change of current I (from Lenz's law), as illustrated in Figure 3.94.

Test your knowledge 3.17

1. Give three examples of magnetic materials.
2. From which pole do the magnetic flux lines come from?
3. If, when using Fleming's left-hand motor rule, your thumb is pointing upwards, sketch the direction of the magnetic field and current flow.
4. What are the units of flux *and* flux density?
5. How may the torque produced on the coil of an electric motor be found?
6. What is a commutator used for?
7. What does the deflecting force in a moving coil meter depend upon?
8. Explain the requirements necessary to produce electricity.
9. State Lenz's law and explain what it tells us.
10. What does Faraday's law tell us about induced emf?
11. What is mutual inductance and how is it produced?
12. Define the henry.

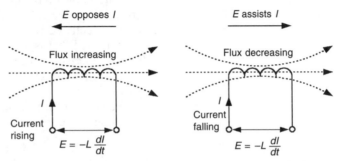

Figure 3.94 *Self-Inductance*

This effect is called *self-inductance*, which has the symbol L. Self-inductance L is measured in *henries* and is calculated in the same way as for mutual inductance M. In symbols:

$$E = -L\frac{dI}{dt}$$

A circuit is said to have an inductance L of one henry if one volt is induced by a rate of change of current

$$\frac{dI}{dt}$$

of one ampere per second.

Example 3.51

A coil has a self-inductance of 15 mH and is subject to a rate of change of current of 450 A s^{-1}. What is the back emf?
  All we need do is place the appropriate values in the formula, then:

$$E = -L\frac{dI}{dt} = -\frac{(15 \times 10^{-3})(450)}{1} = -6.75 \text{ V}$$

So the back emf = **6.75 V**.

Problems 3.6

1. (a) Define the unit of electrical energy.
   (b) An electric heater consumes 1 kJ of energy every second. If the heating element has a resistance of 57.6 Ω:

(i) How much current is taken from the supply?

(ii) What is the supply voltage?

2. The heating element in question 1(b) above, with resistance of 57.6 Ω, is made from copper wire of diameter 0.25 mm, calculate its length in metres.

3. (a) State Ohm's law and explain the circumstances under which it may be used.

(b) Three resistors of 5 Ω, 10 Ω and 30 Ω are connected in parallel with one another, then in series with the remaining 17 Ω resistor across a 120 V supply. Determine:

(i) the total current taken from the supply

(ii) the potential difference across the 17 Ω resistor

(iii) the current flowing in each of the paralleled resistors

(iv) the power dissipated as heat through the 17 Ω resistor.

4. A current of 13 A enters a junction where it splits four ways. The current drawn by three of these pathways is 2 A, 4 A and 2.6 A, respectively. Find the current in the fourth pathway.

5. (a) Four cells each of 1.5 V are connected in parallel as a battery. What is the total emf of the battery?

(b) The internal resistance of each cell in the battery identified in (a) is 0.01 Ω. What is the total internal resistance of the battery?

(c) Each cell of this same battery draws a current of 0.2 A. What is the total current drawn by the battery?

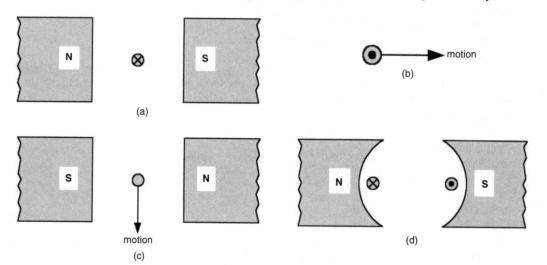

Figure 3.95   *Force on a conductor (a to d)*

6. Study the four diagrams shown in Figure 3.95(a) to (d) and mark on them the direction of motion, field or current as appropriate.

7. Study the six diagrams shown in Figure 3.96(a) to (f) and mark on them the direction of the induced emf.

8. A motor armature conductor of length 10 cm carries 50 A. The flux is at right angles to the conductor and has a flux density of 1.5 T. What is the motor force on the conductor.

9. If an emf of 5 V is induced in a coil when the current in an adjacent coil varies uniformly at the rate of 80 A s^{-1}, what is the value of the mutual inductance of the coil?

10. If the current through a coil of inductance 0.5 H is reduced

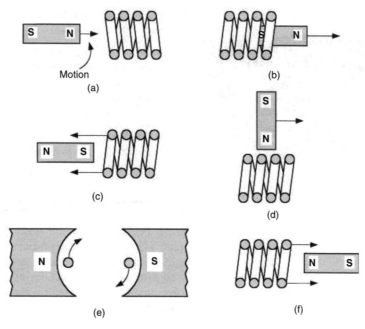

Figure 3.96 *Electromagnetic induction (a to f)*

from 5 A to 2 A in 0.05 seconds, calculate the value of the back emf induced in the coil

ENGINEERING SYSTEMS

Introduction

In this final outcome we look briefly at some of the underlying principles concerned with engineering systems and then apply these principles, in particular, to the study of servo systems.

We start by considering the concept of the *system*, looking in general at the idea of system boundaries, system inputs and outputs and system functions. We then adapt these rather abtract ideas to closed-loop systems and their engineering applications. We also consider the simple analysis of block diagrams for open- and closed-loop systems. Looking in particular at the way we calculate overall system gain through use of transfer functions. We then look at system response and identify the reasons for system damping.

In the final part of our study we will look at the application of closed-loop control to the operation of remote position servo systems and a typical analogue speed control system for a DC electric motor.

What is a system?

Systems are all around us. Indeed, we ourselves are examples of a complex system that in turn contains its own nervous, respiratory and muscular subsystems.

We use the word system so liberally and in such a wide range of contexts that it is difficult to explain what the term means in just a few words. However, at the risk of oversimplification, we can say that every system conforms to the following main points:

(a) it has a function or purpose
(b) it has inputs and outputs
(c) it has a boundary
(d) it comprises a number of smaller components or elements linked together in a particular way.

These fundamental concepts of systems are so important that it is worth expanding on each of them in turn.

Systems have a function or purpose

Some people may argue with the assumption that *every* system must have a function or purpose. They may, for example, ask what is the purpose of the solar system other than, perhaps, providing us with a home! When considering engineering systems, however, we are on pretty safe ground since, by definition, an engineered system *must* have a defined function or purpose.

Systems have a number of inputs and outputs

A system without inputs and outputs cannot interact with its environment and is thus useless! It is worth noting, however, that not all inputs and outputs may be desirable. An example of an undesirable input to a telecommunication system would be noise picked up by the cables and wires along which the signals travel. In effect, this noise presents itself as an unwanted signal superimposed on top of the wanted signal. An example of an undesirable output from a system would be the exhaust gases produced by an internal combustion engine.

Systems have a boundary

Because all systems have a boundary, it is possible to say what is inside the system and what is outside it. This sometimes becomes important when there are a number of systems (and/or subsystems) and they interact with one another.

Sometimes the boundary of a system is not very clearly defined. That said, it is usually possible to construct a diagram for a system showing all of the components and elements within the system and then place this inside a box that contains the system as a whole. We sometimes refer to this as a 'black box', see Figure 3.97.

System components are linked in a particular way

The elements, parts, components or subsystems that make up a system are connected together (i.e. linked) in a particular way.

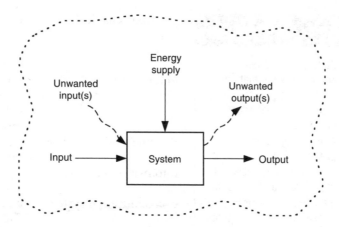

Figure 3.97 *A black box system*

This may sound obvious but it is quite an important point. Consider a box of bicycle parts that can be assembled to produce a fully working bicycle. The box of bicycle parts *does not* constitute a system whereas the fully assembled bicycle *does* constitute a system. Exactly the same physical parts are present in both cases – what *is* important is the way in which the parts are connected together so that they can interact with one another. Figure 3.98 illustrates this point.

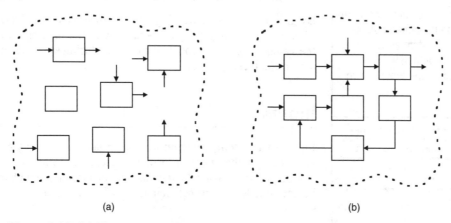

Figure 3.98 *(a) Not a system, (b) a system*

Definition of a system

Systems comprise a number of elements, components or subsystems that are connected together in a particular way. The individual elements of an engineering system interact together to satisfy a particular functional requirement.

Types of system

Various types of system are used in engineering. Listed below are some of the most common types of system. You will already be familiar with some of them but others may be completely new to you.

Electromechanical systems	e.g. a vehicle electrical system comprising battery, starter motor, ignition coil, contact breaker, distributor, etc.
Fluidic systems	e.g. a vehicle braking system comprising foot operated lever, master cylinder, slave cylinder, piping, fluid reservoir, etc.
Electrochemical systems	e.g. a cell that uses gas as a fuel to produce electricity, pure water and heat.
Information systems	e.g. a computerized airport flight arrival system.
Communication systems	e.g. a local area network comprising file server, coaxial cable, network adapters, several computers and a laser printer
Control systems	e.g. a microcomputer-based controller that regulates the flow and temperature of material used in a diecasting process.
Transport systems	e.g. an overhead conveyor for transporting gravel from a quarry to a nearby processing site.

Function and purpose

By definition, an engineering system must have a defined purpose. This purpose is normally associated with the principal function of the system (e.g. the purpose of a manufacturing system is to produce a manufactured product).

Inputs

Inputs to a system may comprise raw materials, energy and control values set by the operator or manufacturer. In practical, real-world systems we must also take into account disturbances (such as changes in ambient temperature) that may affect the performance of the system.

Outputs

Outputs from a system can comprise finished products, processed materials, converted energy, etc. We also need to give some consideration to the by-products produced by the system, such as waste heat, toxic materials, inert materials, chemically active materials, radiation, etc.

System components

Most systems comprise several different types of component. Some components provide the main function of a system (e.g. a boiler in

240 V AC mains electricity to operate the gas ignition system and water pump

Outputs
Heat from radiators
Waste heat and combustion products

Functional components
Gas fired boiler
Gas ignition system
Room thermostat (adjustable)
Water pump
Water tank
Radiators
Gas valve (on/off)
Water valve (on/off)
Control panel

Sketch a diagram showing how the system components are connected together. Label your diagram clearly. (Hint: Take a look at your central heating system at home if you are not sure about how the system components are linked together.)

a central heating system) while others monitor the output of the system (e.g. a room thermostat in a central heating system). Other components are used to set or determine the required output of the system (e.g. the human operator who interacts with the system through a programmable controller).

The following categories can be applied to system components:

Functional components	Essentially, these are the components that do the work of the system.
Performance checking components	These components ascertain the level of performance of the system. As such, they measure or in some way respond to the output produced by the system.
Controlling components	These components are responsible for determining the level of performance of the system. They establish the desired output from the system and, in complex systems, manage the overall operation of the system.

As an example, a satellite positioning system might comprise the following:

Directional thrusters (functional components)
Fuel tank (functional component)
Control valve (functional component)
Earth-link antenna (functional component)
Command receiver/decoder (functional component)
Position computer (performance checking component)
Ground command station (controlling component).

Controlling engineering systems

Now that you have some idea about the general make-up of systems, it is time to consider how they are controlled. All modern engineering systems include a certain aspect of control, at some point. When we *control systems*, we are ensuring that they *behave in a desired way*. So, for example, if you drive a car you need to ensure that the car engine produces the right amount of *torque* to overcome variations in the terrain, such as going up hill. The torque needs to be varied not only to meet changing loads, but also to maintain the appropriate speed of rotation of the driving wheels.

Control systems may be simple as in the case of a light being controlled by a switch, or complex such as an automatic instrument landing system for a modern passenger aircraft. In the simple case, the fact that the light did or did not illuminate would have been observed by the operator and if a fault had been detected then it would need to be remedied by the operator. In the case of the complex system the tendency of the aircraft to move off the correct glide path, as the aircraft is landing, is automatically corrected by the system and requires no operator intervention. The simple system

Key point

Open-loop systems have no automatic control of system performance and generally depend on human operator intervention for their control.

requiring operator intervention is often referred to as *open loop*; the complex system, which automatically corrects for errors, is commonly known as *closed loop*.

The *type of system control method* used is dependent on the type of control strategy being employed, as well as the output being controlled. Thus different control methods are appropriate to different types of system and the overall control strategy can be based on *analogue or digital techniques* and may also be classed as either *sequential* or *combinational*.

Analogue control

Analogue control involves the use of *signals* and quantities that are *continuously variable*. Within analogue control systems, signals may, for example, be represented by voltage, current, pressure, temperature, etc. that can take any value between two set limits. Figure 3.99 shows how the output of a typical analogue system varies with time.

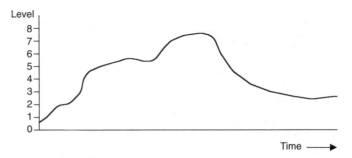

Figure 3.99 *Output of analogue system shown against time*

The majority of analogue control systems use *electrical signalling*, where the signals are *modified or conditioned* using *operational amplifiers* (see Figure 3.100). These devices are capable of performing *mathematical operations* such as addition, subtraction, multiplication, division, integration and differentiation!

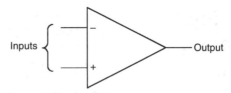

Figure 3.100 *An operational amplifier*

Key point

Analogue signals are continuously variable.

The exact way in which operational amplifiers perform these mathematical operations is not required at this stage in your learning. However, you will see how we use them when we consider closed-loop control, a little later.

Digital control

Digital control involves the use of signals and quantities that vary in *discrete steps*. Values that fall between two adjacent steps must take one or other value as intermediate values are disallowed! Figure 3.101 shows how the output of a typical digital system varies with time.

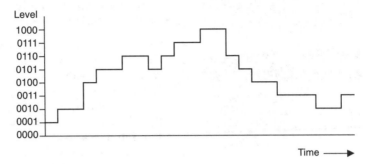

Figure 3.101 *Output of a digital system shown against time*

Digital control systems are usually based on electronic digital logic devices (see Figure 3.102) or microprocessor-based computer systems.

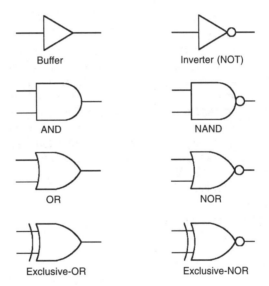

Figure 3.102 *A selection of digital logic devices*

Values represented within a digital system are expressed in *binary coded form* using a number of signal lines. The voltage on each line can be either 'high' (representing logic 1) or 'low' (representing logic 0). The more signal lines, the greater the resolution of the system. For example, with just two signal lines it is possible to represent a number using only two binary digits (or 'bits'). Since each bit can be either 0 or 1 it is possible to represent only four different values (00, 01, 10 and 11) using this system. With three signal lines we can represent numbers using three bits and eight different values are possible (000, 001, 010, 011, 100, 101, 110 and 111).

The relationship between the number of bits, n, and the number of different values possible, m, is given by $m = 2^n$. So, in an 8-bit system the number of different discrete states is given by $m = 2^8 = 256$.

Sequential control systems

Many systems are required to perform a series of operations in a set order. For example, the ignition system of a gas boiler may require the following sequence of operations:

1. Operator's start button pressed.
2. Fan motor operates.
3. Delay 60 seconds.
4. Open gas supply valve.
5. Igniter operates for 2 seconds.
6. If ignition fails, close gas supply valve, delay 60 seconds, then stop fan motor.
7. If ignition succeeds, boiler will continue to operate until either stop switch operates or flame fails.

The components of simple sequential systems often include timers, relays, counters, etc.; however, digital logic and microprocessor-based controllers are used on more complex systems.

Combinational control systems

Combinational control systems take several inputs and perform comparisons on a continuous basis. In effect, everything happens at the same time – there are no delays or predetermined sequences that would be associated with sequential controllers. The aircraft instrument landing system (ILS) we considered earlier, for example, makes continuous comparisons of an aircraft's position relative to the ILS radio beam (Figure 3.103). Where any deviation is detected, appropriate correction is applied to the aircraft's flight controls.

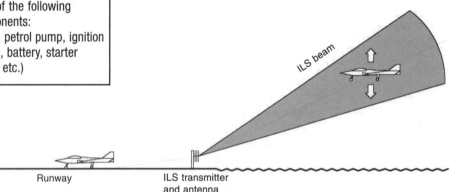

Figure 3.103 *An aircraft instrument landing system (ILS)*

Closed-loop control

We looked earlier at a simple control system, the light switch, and a complex system, the aircraft ILS system, indicating that the former was open loop and the latter closed loop. We need to define and compare open-loop and closed-loop control systems before considering the closed-loop system in detail.

Open-loop control

In a system that employs open-loop control, the value of the input variable is set to a given value in the expectation that the output will reach this desired value. In such a system there is no *automatic* comparison of the actual output value with the desired output value in order to compensate for any differences. Thus *open-loop control* may be regarded as: *a system that needs human operator intervention to monitor performance and to modify matters if the outcome is not correct*.

A simple example of an open-loop control method is the manual adjustment of the regulator that controls the flow of gas to a burner on the hob of a gas cooker. This adjustment is carried out in the expectation that food will be raised to the correct temperature in a given time and without burning. Other than the occasional watchful eye of the chef, there is no means of automatically regulating the gas flow in response to the actual temperature of the food.

Consider another example, the control of the motor vehicle (Figure 3.104), we mentioned earlier.

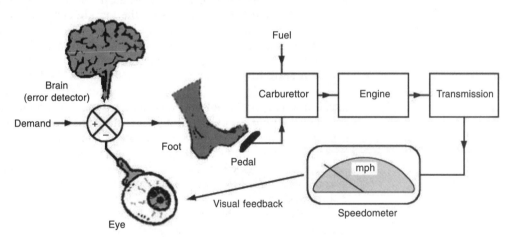

Figure 3.104 *Control of motor vehicle speed, using human operator*

Let us assume that we wish to drive the vehicle at 30 mph. Starting from rest, with the engine running, we need to press down on the accelerator pedal to open the throttle, this in turn increases the amount of fuel charge entering the carburettor and so increases the available power being delivered by the engine. Now, against a given load (weight of the car, terrain, etc.) this increase in power increases the torque to the driving wheels which, acting against frictional resistance, provide the driving force necessary to accelerate the vehicle up to the desired speed. We, as the driver, are now required to monitor the speed of the vehicle and correct matters accordingly, to ensure that we travel at the desired speed of 30 mph. The way in which we correct matters is to continually monitor our speed using the speedometer, while at the same time adjusting the throttle setting, in order to keep us at 30 mph.

The finesse with which the speed may be controlled depends on two main factors. One is the accuracy of reading the speedometer and the other the accuracy of setting the throttle. There is always bound to be a time delay between the observed effect on the speed and the measures taken to remedy the disturbance. Our average thinking time is around 0.2 of a second, this is why we always need to allow a thinking distance as well as a stopping distance between vehicles!

Now, on motor vehicles, this is the normal way of watching and controlling our speed. What if we adopt the same system for controlling the flight speed of, say, a supersonic jet fighter? Imagine two such aircraft approaching each other at, say, 1600 km/hr, that is a relative closing speed of 3200 km/hr or approximately 890 m s^{-1}. By the time we have thought about avoiding action, we have closed by another 200 m, never mind the time it takes for the aircrafts' flight controls to react! We therefore need another form of control, which is faster and more accurate and does not necessarily depend on human operator intervention. For this function and many other engineering functions, we use *closed-loop* control.

Nature of closed-loop control

Clearly, open-loop control has some significant disadvantages. What is required is some means of closing the loop in order to make a continuous automatic comparison of the actual value of the output compared with the setting of the input control variable. The speed at which this is done is also of some importance.

In our cooker example, the chef actually closes the loop on an intermittent basis. In effect, the gas cooker relies on human intervention in order to ensure consistency of the food produced. If our cooking requires only boiling water, this intervention can be kept to a minimum; however, for 'haute cuisine' we require the constant supervision of a skilled human operator!

All practical engineering systems make use of closed-loop control. In some cases, the loop might be closed by a human operator who determines the deviation between the desired and actual output. In most cases, however, the action of the system is made *fully automatic* and no human intervention is necessary other than initially setting the desired value of the output.

If we again consider the example of controlling the speed of the car, this could be done *automatically* by controlling the engine rpm, for varying load conditions, thus producing the correct amount of torque at the driving wheels to maintain the *desired* speed. In this case our automatic closed-loop control system for the car will need to be modified so that we are able to *sense* and *control* engine rpm (Figure 3.105).

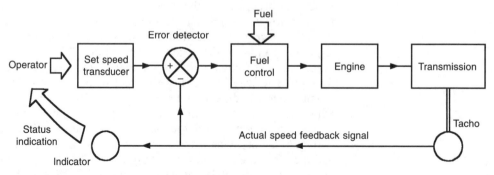

Figure 3.105 *Closed-loop control of engine rpm*

So the demanded speed signal for the car in this system is set using an *input transducer*, such as a potentiometer (transducers and other system components are explained next, when we generalize our closed-loop system). The actual speed achieved by the car (the load) is monitored by a *feedback transducer*. In this case, a tachogenerator is used that converts engine rpm into a corresponding voltage output. The *error detector* is a device which produces a difference or error signal to the throttle (a summing operational amplifier could achieve this). The *error signal* is equal to the demand signal minus the feedback signal. As the speed of the car (the actual speed) approaches the demand speed (set by the input transducer) the signal to the throttle becomes smaller and to provide greater sensitivity and accuracy of the set speed an amplifier is included in the system (the *error signal amplifier*). Remember it is the error between the desired and actual values that continues to drive the system (*error actuated*). Once the desired value is equal to the actual value (of speed in this case), there is no error and the system remains at this value, unless subject to an external disturbance.

From the outline described above, we may deduce the general schematic (Figure 3.106) required for any closed-loop control system.

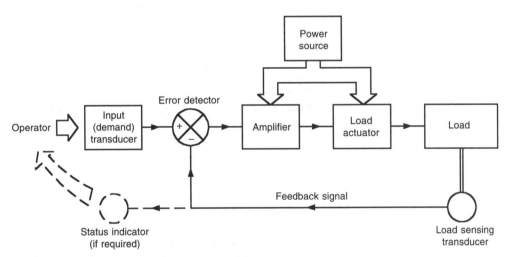

Figure 3.106 *Generalized schematic diagram of a closed-loop control system*

Any variation in the performance of the load, not called for by the operator, produces an automatic change in the error signal which attempts to counteract the load performance variation. How quickly corrections are made to counteract the load performance depends to a large extent on the system *gain*. In order to appreciate control systems more fully we need to consider the nature and function of transducers, amplifiers (signal conditioners), error detectors and power actuation components that go to make up closed-loop control systems.

Transducers

In its simplest sense, a *transducer* is essentially an *energy converter*. Although you might not realize it, transducers are all around us! For example, the light switch we mentioned earlier converts position into a voltage. The tachogenerator mentioned earlier, as a type of transducer, converts rotational velocity into a voltage. Thus transducers may be simple or complex but essentially they all perform the same function, that of energy conversion. More complex transducers may be composed of two components: the primary component or *sensor* and the secondary component or *conversion/ control element*. In this sense, any but the simplest of transducers are really *systems* in their own right.

Consider Figure 3.107, which shows a Bourdon pressure gauge.

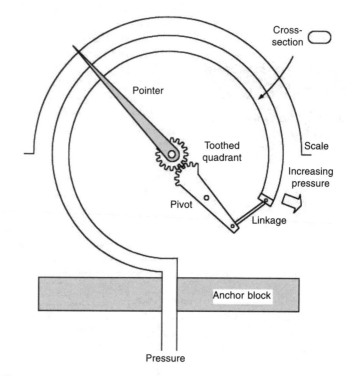

Figure 3.107 *Bourdon pressure gauge*

This instrument is essentially a *transducer system*. It consists of a flattened oval cross-section hollow tube, bent round in a circular manner; this is the *sensing element* that senses fluid pressure, the *primary signal*. This fluid pressure causes the tube to straighten, thus the Bourdon tube converts fluid pressure into movement, so it is in its own right, a transducer; in this system it is used as the *sensing element*. The *secondary signal* (linear movement) from the tube is converted by the *gearing* into rotary movement at the *pointer*. Thus the gearing acts as the *conversion/control element* of the overall transducer system. The final output is the movement of the pointer across a scale, which enables us to read off the pressure.

We can represent any transducer (Figure 3.108) by a generalized flow diagram of the transduction process described above for our Bourdon pressure gauge example.

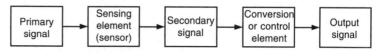

Figure 3.108 *Transducer flow diagram*

Transducers may be grouped according to the type of signal received by the sensing element and given out by the conversion or control element. The most frequently used transducers are those that convert any physical, mechanical, thermal, chemical or radiation input to a voltage or current output. That is, any transducer that measures force, pressure, displacement, vibration, light, concentration, temperature, etc., then converts these input signals into an electrical voltage output. A few examples of these transducers, together with others that do not have a voltage output, are tabulated below.

Devices	Input quantity sensed	output signal
Switch	Position	Voltage
Potentiometer	Position	Voltage
Linear variable differential transformer (LVDT)	Position	Voltage
Tachogenerator	Rotational speed	Voltage
Thermistor	Temperature	Voltage
Thermocouple	Temperature	Voltage
Accelerometer	Acceleration	Voltage
Photodiode	Light	Voltage
Photoelectric cell	Light	Voltage
Electric motor	Voltage	Rotational movement
Voltmeter	Voltage	Movement of needle
Lamp	Voltage	Light
Bourdon gauge	Fluid pressure	Movement of needle
Hydraulic motor	Fluid flow	Rotational movement
Hydraulic pump	Rotational movement	Fluid flow
Bimetallic strip	Heat	Angular movement
Radiation pyrometer	Radiated heat	Voltage
Strain gauge	Displacement	Voltage

As can be seen from the above table, some transducers are relatively simple, others are really systems in their own right, for instance the electric motor and hydraulic pump are *energy conversion devices* and therefore *transducers*, but physically they are made from many parts, in such a way as to be a *system*.

A *transducer is a device which essentially converts one form of energy into another*. Alternatively a *transducer* may be defined as: *a sensing device that converts physical phenomena and chemical composition into electric, mechanical, pneumatic or hydraulic output signals*.

Error detector or comparison element

An *error detector* or *comparison element monitors output performance against a stored model of what it should be. Deviations from the desired output create an error signal, which drives the system.*

This may sound a little complicated. Look back at our closed-loop version of speed control for the car. The operator (driver) sets the desired output speed of the vehicle by setting the engine rpm, as explained earlier. It is this engine rpm that we are controlling for varying loads at the driving wheels. The engine, in turn, provides the necessary torque to maintain the selected driving speed of the car. So with respect to the *error detector*, our stored model may be set using a *potentiometer* (look back at the input/output we get from this device). By altering our potentiometer setting using, say, a dial, a certain engine rpm and therefore a certain car speed are set by the dial. This produces a corresponding voltage output from the potentiometer that becomes the *set input to the error detector*. At this stage it is worth mentioning that the error detector is often a type of *operational amplifier* that *compares* this set imput (the desired output) with the actual output measured by some form of transducer at the output. In the case of our car, this transducer is likely to be positioned on the gearbox, so that drive speed can be measured using a tachogenerator. The tachogenerator, you will remember, converts rotational velocity into a corresponding voltage output. This voltage signal (*feedback signal*) is *compared in the error detector* with our desired signal and the net difference is the *error* which continues to drive the load (in our case supplying the necessary fuel charge to maintain the desired engine rpm and road speed).

Signal conditioning

In reality the electrical output signal from the error detector is very small and in order to provide sufficient power to drive the load actuator (fuel control unit in our car example), we need to amplify the output from the error detector. This amplification modifies (increases) the signal, so we say that it has been *conditioned* and the general name for this is *signal conditioning*. Any modification of the output signal from the error detector is referred to as signal conditioning. So, for example, converting a DC voltage output signal to an AC voltage output signal is also known as signal conditioning. Also, the signal need not necessarily be electrical. Figure 3.109 shows two forms of mechanical amplifiers, which act as signal conditioners.

> **Key point**
>
> The comparator or error detector compares the set point value (desired value) with the actual output value and amends matters accordingly.

> **Key point**
>
> In real systems, signal conditioners may be mechanical or electrical: amplifiers, filters or converters.

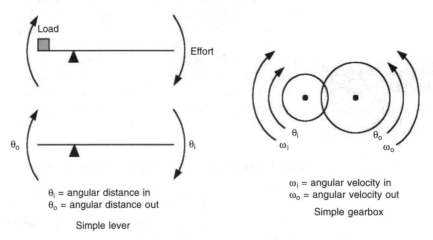

θ_i = angular distance in
θ_o = angular distance out

Simple lever

ω_i = angular velocity in
ω_o = angular velocity out

Simple gearbox

Figure 3.109 *Mechanical amplifiers (signal conditioners)*

In the case of the simple lever, this can be used to amplify distance (displacement) or amplify load. In the case of the gearbox, this can reduce/increase speed and displacement, as well as increase/reduce the torque available at the drive from the gearbox. In the case of our Bourdon pressure gauge (see Figure 3.107), the gearing is used to amplify the displacement of the pointer across a pressure scale (not shown).

Load actuation

The load may be driven by electrical, mechanical, hydraulic or pneumatic motors and associated gearing. These machines are collectively known as the *load actuators* or *power source*; they are the muscle of the system and provide the driving power to the *load*. In the case of our car, the power source was an internal combustion engine, where the input is a chemical charge and the output is driving torque, which is *signal conditioned* via a gearbox and drive shafts to the road wheels.

System block diagrams

We have already represented a closed-loop control system in block diagram form (Figure 3.106). However, this particular representation tells us little about the relationship between system inputs, outputs and feedback. The simplest type of open-loop system may be represented diagrammatically using an input, an output and a block to represent the system process. For example, Figure 3.110 illustrates a simple block diagram of a thermocouple open-loop system.

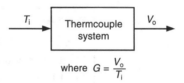

Figure 3.110 *Thermocouple system*

Now we know that for this type of transducer, the input signal is temperature, which is a measure of heat; the output signal is a corresponding voltage. The relationship,

$$\frac{V_o}{T_i}$$

is known as the *gain* (*G*) of the system or as the system *transfer function* (TF). In general terms for any input signal θ_i and output signal θ_o, then:

open-loop system gain $G = \dfrac{\text{output}}{\text{input}} = \dfrac{\theta_o}{\theta_i}$

or the system transfer function

$$\text{TF} = \frac{\theta_o}{\theta_i}$$

So, for example, if in the case of our thermocouple for every 1°C rise in temperature we get a 2 mV increase at the output, then the system gain or transfer function is

$$G = \frac{\theta_o}{\theta_i} = \frac{2\,\text{mV}}{1°C} = 2\,\text{mV}/°C$$

Now we can extend our open-loop system to contain any number of components between the system input and output. When this happens, we simply multiply the individual component values (gains) together (Figure 3.111) to produce the overall system output.

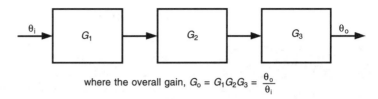

where the overall gain, $G_o = G_1 G_2 G_3 = \dfrac{\theta_o}{\theta_i}$

Figure 3.111 *Overall system gain for components in cascade*

In this particular configuration, where individual system components are connected in series, we refer to them as being connected in *cascade* and, as shown, in this case:

$$G_o = G_1 G_2 G_3 = \frac{\theta_o}{\theta_i}$$

Also these components are said to be in the *feed-forward path* of the system. Symbolically, we may represent feed-forward components by the letter G on system block diagrams. If we were to convert this system to *closed loop*, then we require some percentage of the output to be *monitored* and *fed back* to the error detector for comparison.

Example 3.52

An open-loop system consists of an input potentiometer, an amplifier and a motor. If the gain of these components 2 mV/degree, 1 V/mV and 1 Nm/V, respectively, determine the system transfer function and the value of the output for a given input of 30°. Then the overall system gain G_0 or transfer function

$$TF = \left(\frac{2\ mV}{degree}\right)\left(\frac{1\ V}{mV}\right)\left(\frac{0.1\ Nm}{V}\right) = \frac{0.2\ Nm}{degree}$$

and so for 30° of input,

$$\theta_0 = \left(\frac{0.2\ Nm}{degree}\right)30° = 6\ Nm$$

In other words, the motor produces 6 Nm of torque at the output shaft when the input transducer dial is turned through 30°.

We can represent a closed-loop system in its simplest form by the block diagram shown in Figure 3.112.

In this diagram the *feed-forward path components* are represented by G, as before. The additional components shown are required to convert the system from open loop to *closed loop*. They include the essential *error detector* or comparison element, with output ε, which is the sum of the input signal θ_i (desired output) and a proportion of the output signal $-\gamma_o$, being fed back to the error detector, the negative sign indicating *negative feedback*. In reality, if the feedback signal is a voltage, the *negative symbol* represents an *inverse in polarity* of the voltage, when compared with the input signal to the error detector. Thus the error signal (which drives the system) may be represented symbolically as:

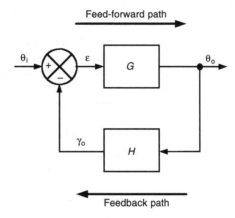

Figure 3.112 *Closed-loop system block diagram*

$$\varepsilon = \theta_i - \gamma_o \qquad (1)$$

Also, in the feed-forward path, the system output θ_o is the product of the error signal ε and the overall gain of the feed-forward components G, that is:

$$\theta_o = G\varepsilon \qquad (2)$$

Now, if we substitute the expression for ε from equation (1) into equation (2), we get:

$$\theta_o = G(\theta_i - \gamma_o)$$

or

$$\theta_o = G\theta_i - G\gamma_o \qquad (3)$$

Now in the *feedback line* we have the component H, which represents the feedback transducer, used to modify (signal condition) the signal from the output of the system, prior to entry into the error detector. For example, a transducer that converts, say, rpm into a corresponding voltage. Then, this modified feedback signal:

$$\gamma_o = H\theta_o$$

and on substitution of this expression for γ_o into equation (3) we get:

$$\theta_o = G\theta_i - GH\theta_o$$

or

$$\theta_o + GH\theta_o = G\theta_i$$

$$(1 + GH)\theta_o = G\theta_i$$

therefore the closed-loop transfer function (CLTF) for a *negative feedback* system is:

$$\frac{\theta_o}{\theta_i} = \frac{G}{1 + GH}$$

Note that in our block diagram of a closed-loop system, the feedback components are represented conventionally by the letter H. The expression for the CLTF also represents the *overall system gain* for this particular type of system.

Example 3.53

In a particular system the forward path gain is 56. Determine the gain if negative feedback is used to convert the system to closed loop, where the feedback component has a gain of 8.

This is a simple example of the use of our CLTF for a negative feedback system, i.e.:

$$\frac{\theta_o}{\theta_i} = \text{closed loop gain} = \frac{G}{1 + GH} = \frac{56}{1 + (56)(8)} = \frac{56}{449} \simeq \mathbf{0.125}$$

Now if *positive feedback* is used, it can be easily shown that

$$\text{CLTF} = \frac{\theta_o}{\theta_i} = \frac{G}{1 - GH}$$

simply by replacing the negative sign in the error detector (hot-cross bun) by a positive sign and following through the analysis, shown above!

Also, if there is no necessity to include a component in the feedback line, then we say that we have a *unity feedback* (because we replace the *H* component(s) by the number one) system.

Thus with unity feedback the CLTF for a negative feedback system is given by:

$$\frac{\theta_o}{\theta_i} = \frac{G}{1 + G}$$

Strictly speaking, we should differentiate between the overall gain of a system and the system transfer function. The TF is the generalized version of the system gain. In other words we tend to use the words 'system gain *G*' when we are finding numerical values. The TF is often an algebraic expression that is found when analysing relatively complex systems.

Example 3.54

1. Determine an expression for the overall transfer function, for the system shown in Figure 3.113.

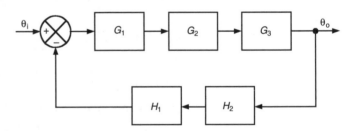

Figure 3.113

2. Determine the overall gain of the system if $G_1 = 20$, $G_2 = 15$, $G_3 = 10$, $H_1 = 0.05$ and $H_2 = 0.05$.

1. From the system block diagram it can be seen that the system is closed loop with negative feedback. Therefore we can use,

$$\frac{\theta_o}{\theta_i} = \frac{G}{1 + GH}$$

Test your knowledge 3.20

1. Detail three advantages of fully automatic closed-loop control systems over their open-loop counterparts.
2. Explain the function of transducers in a typical closed-loop control system.
3. What is the function of the error detector or comparator in a closed-loop control system?
4. In a closed-loop control system, explain how the load is always driven towards its demanded output value.
5. What is the function of the feedback transducer when fitted into a closed-loop control system?
6. Write down the CLTF for a unity positive feedback, closed-loop control system.
7. How may we differentiate between the overall gain of a closed-loop system and the system transfer function?

where $G = G_1G_2G_3$ and $H = H_1H_2$.
Then:

$$\frac{\theta_0}{\theta_i} = \frac{G_1G_2G_3}{1 + G_1G_2G_3H_1H_2}$$

2. To find the *overall system gain*, it is just a question of substituting the appropriate values and completing the arithmetic. Then:

    overall system gain

    $$= \frac{(20)(15)(10)}{1 + (20)(15)(10)(0.05)(0.05)} = \frac{3000}{1 + 7.5} = \mathbf{400}$$

Closed-loop system response

In a perfect system, the output value, θ_o, will respond instantaneously to a change in the input, θ_i. There will be no delay when changing from one value to another and no time required for the output to 'settle' to its final value. This ideal state of affairs is illustrated in Figure 3.114(b). In practice, real-world systems take time to reach their final state. Indeed, a very sudden change in output may, in some cases, be undesirable.

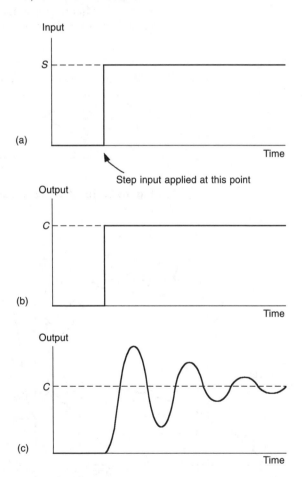

Figure 3.114 *(a) Step input, (b) ideal response to step input, (c) response of real system to step input*

Furthermore, *inertia* is present in many systems. Consider the case of the motor speed control system shown in Figure 3.115 (we study this system in detail later) where the output shaft is connected to a substantial flywheel. The flywheel effectively limits the acceleration of the motor speed when the set point (S) is increased. Furthermore, as the output speed reaches the desired value, the inertia present will keep the speed increasing despite the reduction in voltage (V) applied to the motor. Thus the output shaft speed *overshoots* the desired value before eventually falling back to the required value.

Increasing the *gain* present in the system will have the effect of increasing the acceleration but this, in turn, will also produce a correspondingly greater value of overshoot. Conversely, decreasing the gain will reduce the overshoot but at the expense of slowing down the response. The actual response of the system represents a compromise between speed and an acceptable value of overshoot. Figure 3.114(c) shows the typical response of a system to the step input shown in Figure 3.114(a).

> **Key point**
>
> Increasing the gain of a system reduces response time but increases overshoot.

Second order response

The graph shown in Figure 3.114(c) is known as a '*second order*' *response*. This response has two basic components, a logarithmic growth curve and a damped oscillation (see Figure 3.116).

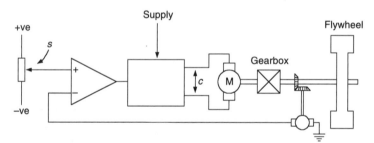

Figure 3.115 *Motor speed control system with inertia*

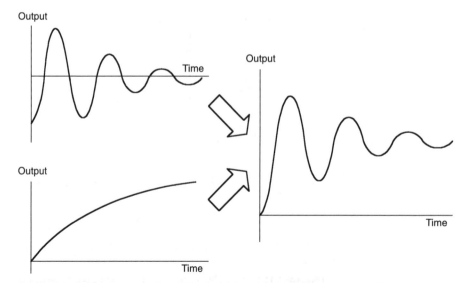

Figure 3.116 *System response resulting from added damped oscillation to a logarithmic growth curve*

The oscillatory component can be reduced (or eliminated) by artificially slowing down the response of the system. This is known as *damping*. The optimum value of damping is that which *just* prevents overshoot. When a system is 'underdamped', some overshoot is still present. Conversely, an 'overdamped' system may take a significantly greater time to respond to a sudden change in input (see Figure 3.117).

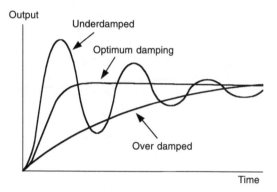

Figure 3.117 *System response with different amounts of damping*

Delay time, settling time and damping factor

The delay time for a system is the time taken for the output to reach 50 per cent of its final (steady-state) value after the application of a step input (see Figure 3.118(a)). The settling time for a system is the time taken for the system to enter and remain within defined tolerance limits for its output. Typical values for tolerance limits are 5 per cent or 10 per cent. In relation to the steady-state step value, the size of the first overshoot is an indicator of the *damping factor*. The smaller the damping factor, the larger the overshoot. Optimum damping is associated with a damping factor of unity.

Natural frequency

The natural frequency f_n of a system is the reciprocal of the time taken t for one cycle of the damped oscillation (see Figure 3.118(b)). Natural frequency is normally specified in Hertz (Hz) or in terms of angular velocity, ω_n, where:

$$\omega_n = 2\pi f_n \quad \text{and} \quad f_n = 1/t$$

Decrement

From Figure 3.118 you should note that each successive cycle of damped oscillation has a smaller amplitude than its predecessor. However, the amount of reduction in amplitude is constant from each peak to the next. This is known as the *decrement* and it can be most easily measured from the ratio of the amplitude of the first undershoot to that of the first overshoot. In each case the amplitudes

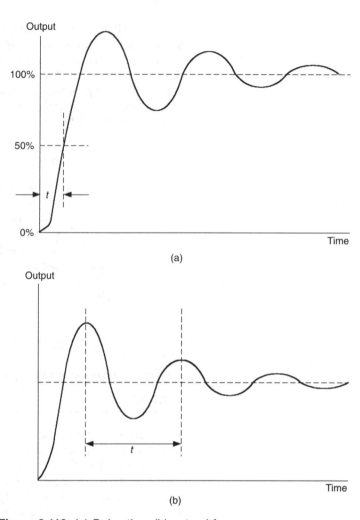

Figure 3.118 *(a) Delay time (b) natural frequency*

are measured either side of the steady-state value. If, for example, the amplitude of the first undershoot is 0.1 while that of the first overshoot is 0.2, the decrement is (0.1/0.2) or 0.5.

Engineering applications of closed-loop control systems

To finish our short study of engineering systems we are going to look at two engineering applications of automatic closed-loop control. Before we do so, let me remind you of some of the reasons for making a system *fully automatic*.

1. Some systems use a very large number of input variables and it may be difficult or impossible for a human operator to keep track of them.
2. Some processes are extremely complex and there may be significant interaction between the input variables.
3. Some systems may have to respond very quickly to changes in variables (human reaction times may just not be fast enough).
4. Some systems require a very high degree of precision (human operators may be unable to work to a sufficiently high degree of accuracy).

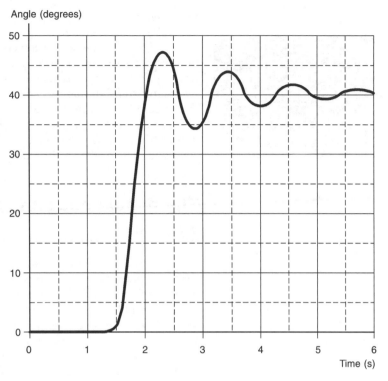

Angle (degrees)

Time (s)

Figure 3.119 *Test your knowledge 3.21*

Test your knowledge 3.21

Figure 3.119 shows the response of a rotary position control system to a sudden change in input. Determine each of the following:

(a) the amplitude of the first overshoot
(b) the decrement
(c) the natural frequency
(d) the 10 per cent settling time.

Key point

Automatic closed-loop control is necessary when the system is complex or requires great precision or is required to respond quickly and accurately.

Thus reasons (1) and (2) could apply to the aircraft independent landing system (ILS), which we considered earlier. There are simply too many variables for the pilot to contend with during take-off and landing of a complex modern airliner. Reason (4) is particularly applicable to modern manufacturing systems, where a high degree of precision is necessary to produce complex components.

Servo systems

Servo systems are closed-loop automatic control systems that are designed to control the position (linear or angular) and velocity (linear or angular) of a load, remotely. Their major engineering use is to overcome the large torques required to move heavy loads through the application of a relatively small input load or torque. Examples of such systems include:

- controlling torque of gearboxes and other transmission components used with heavy machinery
- controlling the movement of an industrial robot
- the automatic piloting of an aircraft
- control of a ship's steering system
- control of a guided missile
- control of a vehicle transmission system
- industrial precision positioning system
- speed control of a process control system.

These systems can include electronic, electromechanical, pneumatic, hydraulic and mechanical devices. The above examples demonstrate the versatility of these systems. We are going to look at two particular examples of such systems.

The first is a remote position control (RPC) system which may be used for such functions as accurately positioning a workpiece prior to a drilling operation or as a powerful gear selector in a large automatic transmission system or even to operate the powered flying controls of a modern aircraft.

The second example is concerned with controlling the speed of an electric motor by automatically controlling the current to the field windings of the motor, under varying torque conditions.

The remote position control (RPC) servo system

For this system we will consider two methods that may be adopted for the remote and accurate positioning of a large workpiece in an automated manufacturing system. The power source for such a system could be electrical, hydraulic or pneumatic. In our system we will be using hydraulic power, which we are able to control both mechanically and electrically.

The first of our systems is illustrated in Figure 3.120 where a *hydromechanical* RPC servo system is shown. The hydraulic fluid flow to the servo valve input is supplied by a constant displacement hydraulic pump and a solenoid operated, closed centre, directional control valve. The displacement of the load is controlled by the operator, via a mechanically operated spool valve. If, for example, the operator moves the spool valve to the right, then fluid is ported into the right-hand chamber via hydraulic line A. This fluid flow reacts against the fixed piston assembly and walls of the cylinder body. At the same time, hydraulic fluid in the left-hand oil chamber is ported back to return, via line B. The whole cylinder body now moves to the right, this movement of the cylinder body, over the now stationary spool valve, gradually shuts off the hydraulic system supply and return lines, creating a hydraulic lock, bringing the cylinder body and thus the load (workpiece) to rest. If the operator mechanically moves the spool valve to the left, the cylinder body moves the workpiece to the left, until follow-up action again causes the movement to cease.

The speed of operation is dependent on the amount of displacement of the spool valve pistons relative to the cylinder body. Thus the workpiece can be moved in both directions, at the required feed rate, dependent on the movement of the spool valve, initiated by the operator, who may be remote from the workstation.

A similar system to that shown in Figure 3.120 could also be used to operate the flying controls of an aircraft where the input to the spool valve is controlled by the pilot.

A rather more sophisticated electrohydraulic servo system could be used for workpiece positioning. Such a system is illustrated in Figure 3.121, where, in this case, the servo valve (spool valve) is electrically actuated using an appropriate motor and the follow-up action is electrically sensed using a pair of electrical potentiometers.

The hydraulic power supply and hydraulic actuator are identical to those shown before in Figure 3.120. The difference lies in the

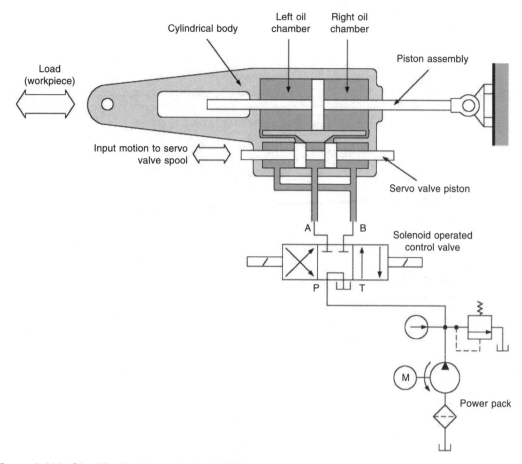

Figure 3.120 *Simplified hydromechanical RPC servo system*

way the servo spool valve is actuated. In this case electrical positioning of the valve spool is set via a rotary potentiometer, which supplies the electrical input voltage via an amplifier to the spool motor. The follow-up signal is provided by the linear potentiometer, which acts as the feedback transducer. There is a separate electrical power supply for the electrical position control system.

There are several alternative ways in which to control the position of a load remotely, including an electric–electric system or electropneumatic system, where the power for the heavy duty position actuator could be supplied by electrical or pneumatic means. The advantages of hydraulic actuation include available actuator power, speed of response, robustness and sensitivity. Disadvantages include possible leakage of hydraulic fluid and associated contamination, component weight and expense.

Electric motor speed control

Figure 3.122 shows a schematic diagram of a motor speed control system, which you met earlier. Here the closed-loop servo system provides speed control for the DC motor *M*. The actual motor speed is sensed by means of a small DC tachogenerator, *G*, coupled

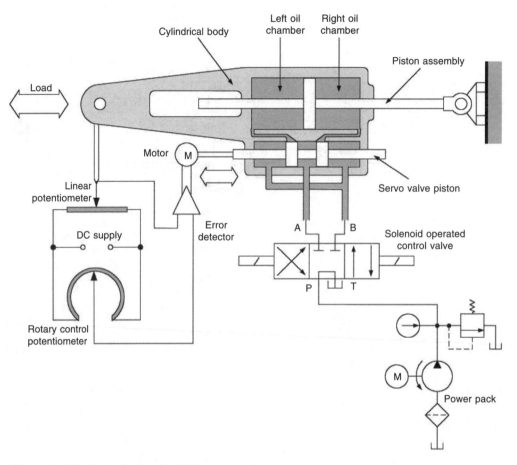

Figure 3.121 *Electrohydraulic RPC servo system*

to the output shaft. The voltage produced by the tachogenerator is compared with that produced at the slider of a potentiometer, *R*, which is used to set the desired speed. The comparison of the two voltages (i.e. that of the tachogenerator with that corresponding to the desired value or *set point*) is performed by an operational amplifier connected as the error detector or comparator. The output of the comparator stage is applied to a power amplifier that supplies current to the DC motor. Energy is derived from a mains-powered DC power supply comprising transformer, rectifier and smoothing circuits.

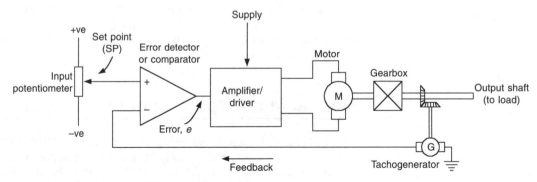

Figure 3.122 *Motor speed control system*

How exactly does the variation in voltage to the motor control its speed? When you studied *motor force*, you may well remember that it depends on the electromagnetic field strength β, the current in the armature windings *I* and the length of the conductor *L* in the magnetic field. This force, acting at a distance from the centre of rotation of the armature, produced a *driving torque*, i.e. the *output* from the motor. Therefore, this torque may be varied as the force is varied within the motor. When the motor operates against varying loads, the amount of torque required to sustain a particular speed (rpm) also needs to be varied. This is achieved by varying the voltage and therefore current to the electromagnetic *field* windings. The higher the current in these windings the greater the field strength β and so the greater the motor force and torque, from $F = βIL$.

A simplified schematic diagram of this control system is shown in Figure 3.123. Note, once again, that the error signal from the comparator requires amplification (signal conditioning) in order to supply sufficient change in current to alter the field strength.

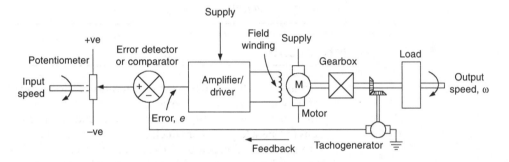

Figure 3.123 *Motor speed control using variation in field strength*

Example 3.55 A digital closed-loop control system

Figure 3.124 shows a microprocessor-based digital, closed-loop control system developed by Lotus. This system provides active ride suspension control for a high-performance road vehicle and it avoids the need for the road springs, dampers and anti-roll bars that are found in a conventional vehicle. The system allows the car to be 'self-levelling' and it can eliminate roll when the vehicle corners at both low and high speed. The overall ride can be set to 'hard' or 'soft' to suit the driver and also to cater for a variety of load conditions. The three principal input signals (all in analogue form) are derived from a load cell (which forms part of the suspension unit), an accelerometer and a linear displacement (position) transducer attached to each wheel. These analogue signals are converted to digital signals that can be processed by the microprocessor by means of analogue to digital converters (ADC). The 8-bit digital signals produced by the ADC are connected directly to the microprocessor's 8-bit data bus.

The two main output signals are used to drive two servo control valves that regulate the supply of hydraulic fluid to the hydraulic ram arrangement. The supply of hydraulic fluid is powered by a small engine-driven pump.

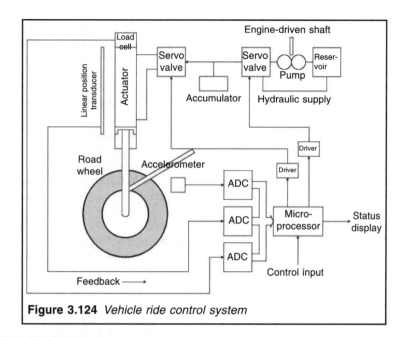

Figure 3.124 *Vehicle ride control system*

Problems

1. Explain the difference between the terms open loop and closed loop, when applied to a system.

2. An automatic telescope, used for tracking and monitoring solar activity, has a large motor which drives the telescope mechanism through a reduction gearbox. The motor is driven by an amplifier, the signal to which is a voltage proportional to the error between the desired and actual positions. Sketch a block diagram to represent the system and explain its operation.

3. Classify the following list of system components under the headings: sensor or transducer, controlling device, signal conditioning device, actuation device or output device:

    a rotary switch
    an actuator
    a gearbox
    a voltage regulator
    a motor
    a pressure relief valve
    a speed governor
    an amplifier
    a visual display unit (VDU).

4. A guided missile direction control system operates as follows. The directions of the missile and the target are measured using the missile's gyroscope. The difference between these two measurements is fed, as an electrical signal, to a comparator. The error signal from the comparator is amplified and used to drive a servo motor which deflects the jet efflux from the rocket, so altering the rocket's trajectory and guiding it towards the target. Sketch a block diagram of this system.

5. Sketch block diagrams of each of the following systems, clearly identify the system's inputs, outputs, signal conditioning and actuation devices:

a petrol engine ignition system
a hi-fi system.

6. The following data refer to measurements made on a motor speed control system:

Time (s)	0	1	2	3	4	5
Output (rev/sec)	0	2.1	11.0	19.5	25.0	28.6

	6	7	8	9	10	11
	29.8	28.8	24.5	20.5	17.0	15.5

	12	13	14	15	16	17
	14.8	15.6	16.9	19.1	21.0	21.9

	18	19	20
	22.8	21.9	21.0

Plot the response of this system and use it to determine:

(a) the final (steady-state) value
(b) the delay time
(c) the 15 per cent settling time
(d) the natural frequency.

7. An amplifier has an open-loop gain of 200 and is used in a circuit in which 5 per cent of the output is fed back as negative feedback. Determine the overall voltage gain with negative feedback applied.

8. The forward path gain of an open-loop system is 100. If the system is made into a closed-loop system with negative feedback gain of 2 determine the percentage reduction in the overall gain.

9. Investigate the following systems and in particular determine how these systems are controlled during their operation:

(a) a water-level control system
(b) a milk bottle filling, conveyor system
(c) a fluid flow, measurement and display system
(d) a digital velocity measuring system.

10. Explain the nature of servo system damping, including the need for compromise with respect to the 'degree' of damping.

This unit aims to provide you with a sound foundation in mathematical principles, which will enable you to solve mathematical, scientific and associated engineering problems at the technician level.

We start by revising the *fundamental principles* of number and by considering the use of letters to represent numbers and to aid our understanding of the arithmetic operations that we are able to perform on numbers. Powers and exponents of numbers are considered, together with a coverage of the manipulation of fractions, without the use of a calculator! Estimation techniques involving various forms of number are also covered in this introductory section.

Next we move on to consider the syllabus itself, where we will study the four major outcomes which make up the unit – algebra, trigonometry, statistics and calculus arithmetic.

In our study of algebra we will determine and use the fundamental laws, which are needed to correctly manipulate algebraic and logarithmic expressions, functions and formulae. We will also use graphs to represent algebraic and logarithmic functions and to solve simple equations. Our study of trigonometry will include the use of trigonometric ratios to solve problems involving areas, volumes and measurement. In addition we will again use graphs of sinusoidal functions to illustrate and solve problems that are particularly relevant to the solution of trigonometric function. Next, we consider and use a variety of statistical methods to gather, manipulate and display scientific and engineering data. We will then consider the ways in which the elementary rules of calculus arithmetic may be used to solve problems that involve simple differentiation and integration of algebraic and trigonometric functions. Finally, we will use the calculus to solve some elementary engineering problems, involving rates of change and summation.

In order to aid your understanding of mathematics, you will find numerous fully worked examples and *Test your knowledge* exercises spread throughout this unit.

Fundamentals

Introduction

If you feel that you already fully understand the ideas on numbers and symbols presented below, then go directly to *Test your knowledge 4.1* and answer *all* questions. Remember to check your answers in the back of the book and revise any areas in this section that you get wrong! Please remember that *all* the material contained in this section on *fundamentals* is considered *essential prerequisite knowledge* for your study of the unit outcomes that follow. In particular, the sections on: *factors*, *products*, *powers*, *indices* and *exponents* need to be thoroughly understood, as they form an *essential part of the outcome on algebra*. Therefore, you are strongly advised to work your way carefully through *all* the ideas and principles presented here, before moving on to the unit outcomes.

Numbers and symbols

It is generally believed that our present number system began with the use of the *natural numbers*, such as 1, 2, 3, 4, . . . , etc. These whole numbers, known as the *positive integers*, were used primarily for counting. However, as time went on, it became apparent that whole numbers could not be used for defining certain mathematical quantities. For example, a period in time might be between 3 and 4 days or the area of a field might be between 2 or 3 acres (or whatever unit of measure was used at the time). So the *positive fractions* were introduced, for example $^1/_2$, $^1/_4$, $^3/_4$. These two groups of numbers, the positive integers and the positive fractions, constitute what we call the *positive rational numbers*. Thus, 711 is an integer or whole number, $^1/_4$ is a positive fraction and $234^3/_5$ is a rational number. In fact a *rational number is any number that can be expressed as the quotient of two integers*, that is any number that can be written in the form *a/b* where *a* and *b* represent any integers. Thus $^2/_5$, $^8/_9$ and 1 are all rational numbers. The number 1 can be represented by the quotient $^1/_1 = 1$, in fact *any number divided by itself must always be equal to 1*.

The natural numbers are *positive* integers, but suppose we wish to subtract a larger natural number from a smaller natural number. For example, 10 subtracted from 7, we obviously obtain a number which is *less than zero*, i.e. $7 - 10 = -3$. So our idea of numbers must be enlarged to include numbers less than zero, called *negative numbers*. The number zero (0) is unique. It is not a natural number because all natural numbers represent positive integer values, that is numbers above zero, and quite clearly from what has been said, it is not a negative number either. It sits uniquely on its own and must be added to our number collection.

So to the natural numbers (positive integers) we have added negative integers, the concept of zero, positive rational numbers and negative natural numbers. What about numbers like $\sqrt{2}$? This is *not* a rational number because it cannot be represented by the quotient of two integers. So yet another class of number needs to be included, the *irrational* or non-rational numbers. Together all the above kinds of number constitute the broad class of numbers known as *real numbers*. They include positive and negative terminating and non-terminating decimals (e.g. $\pm^1/_9 = \pm0.1111$, . . . , 0.48299999, ±2.5, 1.73205, . . .). The real numbers are so called to distinguish them from others such as *imaginary* or *complex* numbers, the latter may be made up of both real and imaginary number parts. Complex numbers will not be considered during our study of technician mathematics.

Although we have mentioned negative numbers, we have not considered their arithmetic manipulation. All positive and negative numbers are referred to as *signed numbers* and they obey the arithmetic laws of sign. Before we consider these laws, let us first consider what we mean by signed numbers.

Conventional representation of signed numbers is shown below, with zero at the midpoint. Positive numbers are conventionally shown to the right of zero and negative numbers to the left.

−12 −11 −10 −9 −8 −7 −6 −5 −4 −3 −2 −1 0 +1 +2 +3 +4 +5 +6 +7 +8 +9 +10 +11 +12

The number of units a point is from zero, regardless of its direction, is called the *absolute value* of the number corresponding to the point on the above number system when points are drawn to scale. Thus the absolute value of a positive number, or of zero, is the number itself. While the absolute value of a negative number is the number with its sign changed. For example, the absolute value of +10 is 10 and the absolute value of −10 is also 10. Now the absolute value of any number n is represented by the symbol $|n|$. Thus $|+24|$ means the absolute value of +24. Which is larger, $|+3|$ or $|−14|$?

I hope you said $|−14|$ because its absolute value is 14, while that of $|+3|$ is 3 and of course 14 is larger than 3. We are now ready to consider the laws of signs.

The laws of signs

You are probably already familiar with these laws, here they are:

First law: *To add two numbers with like signs, add their absolute values and prefix their common sign to the result.* This law works for ordinary arithmetic numbers and simply defines what we have always done in arithmetic addition.

For example: 3 + 4 = 7 or in full (+3) + (+4) = +7

After the introduction of the negative numbers, the unsigned arithmetic numbers became the positive numbers, as illustrated above. So now all numbers may be considered either positive or negative, and the laws of signs apply to them all.

Does the above law apply to the addition of two negative numbers? From ordinary arithmetic we know that (−7) + (−5) = −12. This again obeys the first law of signs, because we add their *absolute value* and prefix their common sign.

Second law: *To add two signed numbers with unlike signs, subtract the smaller absolute value from the larger and prefix the sign of the number with the larger absolute value to the results.*

So following this rule, we get, for example:

5 + (−3) = 2; −12 +9 = −3; 6 + (−11) = −5 and so on.

The numbers written without signs are, of course, positive numbers. Notice that brackets have been removed when not necessary.

Third law: *To subtract one signed number from another, change the sign of the number to be subtracted and follow the rules for addition.*

For example, if we subtract 5 from –3, we get –3 – (+5) = –3 + (–5) = –8.

Now what about the multiplication and division of negative and positive numbers; so as not to labour the point the rules for these operations are combined in our fourth and final law.

Fourth law: *To multiply (or divide) one signed number by another, multiply (or divide) their absolute values; then, if the numbers have like signs, prefix the plus sign to the result; if they have unlike signs, prefix the minus sign to the result.*

Therefore applying this rule, to the multiplication of two positive numbers, then for example, 3 × 4 = 12, 12 × 8 = 96, . . . , and so on, which of course is simple arithmetic. Now applying the rule to the multiplication of mixed sign numbers we get, for example, –3 × 4 = –12, 12 × (–8) = –96, . . . , and so on. We can show, equally well, that the above rule yields similar results for division.

Key point

(+)(+) = + and (–)(–) = + and (–)(+) = – or (+)(–) = –.

Example 4.1

Apply the fourth law to the following arithmetic problems and determine the arithmetic result.

(a) (–4)(–3)(–7) = ?
(b) 14/–2 = ?
(c) 5(–6)(–2) = ?
(d) –22/–11 = ?

(a) In this example we apply the fourth law twice. (–4)(–3) = 12 (like signs) and so 12(–7) = –84.
(b) 14/–2 applying the third law for unlike signs immediately gives –7, the correct result.
(c) Again applying the third law twice. 5(–6) = –30 (unlike signs) and (–30)(–2) = 60.
(d) –22/–11 applying the third law for like sign gives 2, the correct result.

The use of symbols

We earlier introduced the concept of *symbols* to represent numbers when we defined rational numbers where the *letters a and b* were used to represent *any* integer. Look at the symbols below. Do they represent the same number?

IX; 9; nine; $+\sqrt{81}$

I hope you answered yes, since each expression is a perfectly valid way of representing the positive integer 9. In *algebra* we use letters to represent Arabic numerals; such numbers are called *general numbers* or *literal numbers*, as distinguished from *explicit numbers* like 1, 2, 3, etc. Thus a literal number is simply a number represented by a letter, instead of a numeral. Literal numbers are used to state algebraic rules, laws and formulae; these statements being made in mathematical sentences are called *equations*.

If a is a positive integer and b is 1, what is a/b? I hope you were able to see that $a/b = a$. Any number divided by 1 is always itself. Thus: $a/1 = a$, $c/1 = c$, $45.6/1 = 45.6$.

Suppose a is again any positive integer, but b is 0. What is the value of *a/b*? What we are asking is what is the value of any positive integer divided by zero? Well, the answer is that we really do not know! *The value of the quotient a/b, if b = 0, is not defined in mathematics.* This is because there is no such quotient that meets the conditions required of quotients. For example, you know that to check the accuracy of a division problem, you can multiply the quotient by the *divisor* to get the *dividend*. Where, for example, if 21/7 = 3 then 7 is the divisor, 21 is the dividend and 3 is the quotient and so 3 × 7 = 21, as expected. So if 17/0 were equal to 17, then 17 × 0 should again equal 17 but it does not! Or if 17/0 were equal to zero, then 0 × 0 should equal 17 but again it does not. *Any number multiplied by zero is always zero.* Therefore, division of any number by zero (as well as zero divided by zero) is excluded from mathematics. *If b = 0, or if both a and b are zero, then a/b is meaningless.*

When multiplying *literal numbers* together we try to avoid the multiplication sign (×), this is because it can be easily mistaken for the letter x. Thus instead of writing $a \times b$ for the product of two general numbers, we write $a \cdot b$ (the dot notation for multiplication) or more usually just ab to indicate the product of two general numbers a and b.

> **Key point**
>
> Division by zero (0) is meaningless and not allowed.

Example 4.2

If we let the letter *n* stand for any real number, what does each of the following expressions equal?

(a) $n/n = ?$
(b) $n \cdot 0 = ?$
(c) $n \cdot 1 = ?$
(d) $n + 0 = ?$
(e) $n - 0 = ?$
(f) $n - n = ?$
(g) $n/0 = ?$

(a) $n/n = 1$,   that is any number divided by itself is equal to 1.
(b) $n \cdot 0 = 1$,   any number multiplied by zero is itself zero.
(c) $n \cdot 1 = n$,   any number multiplied *or* divided by 1 is itself.
(d) $n + 0 = n$,   the addition of zero to any number will not alter that number.
(e) $n - 0 = n$,   the subtraction of zero from any number will not alter that number.
(f) $n - n = 0$,   subtraction of any number from itself will always equal zero.
(g) $n/0$,   division by zero is not defined in mathematics.

> **Key point**
>
> When any number is divided by itself the result is always 1.0.

The commutative, associative and distributive laws

We all know that 6 × 5 = 30 and that 5 × 6 = 30, so is it true that when multiplying any two numbers together, the result is the same no matter what the order? The answer is yes. The above relationship may be stated as:

> *The product of two real numbers is the same no matter in what order they are multiplied, that is, ab = ba. This is known as the commutative law of multiplication.*

Key point

The multiplication of real numbers can be carried out in any order. The result is always the same.

Key point

The addition of numbers is always the same no matter in what order they are taken.

If three or more real numbers are mulitiplied together, the order in which they are multiplied still makes no difference to the product. For example, $3 \times 4 \times 5 = 60$ and $5 \times 3 \times 4 = 60$. This relationship may be stated formally as:

> *The product of three or more numbers is the same no matter in what manner they are grouped, that is, a(bc) = (ab)c. This is known as the associative law of multiplication.*

These laws may seem ridiculously simple, yet they form the basis of many algebraic techniques, you will be using latter!

We also have commutative and associative laws for addition of numbers, which by now will be quite obvious to you. Here they are:

- The sum of two numbers is the same no matter in what order they are added, that is, $a + b = b + a$. This is known as the *commutative law of addition*.
- The sum of three or more numbers is the same no matter in what manner they are grouped, that is, $(a + b) + c = a + (b + c)$. This is known as the *associative law of addition*.

You may be wondering where the laws are for subtraction. Well, you have already covered these when we considered the laws of signs. In other words the above laws are valid whether or not the number is positive or negative. So, for example, $-8 + (16 - 5) = 3$ and $(-8 + 16) - 5 = 3$.

In order to complete our laws we need to consider the following problem: $4(5 + 6) = ?$. We may solve this problem in one of two ways; first, by adding the numbers inside the brackets and then multiplying the result by 4, this gives: $4(11) = 44$. Alternatively we may multiply out the bracket as follows: $(4 \times 5) + (4 \times 6) = 20 + 24 = 44$. Thus whichever method we choose, the arithmetic result is the same. This result is true in all cases, no matter how many numbers are contained within the brackets!

So, in general, using literal numbers we have:

$$a(b + c) = ab + ac$$

This is the *distributive law*. In words it is rather complicated:

> *The product of a number by the sum of two or more numbers is equal to the sum of the products of the first number by each of the numbers of the sum.*

Now, perhaps you can see the power of algebra in representing this law; it is a lot easier to remember than the wordy explanation!

Remember that the distributive law is valid no matter how many numbers are contained in the brackets, and no matter whether the sign connecting them is a plus or minus. As you will see later, this law is one of the most useful and convenient rules for manipulating formulae and solving algebraic equations.

Example 4.3

If $a = 4$, $b = 3$, and $c = 7$, does

$$a(b - c) = ab - ac$$

The above expression is just the distributive law, with the sign of one number within the bracket changed. This of course is valid since the sign connecting the numbers within the bracket may be a plus or minus. Nevertheless we will substitute the arithmetic values in order to check the validity of the expression.

Then:

$$4(3 − 7) = 4(3) − 4(7)$$

$$4(−4) = 12 − 28$$

$$−16 = −16$$

So our law works irrespective of whether the sign joining the numbers is positive or negative.

You should now attempt Test your knowledge 4.1, *without* the aid of a calculator!

Factors, powers and exponents

Factors

When two or more numbers are multiplied together, each of them, or the product of any number of them (apart from them all), is a factor of the product. This applies to explicit arithmetic numbers and to literal numbers.

So, for example, if we multiply the numbers 2 and 6, we get 2 × 6 = 12, thus 2 and 6 are factors of the number 12. However, the number 12 has more than one set of factors, 3 × 4 = 12, so 3 and 4 are also factors of the number 12. We can also multiply 2 × 2 × 3 to get 12. So the numbers 2, 2 and 3 are yet another set of factors of the number 12. Finally you will remember that any number *n* multiplied by 1 is itself, or $n \times 1 = n$. So every number has itself and 1 as factors. 1 and *n* are considered *trivial factors* and when asked to find the factors of an explicit or literal number, we will exclude the number itself and 1.

Example 4.4

Find the factors of:

(a) 8
(b) *xy*
(c) 24
(d) *abc*
(e) −*m*.

(a) Apart from the trivial factors 1 and 8, which we agreed to ignore, the number 8 has only the factors 2 and 4 since; 2 × 4 = 8; remember that these factors can be presented in reverse order, 4 × 2 = 8, but 2 and 4 are still the only factors.

(b) Similarly the literal number *xy* can only have the factors *x* and *y*, if we ignore the trivial factors. Thus the numbers *x* and *y* multiplied together to form the product *xy* are factors of that product.

(c) The number 24 has several sets of factors, with varying numbers in each set. First, we find the number of sets with two factors, these are:

$24 = 6 \times 4$
$24 = 8 \times 3$
$24 = 12 \times 2$.

More than two factors:

$24 = 2 \times 2 \times 6$
$24 = 4 \times 3 \times 2$
$24 = 2 \times 2 \times 2 \times 3$.

However, if we look closely we see that the number 24 has only six *different* factors: 12, 8, 6, 4, 3 and 2.

(d) So what about the factors in the number *abc*? Well, I hope you can see that the product of each individual factor *a*, *b* and *c* constitutes one set of factors. Also *ab* and *c*, *a* and *bc* and *b* and *ac* form a further three sets. So extracting the different factors from these sets we have: *a*, *b*, *c*, *ab*, *ac* and *bc* as the six factors of the number *abc*.

(e) We have two sets of factors here 1 and –*n*, which is the trivial factor, but also the set *n* and –1; notice the subtlety with the sign change. When dealing with minus numbers, any two factors must have opposite signs.

Powers and exponents

When a number is the product of the same factor multiplied by itself, this number is called a *power* of the factor. For example, we know that $3 \times 3 = 9$. Therefore we can say that 9 is a power of 3. To be precise, it is the second power of 3, because two 3s are multiplied together to produce 9. Similarly 16 is the second power of 4. We may use literal terminology to generalize the relationship between powers and factors.

So the second power of *a* means $a \times a$ (or *aa*), this is written as a^2, where *a* is known as the *base* (factor) and 2 is the *exponent* (or index). Thus writing the number 9 in exponent form we get $9 = 3^2$ where *9 is the second power, 3 is the base (factor) and 2 is the exponent (index)*.

The above ideal can be extended to write arithmetic numbers in exponent or index form. For example, $5^2 = 25$, $9^2 = 81$ and $3^3 = 27$. Notice that the second power of 5 gives the number 25 or $5 \times 5 = 25$, similarly 3^3 means the third power of 3, literally $3 \times 3 \times 3 = 27$. The idea of powers and exponents (indices) can be extended to literal numbers. For example: $a \cdot a \cdot a \cdot a \cdot a$ or a^5 *or in general a^m where a is the base (factor) and the exponent m (or index) is any positive integer. a^m means a used as a factor m times and is read as* the '*m*th power of *a*'. Note, that since any number used as a factor once would simply be the number itself, the index (exponent) is not usually written, in other words *a* means a^1.

Now, providing the base of two or more numbers expressed in index (exponent) form is the same, we can perform multiplication and division on these numbers, by adding or subtracting the *indices* accordingly.

We will from now on refer to the exponent of a number as its index, in order to avoid confusion with particular functions, such as the exponential function, which we study latter.

Consider the following literal numbers in index form:

$$x^2 \cdot x^2 = (x \cdot x)(x \cdot x) = x \cdot x \cdot x \cdot x = x^4$$

$$x^2 \cdot x^4 = (x \cdot x)(x \cdot x \cdot x \cdot x) = x \cdot x \cdot x \cdot x \cdot x \cdot x = x^6$$

$$\frac{x^2}{x^2} = \frac{x \cdot x}{x \cdot x} = x^0 = 1$$

$$\frac{x^2}{x^4} = \frac{x \cdot x}{x \cdot x \cdot x \cdot x} = \frac{1}{x \cdot x} = x^{-2}$$

What you are looking for is a pattern between the first two literal numbers which involve multiplication and the second two which involve division.

For multiplication of numbers with the same base, we add the indices, and for division of numbers with the same base, we subtract the indices in the *denominator* (below the line) from those in the *numerator* (above the line). Remember also that the base number $x = x^1$.

We will now generalize our observations and so formulate the *laws of indices*.

The laws of indices

In the following laws a is the common *base*, m and n are the *indices* (exponents). Each law has an example of its use alongside.

1. $a^m \times a^n = a^{m+n}$ $2^2 \times 2^4 = 2^{2+4} = 2^6 = 64$

2. $\dfrac{a^m}{a^n} = a^{m-n}$ $\dfrac{3^4}{3^2} = 3^{4-2} = 3^2 = 9$

3. $(a^m)^n = a^{mn}$ $(2^2)^3 = 2^{2\times3} = 2^6 = 64$

4. $a^0 = 1$ Any number raised to the power 0 is always 1

5. $a^{\frac{m}{n}} = \sqrt[n]{a^m}$ $27^{\frac{4}{3}} = \sqrt[3]{27^4} = 3^4 = 81$

6. $a^{-n} = \dfrac{1}{a^n}$ $6^{-2} = \dfrac{1}{6^2} = \dfrac{1}{36}$

We need to study these laws carefully in order to understand the significance of each.

Law 1 you have already meet, it enables us to *multiply numbers* given in index form that have a common base. In the example the common base is 2, the first number raises this base (factor) to the power 2 and the second raises the same base to the power 3. In order to find the result we simply *add* the indices.

Law 2 we have again used when *dividing numbers* with a common base, in this case the base is 3. Note that since division is the opposite arithmetic operation to multiplication, it follows that we should perform the opposite arithmetic operation on the indices, that of *subtraction*. Remember we always subtract the index in the denominator from the index in the numerator.

Law 3 is concerned with raising the powers of numbers. Do not mix this law up with law 1. When *raising powers of numbers* in index form, we *multiply* the indices.

Law 4 you have also met; this law simply states that *any number raised to the power 0 is always 1.* Knowing that any number divided by itself is also 1, we can use this fact to show that a number raised to the power 0 is also 1. What we need to do is use the second law concerning the division of numbers in index form.

We know that $\frac{9}{9} = 1$ or $\frac{3^2}{3^2} = 3^{2-2} = 3^0 = 1$ which shows that $3^0 = 1$ and in fact because we have used the second law of indices, this must be true in all cases.

<table>
<tr><td>

Key point

When you move a number in index form above or below the fraction line always change the sign of the index.

</td><td>

Law 5 is a rather complicated looking law, which enables us to find the decimal equivalent of a number in index form, where the index is a fraction. All that you need to remember is that the index number above the fraction line is raised to that power and the index number below the fraction line has that number root.

So for the number, $8^{\frac{2}{3}}$ then we raise 8 to the power 2 and then take the cube root of the result. It does not matter in which order we perform these operations. So we could have just as easily taken the cube root of 8 and then raised it to the power 2.

</td></tr>
</table>

Law 6 is very useful when you wish to convert the division of a number to multiplication. In other words bring a number from underneath the division line to the top of the division line. *As the number crosses the line we change the sign of its index.* This is illustrated in the example which accompanies this law.

The following examples, further illustrate the use of the above laws, when evaluating or simplifying expressions that involve numbers and symbols.

Example 4.5

Evaluate the following expressions:

(a) $\dfrac{3^2 \times 3^3 \times 3}{3^4}$

(b) $(6)(2x^0)$

(c) $36^{-\frac{1}{2}}$

(d) $16^{-\frac{3}{4}}$

(e) $\dfrac{(2^3)^2 (3^2)^3}{(3^4)}$.

(a) $\dfrac{3^2 \times 3^3 \times 3}{3^4} = \dfrac{3^{2+3+1}}{3^4}$ (law 1) $\dfrac{3^6}{3^4} = 3^{6-4}$ (law 2) $= 3^2 = 9$

(b) $(6)(2x^0) = (6)(2) = 12$ remembering that $x^0 = 1$ (law 4)

(c) $36^{-\frac{1}{2}} = \dfrac{1}{36^{\frac{1}{2}}} =$ (law 6) $= \dfrac{1}{\sqrt{36}}$ (law 5)

$= \pm\dfrac{1}{6}$ (note $\pm$ square root)

(d) $16^{-\frac{3}{4}} = \dfrac{1}{16^{\frac{3}{4}}}$ (law 6) $= \dfrac{1}{\sqrt[4]{16^3}}$ (law 5) $= \dfrac{1}{2^3} = \dfrac{1}{8}$

(e) $\dfrac{(2^3)^2\,(3^2)3}{3^4} = \dfrac{(2^{3\times2})(3^{2+1})}{3^4}$ (law 3) $= \dfrac{2^6 \times 3^3}{3^4} = 2^6 \times 3^{3-4}$ (law 2)

$= 2^6 \times 3^{-1} = 64 \times \dfrac{1}{3}$ (law 6) $= \dfrac{64}{3}$

Example 4.6

Simplify the following expressions:

(a) $\dfrac{12x^3y^2}{4x^2y}$

(b) $\left(\dfrac{a^3b^2c^4}{a^4bc}\right)\left(\dfrac{a^2}{c^2}\right)$

(c) $[(b^3c^2)(ab^3c^2)(a^0)]^2$

(a) $\dfrac{12x^3y^2}{4x^2y} = 3x^{3-2}y^{2-1}$

(rule 2 and simple division of integers) $= 3xy$

(b) $\left(\dfrac{a^3b^2c^4}{a^4bc}\right)\left(\dfrac{a^2}{c^2}\right) = a^{3+2-4}b^{2-1}c^{4-1-2} = abc$

(rules 1 and 2 and operating on like bases)

Note also in the above problem that there was no real need for the second set of brackets, since all numbers were multiplied together.

(c) $[(b^3c^2)(ab^3c^2)(a^0)]^2 = [(b^3c^2)(ab^3c^2)(1)]^2$ (rule 4)

$= [ab^{3+3}c^{2+2}]^2$ (rule 1) $= [ab^6c^4]^2 = a^2b^{12}c^8$ (rule 3)

Powers of ten and estimation techniques

The powers of ten are sometimes called 'the technician's shorthand'. They enable very large and very small numbers to be expressed in simple terms. You may have wondered why, in our study of numbers, we have not mentioned *decimal numbers* before now. Well the reason is simple, these are the numbers you are most familiar with; they may be rational, irrational or real numbers. Other numbers such as the positive and negative integers are a subset of real numbers. The exception are the complex numbers; these are not a subset of the real numbers and do not form part of our study in this course.

Essentially then, decimal numbers may be expressed in index form, using the powers of ten. For example:

1 000 000	$= 1 \times 10^6$
100 000	$= 1 \times 10^5$
10 000	$= 1 \times 10^4$
1000	$= 1 \times 10^3$
100	$= 1 \times 10^2$

$$10 \qquad\qquad\qquad = 1 \times 10^1$$
$$0 \qquad\qquad\qquad\quad = 0$$
$$1/10 = 0.1 \qquad\qquad = 1 \times 10^{-1}$$
$$1/100 = 0.01 \qquad\quad = 1 \times 10^{-2}$$
$$1/1000 = 0.001 \qquad = 1 \times 10^{-3}$$
$$1/10\ 000 = 0.0001 \qquad = 1 \times 10^{-4}$$
$$1/100\ 000 = 0.00001 \quad = 1 \times 10^{-5}$$
$$1/1\ 000\ 000 = 0.000001 = 1 \times 10^{-6}$$

I am sure you are familiar with the above shorthand way of representing numbers. We show, for example, the number one million (1 000 000) as 1×10^6, that is 1 multiplied by 10 six times. The *exponent (index)* of 10 is 6, thus the number is in exponent or *exponential form*, the *exp* button on your calculator!

Notice we multiply all the numbers represented in this manner by the number 1. This is because we are representing one million, one hundred thousand, one-tenth, etc. When using your calculator *always input the multiplier* (one in this case) *then press the exp button and input the index (exponent) number, remembering to operate the +/– button if you are imputing a negative index.*

When representing decimal numbers in index (exponent) form, the multiplier *is always a number which is ≥ 1.0 or < 10*, that is a number greater than or equal to (≥ 1.0) one or less than (< 10) ten.

So, for example, the decimal number $8762.0 = 8.762 \times 10^3$ in index form. Notice with this *number, greater than 1.0, we displace the decimal point* three (3) places *to the left*, that is three powers of ten. Numbers rearranged in this way, using powers of ten, are said to be in *index form* or *exponent form* or *standard form*, dependent on the literature you read.

What about the decimal number 0.000245? Well, I hope you can see that in order to obtain a multiplier that is greater than or equal to one and less than 10, we need to *displace the decimal point* four (4) places *to the right*. Note that the zero in front of the decimal point is placed there to indicate that a whole number has not been omitted. Therefore the number in index form now becomes 2.45×10^{-4}. Notice *that for numbers less than 1.0, we use a negative index*. In other words *all decimal fractions represented in index form have a negative index and all numbers greater than 1.0, represented in this way, have a positive index.*

Every step in our argument up till now has been perfectly logical, but how would we deal with a mixed whole number and decimal number such as 8762.87412355? Well, again to represent this number exactly, in index form, we proceed in the same manner as when dealing with just the whole number. So displacing the decimal point three places to the left to obtain our multiplier gives $8.76287412355 \times 10^3$. This is all very well but one of the important reasons for dealing with numbers in index form is that the manipulation should be easier! In the above example we still have 12 numbers to contend with plus the powers of ten.

In most areas of engineering, there is little need to work to so many places of decimals. In the above example for the original number we have eight decimal place accuracy; this is unlikely to be needed unless we are dealing with a subject like rocket science or astrophysics! So this leads us in to the very important skill of being able to provide *approximations or estimates to a stated degree of accuracy.*

Key point

Decimal numbers greater or equal to 1 (≥ 1) written in index form (standard form) always have a positive exponent (index).

Key point

Decimal fractions, written in index form, always have a negative index.

Example 4.7

For the numbers:

(a) 8762.87412355

(b) 0.0000000234876.

(i) Convert these numbers into *standard form* with 3 decimal place accuracy.

(ii) Write down these numbers in *decimal form*, correct to 2 significant figures.

(a)(i) We have already converted this number into standard form, it is:

$8.76287412355 \times 10^3$. Now looking at the decimal places for the stated accuracy we must consider the first four places 8.7628 and since the last *significant figure* is 8, in this case (*greater than 5*) *we round up* to give the required answer as **8.763×10^3**.

(b)(i) $0.0000000234876 = 2.34876 \times 10^{-7}$ and now following the same argument as above this number is to three decimal places $= \mathbf{2.349 \times 10^{-7}}$.

(a)(ii) For the number 8762.87412355 the 2 required significant figures are to the left of the decimal place. So we are concerned with the whole number 8762 and the first two figures are of primary concern. Again to find our approximation we need first to consider the three figures 876, again since 6 is above halfway between 1 and 10, then we round up to give the required answer **8800**.

Note that we had to add two zeros to the left of the decimal point. This should be obvious when you consider that all we have been asked to do is approximate the number 8762 to within two significant figures.

(b)(ii) For the number 0.0000000234876 the significant figures are any integers to the right of the decimal point and the zeros. So in this case the number to the required number of significant figures is **0.000000023**.

We are now in a position to be able to determine estimates, not just for single numbers but also for expressions involving several numbers. The easiest way of achieving these estimates is to place all numbers involved into standard form and then determine the estimate to the correct degree of accuracy. You may wonder why we do not simply use our calculators and determine values to 8 decimal place accuracy. Well, you need only press one button incorrectly on your calculator to produce an incorrect answer, but how will you know if your answer is incorrect, if you are unable to obtain a rough estimate of what is the correct answer? This is where the use of estimation techniques proves to be most useful; these techniques are best illustrated by example.

Example 4.8

(a) Determine an estimate for $3.27 \times 10.2 \times 0.124$ correct to 1 significant figure.

(b) Simplify $\dfrac{3177.8256 \times 0.000314}{(154\,025)^2}$ giving your answer correct to 2 significant figures.

Test your knowledge 4.2

1. Find the factors (other than the trivial factors) of:
   (a) 16
   (b) n^2
   (c) $wxyz$.
2. Find the common factors in the expression $ab^2c^2 + a^3b^2c^2 + ab^2c$
3. Simplify:
   (a) $\dfrac{1}{2^3} \times 2^7 \times \dfrac{1}{2^{-5}} \times 2^{-4}$
   (b) $\left(\dfrac{16}{81}\right)^{\frac{3}{4}}$
   (c) $\dfrac{b^3 b^{-8} b^2}{b^0 b^{-5}}$.
4. Simplify:
   (a) $(2^2)^3 - 6 \times 3 + 24$
   (b) $\dfrac{1}{2^{-2}} + \dfrac{1}{3^2} - \dfrac{1}{3^{-1}}$.
5. Express the following numbers in normal decimal notation:
   (a) $3 \times 10^{-1} + 5 \times 10^{-2} + 8 \times 10^{-2}$
   (b) $5 \times 10^3 + 81 - 10^0$
6. Express the following numbers in standard form:
   (a) 318.62
   (b) 0.00004702
   (c) 51 292 000 000
   (d) −0.00041045
7. Round off the following numbers correct to 3 significant figures:
   (a) 2.713
   (b) 0.0001267
   (c) 5.435×10^4
8. Evaluate:
   (a) $(81.7251 \times 20.739)2 - 52\,982$
   (b) $\dfrac{(56.739721)^2 \times 0.0997}{(1.9787 \times 10^3)^2}$

   correct to 2 significant figures, expressing your answers in standard form.

(a) You might be able to provide an estimate for this calculation without converting to standard form. For the sake of completeness and to illustrate an important point we will solve this problem using the complete process.

First we convert all numbers to standard form; this gives:

$(3.27 \times 10^0)(1.02 \times 10^1)(1.24 \times 10^{-1})$

Note that $3.27 \times 10^0 = 3.27 \times 1 = 3.27$, in other words it is already in standard form!

Now considering each of the multipliers and rounding to *one significant figure* gives:

$(3 \times 10^0)(1 \times 10^1)(1 \times 10^{-1})$

and remembering your first law of indices!

$(3 \times 1 \times 1)(10^{0+1-1}) = (3)(10^0) = 3(1) = \mathbf{3.0}$

You may feel that this is a terribly long winded way to obtain an estimation because the numbers are so simple, but with more complex calculations, the method is very useful indeed.

(b) Following the same procedures as above gives:

$$\frac{(3.1778256 \times 10^3)(3.14 \times 10^{-4})}{(1.54025 \times 10^5)^2} = \frac{(3.2 \times 10^3)(3.1 \times 10^{-4})}{(1.5 \times 10^5)^2}$$

Now again applying the laws of indices and the distributive law of arithmetic we get

$$\frac{(3.2 \times 3.1)(10^{3-4})}{2.25 \times 10^{5 \times 2}} = \frac{(3.2 \times 3.1)10^{-1}}{2.25 \times 10^{10}}$$

$$= \left(\frac{3.2 \times 3.1}{2.25}\right)10^{-11} = \mathbf{4.4 \times 10^{-11}}$$

Note that if you were unable to work out the multiplication and division in your head, then to one significant figure we would have $3 \times 3/2 = 4.5$, very near our approximation using 2 significant figures. The calculator answer to 10 significant figures is $4.206077518 \times 10^{-11}$. The error in this very small number (compared with our estimation) is something like two in one thousand million! Of course the errors for very large numbers, when squared or raised to greater powers, can be significant!

Before leaving the subject of estimation, there is one important convention which you should know. Consider the number 3.7865. If we require an estimate of this number correct to 4 significant figures, what do we write? In this case the last significant figure is a 5, so should we write this number as 3.786 or 3.787, correct to 4 significant figures? The convention states that we *round up when confronted with the number* 5. So the correct answer in this case would be 3.787.

Now try the Test your knowledge 4.2. Remember that even if you are fully familiar with the material in this section, you should check your answers and revise the subject matter for any you get wrong!

Fractions

Before we look at some examples of algebraic manipulation, using the techniques we have just learnt, we need to devote a little time to the study of fractions. In this section we will consider fractions using only *explicit* numbers. Later, in the main syllabus on algebra, we also consider simple fractions using *literal* numbers, that is algebraic fractions. A study of the work that follows should enable you to manipulate simple fractions, without the use of a calculator!

I am often asked why do we need to use fractions at all? Why not use only decimal fractions? Well one very valid reason is that fractions provide *exact* relationships between numbers. For example, the fraction 1/3 is exact, but the decimal fraction equivalent has to be an approximation to a given number of decimals – 0.3333 is correct to 4 decimal places. Thus $1/3 + 1/3 + 1/3 = 1$ but $0.3333 + 0.3333 + 0.3333 = 0.9999$, not quite 1.

A fraction is a *division* of one number by another. Thus the fraction 2/3 means two divided by three. The fraction x/y means the literal number x divided by y. The number above the line is called the *numerator*; the number below the line is the *denominator*, as you learnt before. Thus fractions are represented as:

$$\frac{\text{numerator}}{\text{denominator}}$$

Fractions written in this form, with integers in the numerator and denominator, are often known as *vulgar fractions*, for example $1/2$, $3^1/4$, $3/4$, etc. Whereas fractions written in decimal form 0.5, 3.25, 0.75, 0.333, etc., are known, as their name implies, as *decimal fractions*.

Having defined the vulgar fraction, let us now look at how we multiply, divide, add and subtract these fractions. We start with multiplication, because unlike arithmetic on ordinary numbers, multiplication of fractions is the easiest operation.

Multiplication of fractions

In order to multiply two or more fractions together all that is necessary is to multiply all the numbers in the numerator together and all the numbers in the denominator together, in order to obtain the desired result.

For example: $\dfrac{1}{3} \times \dfrac{2}{3} \times \dfrac{1}{4} = \dfrac{1 \times 2 \times 1}{3 \times 3 \times 4} = \dfrac{2}{36}$ now we are not quite finished, because the fraction 2/36 has numbers in the numerator and denominator which can be further reduced, without affecting the value of the fraction. I hope you can see that if we divide the *numerator and denominator by 2*, we reduce the fraction to 1/18 without affecting the value. Because we have divided the fraction by $2/2 = 1$ the *whole fraction* is unaltered. You can easily check the validity of the process by dividing 1 by 18 and also 2 by 36 on your calculator; in both cases we get the *recurring decimal fraction* 0·055555. Note that the exact value of this fraction cannot be given in decimal form.

Division of fractions

Suppose we wish to divide 1/3 by 2/3, in other words $\dfrac{1/3}{2/3}$. The trick is *to turn the devisor (the fraction doing the dividing) upside down and multiply.* In the above example we get $1/3 \times 3/2$ and we proceed as for multiplication, that is, $\dfrac{1 \times 3}{3 \times 2} = \dfrac{3}{6} = \dfrac{1}{2}$.

Note that again by cancelling numerator and denominator by 3, we get the lowest vulgar fraction. Now again, if you are not convinced that division can be turned into multiplication by using the above method, check on your calculator, or use decimal fractions, to confirm the result.

Addition of fractions

To add fractions, we are required to use some of our previous knowledge concerning *factors*. In particular we need to determine the *lowest common multiple* (LCM) of two or more numbers. That is the smallest possible number that is a common multiple of two or more numbers. For example, 10 is a multiple of 5, 30 is a common multiple of 5 and 3, but 15 is the *lowest* common multiple of 5 and 3. Thus, 15 is the smallest possible number that is exactly divisible by both 5 and 3. What is the lowest common multiple of 2, 3 and 4? One multiple is found simply as $2 \times 3 \times 4 = 24$, but is this the lowest? Well, of course it is not. The number 4 is exactly divisible by 2, so is the number 24, to give 12, so is this the lowest common multiple? Well, 12 is divisible by the numbers 2, 3 and 4, but the number 6 is not. So 12 *is the lowest common multiple* of the three numbers 2, 3 and 4.

So when adding fractions, why may it be necessary to find the lowest common multiple? We will illustrate the process by example.

> **Key point**
>
> When dividing a fraction by a fraction, turn the bottom fraction upside down (invert) and multiply.

> **Key point**
>
> A common multiple (not necessarily the lowest) can always be found by multiplying all the numbers in the denominator together.

Example 4.9

Add the following fractions:

(a) $\dfrac{1}{3} + \dfrac{1}{4}$

(b) $\dfrac{2}{5} + \dfrac{1}{3} + \dfrac{1}{2}$.

(a)   We first determine the LCM of the numbers in the denominator. In this case the lowest number divisible by both 3 and 4 is 12. So 12 is the LCM.

Now remembering that the whole idea of adding fractions together is to create one fraction as their sum then we place the LCM below the denominators of all the fractions we wish to add. In this case we get:

$$\dfrac{\dfrac{1}{3} + \dfrac{1}{4}}{12}$$

We now divide 3 into 12 to give 4 and then multiply 4 by the number in the numerator of the fraction 1/3, in this case it is a 1, so $4 \times 1 = 4$, which is the result that will now be placed above the 12. In a similar way we now consider the fraction 1/4 to be added, where 4 into 12 is 3 and $3 \times 1 = 3$. Thus we now have the numbers to be added as:

$$\frac{\frac{1}{3} + \frac{1}{4}}{12} = \frac{4 + 3}{12} = \frac{7}{12}$$

Make sure you follow the rather complex logic to obtain the numbers 4 and 3 above the denominator 12, as shown above. Again just to remind you, let us consider the first fraction to be added, 1/3. We take the denominator of this fraction 3 and divide it into our LCM to give the result 4. We then multiply this result (4 in our case) by the numerator of the fraction 1/3, which gives $4 \times 1 = 4$. This process is then repeated on the second fraction to be added, and so on. We then add the numbers in the numerator to give the required result.

(b) We follow the same process as above to add these three fractions together. The LCM is 30, I hope you can see this. Remember, even if you cannot find the LCM, multiplying all the numbers in the denominator together will always produce a *common multiple*, which can always be used in the denominator of the final fraction.
So we get:

$$\frac{2}{5} + \frac{1}{3} + \frac{1}{2} = \frac{12 + 10 + 15}{30} = \frac{37}{30} = 1\frac{7}{30}$$

Again the number 12 was arrived at by dividing 5 into 30 to give 6 and then multiplying this result by the numerator of the first fraction to give $2 \times 6 = 12$. The numbers 1 and 15 were derived in the same way.

The result of adding the numbers in the numerator of the final fraction gives 37/30, this is known as an *improper fraction*, because it contains a whole integer of 1 or more and a fraction. The final result is found simply by dividing the denominator (30) into the numerator (37) to give 1 and a remainder of 7/30.

Subtraction of fractions

In the case of subtraction of fractions we follow the same procedure as with addition, until we obtain the numbers above the common denominator. At which point we subtract them, rather than add them. So, for example, for the fractions given below, we get:

$$\frac{2}{5} + \frac{1}{3} - \frac{1}{2} = \frac{12 + 10 - 15}{30} = \frac{7}{30}$$

Similarly:

$$\frac{3}{8} - \frac{1}{4} + \frac{1}{2} - \frac{1}{8} = \frac{3 - 2 + 4 - 1}{8} = \frac{4}{8} = \frac{1}{2}$$

Notice that for these fractions the LCM is not just the product of the factors, but is truly the lowest number which is divisible by all the numbers in the devisors of these fractions.

Example 4.10

Simplify the following fractions:

(a) $\dfrac{2}{3} + \dfrac{3}{5} - \dfrac{1}{2}$

(b) $\left(\dfrac{3}{4}\right) \times \left(\dfrac{3}{8} + \dfrac{5}{16} - \dfrac{1}{2}\right)$

(c) $2\dfrac{5}{8} \div \dfrac{7}{16} - \dfrac{3}{8}$

(a) Recognizing that the lowest common multiple is 30 enables us to evaluate this fraction using the rules for addition and subtraction of fractions given before, then:

$$\dfrac{2}{3} + \dfrac{3}{5} - \dfrac{1}{2} = \dfrac{20 + 18 - 15}{30} = \dfrac{21}{30} = \dfrac{7}{10} \quad \text{in its simplest form.}$$

(b) In this example we need to simplify the right-hand bracket *before* we multiply. So we get:

$$\left(\dfrac{3}{4}\right) \times \left(\dfrac{6 + 5 - 8}{16}\right) = \left(\dfrac{3}{4}\right) \times \left(\dfrac{3}{16}\right) = \dfrac{9}{64}$$

(c) This example involves a whole number fraction. To apply the rules, the fraction $2\dfrac{5}{8}$ is best put into improper form, that is $\dfrac{21}{8}$.

Note, to obtain this form we simply multiply the denominator by the whole number and add the existing numerator, i.e. $(2 \times 8) + 5 = 21$, to obtain the new numerator. We next need to apply the rules of arithmetic, *in the correct order*, to solve the fraction. This follows on from the number laws you learnt earlier.

The *arithmetic law of precedence* tells us that we must carry out the operations in the following order:

brackets, of, division, multiplication, addition, subtraction
(you may have remembered this order using the acronym BODMAS).

This tells us (for our example) that we must carry out division before subtraction, there is no choice!

So following the process discussed above, we get:

$$\left(\dfrac{21}{8} \div \dfrac{7}{16} - \dfrac{3}{8}\right) = \left(\dfrac{21}{8} \times \dfrac{16}{7} - \dfrac{3}{8}\right) = \left(\dfrac{6}{1} - \dfrac{3}{8}\right)$$

$$= \left(\dfrac{48 - 3}{8}\right) = \dfrac{45}{8} = 5\dfrac{5}{8}$$

Note that the brackets have been included for clarity.

Key point

Always carry out arithmetic operations in the order of BODMAS – brackets, of, division, multiplication, addition, subtraction.

We finish this section on fundamentals with a brief look at algebraic factorization and algebraic products.

Factorization and products

There are many occasions when we are required to determine the factors and products of algebraic expressions. Literal numbers are used in expressions and formulae to provide a precise, technically accurate way of generalizing laws and statements associated with mathematics, science and engineering, as mentioned previously. When manipulating such expressions, we are often required to multiply them together (determine their *product*) or carry out the reverse process, that of *factorization*. You will see in your later studies that these techniques are very useful when it comes to changing the subject of a particular algebraic formula. In other words when you are required to *transpose a formula* in terms of a particular variable.

We begin by considering the products of some algebraic expressions. Once we are familiar with the way in which these expressions are 'built up' we can then look at the rather more difficult inverse process, that of factorization.

Products

Consider the two factors $(1 + a)$ and $(1 + b)$, noting that each factor consists of a *natural number* and a *literal number*. Suppose we are required to find $(1 + a)(1 + b)$, in other words, their product. Providing we follow a set sequence, obeying the laws of multiplication of arithmetic, then the process is really quite simple!

In order to describe the process accurately, I need to remind you of some basic terminology. In the factor $(1 + a)$ the natural number 1 is considered to be a *constant* because it has no other value; on the other hand the literal number a, can be assigned any number of values, therefore it is referred to as a *variable*. Any number or group of numbers, whether natural or literal, separated by a +, – or = sign is referred to as a *term*. So, for example, the expression $(1 + a)$ has *two* terms.

When multiplying $(1 + a)$ by $(1 + b)$ we start the multiplication process from the left and work to the right, in the same manner as reading a book. We multiply each term in the left-hand bracket by each of the terms in the right-hand bracket, as follows:

$(1 + a)(1 + b) = (1 \cdot 1) + (1 \cdot b) + (a \cdot 1) + (a \cdot b) = 1 + b + a + ab$
$= 1 + a + b + ab$

Notes:

1. That the 'dot' notation for multiplication has been used to avoid confusion with the variable x.
2. It does not matter in which order the factors are multiplied. Refer back to the commutative law of arithmetic, if you do not understand this fact.

Example 4.11

Determine the product of the following algebraic factors:

(a) $(a + b)(a - b)$
(b) $(2a - 3)(a - 1)$
(c) $(abc^3d)(a^2bc^{-1})$.

(a) In this example we proceed in the same manner as we did above, i.e.:

$(a + b)(a - b) = (a \cdot a) + (a)(-b) + (b \cdot a) + (b)(-b)$
$= a^2 + (-ab) + (ba) + (-b^2)$

which by the laws of signs $= a^2 - ab + ba - b^2$ and by the commutative law this can be written as $a^2 - ab + ab - b^2$ or $(a + b)(a - b) = \mathbf{a^2 - b^2}$. I hope you have followed this process and recognize the notation for multiplying two bracketed terms.

The product $a^2 - b^2$ is a special case and is known as *the product of two squares*. This enables you to write down the product of any two factors that take the form $(x + y)(x - y)$ as equal to $x^2 - y^2$, where x and y are any two variables.

(b) Again for these factors, we follow the process where we get:

$(2a - 3)(a - 1) = 2a \cdot a + (2a)(-1) + (-3)(a) + (-3)(-1)$

$= 2a^2 - 2a - 3a + 3$ and so, $(2a - 3)(a - 1) = \mathbf{2a^2 - 5a + 3}$

(c) In this case we simply multiply together *like variables*, using the *laws of indices*! So we get:

$(abc^3d)(a^2bc^{-1}) = (a^1 \cdot a^2)(b^1 \cdot b^1)(c^1 \cdot c^{-1})(d^1)$

$= (a^{1+2})(b^{1+1})(c^{3-1})(d^1) = \mathbf{a^3 b^2 c^2 d}$

Note that the brackets in the above solution have been included only for clarity, they are not required for any other purpose.

I hope you are getting the idea of how to multiply factors to produce products. So far we have restricted ourselves to just two factors. Can we adopt the process for three or more factors? Well, if you did not know already, you will be pleased to know that we can!

Example 4.12

Simplify the following:

(a) $(x + y)(x + y)(x - y)$
(b) $(a + b)(a^2 - ab + b^2)$.

(a) This expression may be simplified by multiplying out the brackets and collecting like terms. I hope you recognize the fact that the product of $(x + y)(x - y)$ is $x^2 - y^2$. Then all we need do is multiply this product by the remaining factor, we get:

$(x + y)(x^2 - y^2) = \mathbf{x^3 - xy^2 + x^2y - y^3}$

Note the convention of putting the variables in alphabetical order and the fact that it does not matter in what order we multiply the factors, the result will be the same.

(b) This is a straightforward product where:

$(a + b)(a^2 - ab + b^2) = a^3 - a^2b + ab^2 + a^2b - ab^2 + b^3$

$= \mathbf{a^3 - b^3}$

Notice that there are six terms resulting from the necessary six multiplications. When we collect like terms and add we are left with the product known as the *difference between cubes*.

Factorization

Factorizing is the process of finding two or more factors which, when multiplied together, will result in the given expression. Therefore, factorizing is really the opposite of multiplication or finding the product. It was for this reason that we first considered the simpler process of finding the product.

Thus, for example, $x(y + z) = xy + xz$. This product resulted from the multiplication of the two factors x and $(y + z)$. If we now unpick the product you should be able to see that x is a *common factor* that appears in *both terms* of the product.

What about the expression $x^2 - 16$? I hope you are able to recognize the fact that this expression is an example of the *difference between two squares*. Therefore we can write down the factors immediately as: $(x + 4)$ and $(x - 4)$; look back at Example 4.11(a) if you are unsure. We can check the validity of our factors by multiplying and checking that the product we get is identical to the original expression that we were required to factorize, i.e.:

$$(x + 4)(x - 4) = x^2 - 4x + 4x - 16 = x^2 - 16$$

as required.

Suppose you are asked to factorize the expression $a^2 - 6a + 9$, how do we go about it? Well, a good place to start is with the term involving the highest power of the variable, that is a^2. Remember that *convention* dictates we lay out our expression in descending powers of the unknown, starting with the highest power positioned at the extreme left-hand side of the expression. a can only have factors of itself and 1 or a and a, therefore ignoring the trivial factors, $a^2 = a \cdot a$. At the other end of the expression we have the natural number *9*, this has the trivial factors *1* and *9* or the factors *3* and *3* or *–3* and *–3*. Note the importance of considering the *negative* case, where from the laws of signs $(-3)(-3) = 9$. So now we have several sets of factors we can try, these are:

(a) $(a + 3)(a + 3)$ or
(b) $(a - 3)(a - 3)$ or
(c) $(a + 3)(a - 3)$.

Now we could try multiplying up each set of factors until we obtained the required result, that is determine the factors by *trial and error*. This does become rather tedious when there are a significant number of possibilities. So, before resorting to this method, we need to see if we can eliminate some combinations of factors by applying one or two simple rules.

I hope you can see why the factors $(a + 3)(a - 3)$ can be immediately excluded. These are the factors for the difference between squares, which is not the original expression we needed to factorize.

What about the factors $(a + 3)(a + 3)$? Both factors contain only positive terms, therefore any of their products must also be positive by the laws of signs! In our expression $a^2 - 6a + 9$ there is a *minus sign*, so again this set of factors may be eliminated. This leaves us with the factors $(a - 3)(a - 3)$ and on multiplication we get:

$$(a - 3)(a - 3) = a^2 - 3a - 3a + 9 = a^2 - 6a + 9$$

giving us the correct result.

You may have noticed that we left out the sets of factors $(a - 1)(a - 9)$, $(a - 1)(a + 9)$, $(a + 1)(a - 9)$ and $(a + 1)(a + 9)$ from our original group of possibles! Well, in the case of $(a + 1)(a + 9)$, this would be eliminated using the laws of signs, but what about the rest?

There is one more very useful technique we can employ when

considering just two factors. This technique enables us to check the accuracy of our factors by determining the middle term of the expression we are required to factorize. So in our case for the expression $a^2 - 6a + 9$, then $-6a$ is the middle term.

The middle term is derived from our chosen factors by *multiplying the outer terms, multiplying the inner terms and adding.*

So in the case of the correct factors $(a - 3)(a - 3)$, the outer terms are a and -3, which on multiplication $(a)(-3) = -3a$ and similarly the inner terms $(-3)(a) = -3a$ and so their sum $= -3a + (-3a) = -6a$, as required.

If we try this technique to any of the above factors involving *1* and *9*, we will see that they can be quickly eliminated. For example, $(a - 1)(a - 9)$ has an outer product of $(a)(-9) = -9a$ and an inner product of $(-1)(a) = -a$, which when added $= -9a - a = -10a$, which of course is incorrect.

Example 4.13

Factorize the expressions:

(a) $x^2 + 2x - 8$
(b) $12x^2 - 10x - 12$.

(a) To determine the factors for this expression we follow the same procedure as detailed above.
First, we consider the factors for the outer term x^2 (apart from the trivial factors), we have $x^2 = x \cdot x$ and the factors of -8 are $(2)(4)$ or $(-2)(4)$ or $(4)(-2)$ or $(1)(8)$ or $(-1)(8)$ or $(8)(-1)$. So by considering only outer and inner terms we have the following possible combination of factors:

$(x + 2)(x + 4)$, $(x + 2)(x - 4)$, $(x - 2)(x - 4)$ and

$(x + 1)(x + 8)$, $(x + 1)(x - 8)$, $(x - 1)(x + 8)$

Now we eliminate the sets of factors that have only positive terms (by the law of signs). This leaves $(x + 2)(x - 4)$, $(x - 2)(x + 4)$, $(x + 1)(x - 8)$ and $(x - 1)(x + 8)$. *The last two sets of factors can be eliminated by applying the outer and inner term rule.* If you apply this rule, neither of these sets gives the correct middle term. We are therefore left with the two sets of factors:

$(x + 2)(x - 4)$ or $(x - 2)(x + 4)$

So let's try $(x + 2)(x - 4)$. Applying the outer and inner term rule we get $(x)(-4) = -4x$ and $(2)(x) = 2x$, which on addition gives $-2x$, but we require $+2x$, so these are not the factors. So finally we try $(x - 2)(x + 4)$, where on application of the rule we get $(x)(4) = 4x$ and $(-2)(x) = -2x$, which on addition gives $4x - 2x = 2x$, as required. Where the factors of the expression $x^2 + 2x - 8$ are $(x - 2)(x + 4)$.

(b) For the expression $12x^2 - 10x - 12$, we have the added complication of several possibilities for the term involving the square of the variable x, i.e. $12x^2$. This term could be the product of the factors $(x)(12x)$ or $(2x)(6x)$ or $(3x)(4x)$ and the right-hand term could be the product of the factors $(-1)(12)$ or $(1)(-12)$ or $(-2)(6)$ or $(2)(-6)$ or $(-3)(4)$ or $(3)(-4)$. By the *rule of signs*, no set of factors can have all positive terms, so these can be eliminated from the possible solutions. This leaves us with:

Test your knowledge 4.3

1. Simplify the following fractions

   (a) $\dfrac{3}{16} \times \dfrac{8}{15}$

   (b) $\dfrac{3}{5} \div \dfrac{9}{125}$

   (c) $\dfrac{1}{4}$ of $\dfrac{18}{5}$.

2. Simplify:

   (a) $(a^2 b^3 c)(a^3 b^{-4} c^2 d)$

   (b) $(12x^2 - 2)(2xy^2)$.

3. Simplify the following fractions:

   (a) $\dfrac{2}{9} + \dfrac{15}{9} - \dfrac{2}{3}$

   (b) $3\dfrac{2}{3} - 2\dfrac{1}{5} + 1\dfrac{5}{6}$

   (c) $\dfrac{17}{7} - \dfrac{3}{14} \times 2$.

4. Reduce the following fractions to their lowest terms:

   (a) $\dfrac{21a^3 b^4}{28a^9 b^2}$

   (b) $\dfrac{abc}{d} \div \dfrac{abc}{d^2}$.

5. Determine the product of the following:

   (a) $(3a - 1)(2a + 2)$
   (b) $(2 - x^2)(2 + x^2)$
   (c) $ab(3a - 2b)(a + b)$
   (d) $(s - t)(s^2 + st + t^2)$.

6. Factorize the following expressions:

   (a) $x^2 + 2x - 3$
   (b) $a^2 - 3a - 18$
   (c) $4p^2 + 14p + 12$.

7. Find *all* factors of the expressions:

   (a) $3x^3 + 27x^2 + 42x$
   (b) $9y^2 x^2 + 27x^3 y^3 - 6xy$.

8. Evaluate:

   (a) $a^2 + 0.5a + 0.06$, when $a = -0.3$
   (b) $(x - y)(x^2 + xy + y^2)$, when $x = 0.7$, $y = 0.4$.

---

set 1 $(3x + 1)(x - 12)$, $(3x - 1)(x + 12)$, $(x - 1)(3x + 12)$ or $(x + 1)(3x - 12)$

set 2 $(3x + 2)(x - 6)$, $(3x - 2)(x + 6)$, $(x + 2)(3x - 6)$, $(x - 2)(3x + 6)$

set 3 $(3x + 3)(x - 4)$, $(3x - 3)(x + 4)$, $(x + 3)(3x - 4)$, $(x - 3)(3x + 4)$

The choice of possible solution does seem to be getting complicated! However, if we apply the *multiplication of outer terms, multiplication of inner terms rule* to sets 1 and 3, they are quickly eliminated leaving us with just set 2. Application of the rule, once more, to the factors in set 2 gives us our required solution. Where the factors of the expression $12x^2 - 10x - 12$ are **$(3x + 2)(4x - 6)$**.

Example 4.14

Factorize the expression $3x^3 - 18x^2 + 27x$.

We are now dealing with an unknown variable x raised to the third power! Do not worry, in this particular case the trick is to recognize the common factor. If we first consider the integers that multiply the variable we have:

$$3x^3 - 18x^2 + 27x$$

All these numbers are divisible by 3, therefore *3* is a common factor. Also, in a similar manner, the variable itself has a common factor, since all are divisible by x.

So all we need do is remove these common factors to produce the expression:

$$3x(x^2 - 6x + 9)$$

Note that on multiplication you will obtain the original expression, so that $3x$ and $x^2 - 6x + 9$ must be factors.

This expression now has one factor where the greatest power of the unknown is 2. This factor can itself be broken down into two *linear factors* (i.e. where the unknown is raised to the power 1) using the techniques described before. Where the factors of the expression $3x^3 - 18x + 27x$ are **$(3x)(x - 3)(x - 3)$**.

Finally, before we leave our study of factorization, some common algebraic expressions are tabulated (Table 4.1) in general form, with their factors.

Table 4.1 *Common algebraic expressions with their factors*

Expression	Factors
1. $xy + xz$	$x(y + z)$
2. $x^2 - y^2$	$(x + y)(x - y)$
3. $x^2 + 2xy + y^2$	$(x + y)^2$
4. $x^2 - 2xy + y^2$	$(x - y)^2$
5. $x^3 + y^3$	$(x + y)(x^2 - xy + y^2)$
6. $x^3 - y^3$	$(x - y)(x^2 + xy + y^2)$

For example, recognizing that $z^3 + 8 = z^3 + 2^3$ then the factors of the expression $z^3 + 8$ are, from expression (5) $= (z + 2)(z^2 - 2z + 4)$, where in this case $z = x$ and $y = 2$.

Algebra

Introduction

In this the first of four outcomes that make up the Mathematics for Technicians unit, you will be introduced to some of the basic rules and laws needed to manipulate algebraic expressions, functions and equations.

Mastery of algebra is vitally important, because without the ability to manipulate equations and functions quickly and efficiently, the remainder of your mathematical and scientific topics becomes that much more difficult.

When students tell me that they are 'unable to differentiate' or 'unable to solve a problem involving trigonometry', the chances are that what they really mean is that they are unable to simplify or transpose the equations associated with these subjects, which require *fluency in the use of algebra*. You will therefore not be surprised to discover that we will spend some time studying the underpinning algebra necessary for the satisfactory resolution of mathematical, scientific and engineering problems.

We start with a continuation of the material you studied previously in the section on *fundamentals*. We look again, but in more detail, at the simplification and expansion of more complex algebraic expressions and the techniques needed to transpose formulae. After the introduction of some graphical methods, we then consider the analytical and graphical solution of linear, simultaneous and quadratic equations.

More advanced manipulation of algebra is then studied, requiring a thorough understanding of indices and logarithms. Finally, after introducing the ideas associated with algebraic, logarithmic and exponential functions, we will analyse such functions using theoretical and graphical methods.

Algebraic operations

We have already looked at the addition, subtraction, multiplication and factorization of algebraic expressions during our study of algebraic fundamentals. As a reminder of some of the techniques and laws you have already covered, make sure you are able to work through the following examples. If in any doubt, refer back to the appropriate topic in the section on *fundamentals*.

Example 4.15

Carry out the appropriate arithmetic operations on the following algebraic expressions:

(a) $3ab + 2ac - 3c + 5ab - 2ac - 4ab + 2c - b$
(b) $3x - 2y \times 4z - 2x$
(c) $(3a^2b^2c^2 + 2abc)(2a^{-1}b^{-1}c^{-1})$
(d) $(3x + 2y)(2x - 3y + 6z)$.

(a) All that is required here, is to add or subtract *like* terms, so we get:

$3ab + 5ab - 4ab + 2ac - 2ac - 3c + 2c - b = \mathbf{4ab - b - c}$

(b) Here you need to remember the laws of precedence we studied previously in *fundamentals;* look back if you are unsure. From this law we carry out multiplication before addition or subtraction. So we get:

$3x - 8yz - 2x = \boldsymbol{x - 8yz}$

(c) With this expression, when multiplying up the brackets, we need to remember the law of indices for multiplication. Using this law we get:

$6a^{2-1}b^{2-1}c^{2-1} + 4a^{1-1}b^{1-1}c^{1-1} = 6a^1b^1c^1 + 4a^0b^0c^0$

$= \boldsymbol{6abc + 4}$

(Do not forget the 4! Remember that any number raised to the power zero is 1 and $4 \times 1 \times 1 \times 1 = 4$).

(d) This is just the multiplication of brackets, where we multiply all terms in the right-hand bracket by both terms in the left-hand bracket. So that before any simplification we should end up with $2 \times 3 = 6$ terms.

$(3x + 2y)(2x - 3y + 6z) = 6x^2 - 9xy + 18xz + 4xy - 6y^2 + 12yz$

(6 terms)

and so after simplification which involves only two terms in this case, we get:

$\boldsymbol{6x^2 - 5xy + 18xz - 6y^2 - 12yz}$

Key point

Remember the *laws of precedence* by the acronym BODMAS – brackets, of, division, multiplication, addition, subtraction.

Example 4.16

Factorize the following algebraic expressions:

(a) $-x^2 + x + 6$
(b) $5x^2y^3 - 40z^3x^2$
(c) $x^2 - 4x - 165$
(d) $8x^6 + 27y^3$.

(a) This is a straightforward example of factorizing a *trinomial* (an algebraic expression of 3 terms, with ascending powers of the unknown).

  We simply follow the rules you studied in the fundamentals section. I will go through the procedure once more to remind you.

  We first consider the left-hand term $-x^2$, which obeying the rules of multiplication must have factors $-x$ and x (ignoring trivial factors). Also for the right-hand term we have 2 and 3 or the trivial factors 1 and 6. Again ignoring the trivial factors, we first try 2 and 3. So we have the following sets of factors:

$(-x + 2)(x + 3)$, $(x +2)(-x + 3)$, $(-x -2)(x - 3)$, $(x - 2)(-x - 3)$.

Now remembering the rule for determining the middle term (addition of outer and inner products), then by trial and error, we eliminate all sets of factors, except the correct solution which is:

$\boldsymbol{(x + 2)(-x + 3)}.$

(b) Here the trick is to recognize the common factor(s) and pull them out behind a bracket. In this case I hope you can see that x^2 is common to both terms as is the number 5. Then we can write the factors as:

$5x^2(y^3 - 8z^3)$

Your answer can always be checked by multiplying up the factors. You should, of course, obtain the original expression providing your factors are correct and your subsequent multiplication is also correct!

(c) With this example, the only difficulty is in recognizing possible factors for the rather large number 165. Well, this is where it is useful to know your 15 times table! With trial and error and use of your calculator you should eventually find that apart from the trivial factors, the numbers 15 and 11 are factors of 165. Also recognizing that $15 - 11 = 4$, we know that some combination of these numbers will produce the required result. Then by obeying the rules of signs you should eventually find the correct factors as:

$(x - 15)(x + 11)$

(d) If you have faithfully completed all the exercises in *Test your knowledge 4.3*, you will have meet this example before! The trick is to recognize that the expression $8x^6 + 27y^3$ may be written as $(2x^2)^3 + (3y)^3$, by application of the laws of indices. Then all that is needed is to apply rule 5 for the sum of two cubes (found in fundamentals Table 4.1) to obtain the required solution. Thus,

$8x^6 + 27y^3 = (2x^2)^3 + (3y)^3 = (2x^2 + 3y)(4x^4 - 6x^2y + 9y^2)$

Where, using rule 5, $2x^2$ is equivalent to x and $3y$ is equivalent to y. Make sure you are able to multiply out the factors to obtain the original expression!

In our study of algebraic operations we have not, so far, considered *division* of algebraic expressions. This is in part due to the fact that division is the inverse arithmetic operation of multiplication, so there are ways in which division may be turned into multiplication using the laws of indices. However, there are occasions when division cannot be avoided. It is therefore useful to master the art of division of both natural numbers as well as literal numbers. To aid your understanding of division of algebraic expressions, we first look at the long division of natural numbers.

Algebraic division

When dividing the number 5184 by 12, you would use your calculator to obtain the result, which of course is 432. I would like to take you back to the time when you were asked to carry out *long division* to obtain this answer! My reason for doing so is quite logical, once you remember this technique using natural numbers, it will be easy to adapt this same technique to the division of literal numbers or to *algebraic expressions*.

We set the above division out, as follows:

$12\overline{)5184}$

We reason that 12 will not go into 5, so we consider the next number, that is 5 and 1 or 51. 12 will go into 51 four (4) times, with 3 left over, then we now have:

$$\begin{array}{r} 4 \\ 12\overline{)5184} \\ \underline{48} \\ 3 \end{array}$$

We now bring down the 8, because 12 does not go into 3, and get 38. 12 will go into 38 three (3) times ($3 \times 12 = 36$), so we put the 3 on top, as we did the 4, then we are left with a remainder of 2. We now have:

$$\begin{array}{r} 43 \\ 12\overline{)5184} \\ \underline{48} \\ 38 \\ \underline{36} \\ 2 \end{array}$$

We continue this process by bringing down the final figure 4, since again 12 will not go into the remainder 2. We get 24 and 12 goes into 24 two times, leaving no remainder. We place the 2 on top, as before, to finish the division. So the completed long division looks like this:

$$\begin{array}{r} 432 \\ 12\overline{)5184} \\ \underline{48} \\ 38 \\ \underline{36} \\ 24 \\ \underline{24} \\ 0 \end{array}$$ leaving a remainder of zero

This division is easily checked by carrying out the inverse arithmetic operation, i.e. $(12 \times 432) = 5184$.

I hope this reminded you of the long division process, which I am sure you are familiar with. We are now going to use this process to carry out *long division of algebra*. This is best illustrated by example.

Example 4.17

Given that $a + b$ is a factor of $a^3 + b^3$, find all remaining factors.

We can approach this problem using long division, since the factors of any expression when multiplied together produce that expression. So we can determine the factors using the inverse of multiplication, i.e. *division*. Now we are dividing by two literal numbers a and b, so starting with the unknown a, we see that a divides into a^3. Think of it as 3 into 27, leaving 9 or 3^2, then a into a^3 is a^2. Another approach is simply to apply the laws of indices $a^3/a^1 = a^2$, thus a^1 and a^2 are factors of a^3. This first part of the division is shown below:

$$\begin{array}{r} a^2 \\ a + b\overline{)a^3 + b^3} \\ \underline{a^3 + a^2b} \\ -a^2b + b^3 \end{array}$$ (after subtraction)

Notice that the second row underneath the division is obtained by multiplying the *divisor* (the expression doing the dividing, $a + b$ in our case) by the *quotient* (the result above the division line, a^2 in our case). The remainder is obtained after subtraction of the second row from the original expression.

Next we need to find a quotient which when multiplied by the divisor gives us $-a^2b$ (the first term in the bottom line). I hope you can see that $-ab$ when multiplied by the first term in the divisor a gives us $-a^2b$, then $-ab$ is the next term in our quotient, as shown below.

$$
\begin{array}{r}
a^2 - ab \\
a + b \overline{\smash{\big)}\ a^3 + b^3} \\
\underline{a^3 + a^2b} \\
-a^2b + b^3 \\
\underline{-a^2b \qquad -ab^2} \\
+ab^2 + b^3 \qquad \text{(again after subtraction)}
\end{array}
$$

Finally we need the next term in our quotient to yield $+ab^2$, when multiplied by the first term of our divisor a. Again, I hope you can see that this is b^2. This completes the division, as shown below:

$$
\begin{array}{r}
a^2 - ab + b^2 \\
a + b \overline{\smash{\big)}\quad a^3 + b^3} \\
\underline{a^3 + a^2b} \\
-a^2b + b^3 \\
\underline{-a^2b \qquad -ab^3} \\
+ab^2 + b^3 \\
\underline{+ab^2 + b^3} \\
0 \quad \text{(after subtraction the remainder is zero)}
\end{array}
$$

Then the factors of the expression $a^3 + b^3$ are $(a + b)$ and $(a^2 - ab + b^2)$.

We know that these two expressions are *factors*, because there is no remainder after division and if we multiply them together we obtain the original expression. Look back at Table 4.1 in the section on fundamentals where we listed these factors, which you may wish to commit to memory!

Key point

In long division of algebra, always line up terms *in order of powers* leaving gaps where appropriate, *before* carrying out subtraction.

The above process may at first appear rather complicated but I hope you can see the pattern and symmetry that exists in the process.

Below is another completed long division shown without explanation. Study it carefully and make sure you can identify the pattern and sequence of events that go to make up the process:

$$
\begin{array}{r}
a^2 + b^2 \\
a^2 - b^2 \overline{\smash{\big)}\ a^4 - b^4} \\
\underline{a^4 - a^2b^2} \\
+a^2b^2 - b^4 \\
\underline{a^2b^2 - b^4} \\
0
\end{array}
$$

Key point

The factors of two squares $x^2 - y^2$ are $(x - y)(x + y)$.

You might have been able to write down the factors of $a^4 - b^4$ straight away, recognizing that it is the difference between two squares, where the factors are themselves literal numbers raised to the power 2.

The need for long division of algebra may occur in your future studies, should you be required to deal with *partial fractions*. It is often useful to be able to simplify rather complex algebraic fractions into their simpler components when trying to *differentiate* or *integrate* them. You will meet *calculus arithmetic* later, where you will be asked to carry out *differentiation* and *integration* of simple functions. You will be pleased to know that in this course you are not required to find *partial fractions*!

So far we have concentrated on long division of algebraic expressions, where the division is exact, but what happens if we are left with a *remainder*? Below is shown the division of two expressions which both yield a remainder:

$$x^2 - 1 \overline{) \begin{array}{l} 1 \\ x^2 + 1 \\ \underline{-(x^2 - 1)} \\ 2 \end{array}}$$

therefore,

$$\frac{x^2 + 1}{x^2 - 1} \equiv 1 + \frac{2}{x^2 - 1}$$

where $\equiv$ means '*always equal to*'

Similarly:

$$x^3 - 3 \overline{) \begin{array}{l} 3 \\ 3x^3 - x^2 + 2 \\ \underline{-(3x^3 - 3x)} \\ -x^2 + 3x + 2 \end{array}}$$

therefore,

$$\frac{3x^3 - x^2 + 2}{x^3 - x} \equiv 3 + \left(\frac{-x^2 + 3x + 2}{3x^3 - x^2 + 2} \right)$$

In both cases, the division has converted an improper fraction into a proper fraction. An *improper algebraic fraction* is one in which the highest power in the numerator *is greater than or equal to* ($\geq$) the highest power in the denominator. Just to make sure you can distinguish between these two types of fraction, let's substitute the natural number 2 for the unknown variable x, in the first of the two examples shown above, i.e.:

$$\frac{x^2 + 1}{x^2 - 1} = \frac{(2)^2 + 1}{(2)^2 - 1} = \frac{5}{3} \quad \text{or} \quad 1\frac{2}{3} \text{ so } \frac{5}{3} \text{ is a fraction in } \textit{improper} \text{ form}$$

and $1\frac{2}{3}$ *is a proper* fraction.

Note also that the proper fraction $1\frac{2}{3}$ *is the same as*

$$1 + \frac{2}{3} \text{ or } \frac{3}{3} + \frac{2}{3} = \frac{5}{3}.$$

With your study of mathematical fundamentals and mastery of the above techniques you should now be ready to tackle problems involving the manipulation and transposition of formulae, which we consider next.

Transposition of formulae

As mentioned earlier formulae provide engineers with a method of writing down some rather complex relationships and ideas in a very precise and elegant way. For example, the formula, $v = u + at$ tells us that the final velocity (v) of, say, a car is equal to its initial velocity (u) plus its acceleration (a), multiplied by the time (t) the car is accelerating. If the car is neither accelerating or decelerating (negative acceleration), then $v = u$ because the acceleration $a = 0$ and $0 \times t = 0$, as you already know. I think you are already beginning to realize that to explain the meaning of one simple formula requires rather a lot of words! It is for this reason that formulae are used rather than just words, to convey engineering concepts.

Terminology

Before considering the techniques needed to manipulate or transpose formulae, we first need to define some important terms, we will use our equation of motion, $v = u + at$, for this purpose.

Term, this is defined as any variable or combination of variables separated by a +, – or = sign. You have already met this definition in our study on fundamentals. Therefore in our formula, according to the definition, there are three (3) terms, they are v, u and at.

Variable, these are *represented by literal numbers which may be assigned various values*. In our case, v, u, a and t are *all variables*. We say that v is a *dependent variable* because its value is determined by the values given to the *independent variables u, a and t*.

Subject, the subject of a formula *sits on its own on one side of the equals sign*. Convention suggests that the subject is placed to the left of the equals sign. In our case v is the subject of our formula. However, the position of the subject, whether to the left or to the right of the equals sign, makes no difference to the sense of a formula. So, $v = u + at$ is identical to $u + at = v$, the subject is simply pivoted about the equals sign.

Transposition of simple formulae

In the following examples we simply apply the basic arithmetic operations of addition, subtraction, multiplication and division to rearrange the subject of a formula, in other words to transpose a formula.

Example 4.18

Transpose the following formula to make the letter in brackets the subject.

1. $a + b = c$ (b)
2. $y - c = z$ (c)
3. $x = yz$ (y)
4. $y = \dfrac{a}{b}$ (b).

1. In this formula we are required to make b the subject, therefore b needs to sit on its own on the left-hand side (LHS) of the equals sign. To achieve this we need to remove a term from the LHS. We ask the question, how is a attached to the LHS? It is in fact *added*, so to remove it to the right-hand side (RHS) of the equals sign we apply the *inverse* arithmetic operation, that is we *subtract* it. To maintain the equality in the formula we need in effect to subtract it from both sides, i.e.

   $a - a + b = c - a$ which of course gives $\boldsymbol{b = c - a}$

   You will remember this operation as: *whatever we do to the LHS of a formula or equation, we must do to the other*, or *when we take any term over the equals sign we change its sign.*

2. Applying the procedure we used in our first example to $y - c = z$, then we subtract y from both sides to give $y - y - c = z - y$, which again gives $-c = z - y$. Now unfortunately in this case we are left with $-c$ on the LHS and we require $+c$, or just c as we normally write it when on its own. Remembering from your study of fundamentals that a minus multiplied by a minus gives a plus and that any number multiplied by one is itself, then:

   $(-1)(-c) = (-1)(z) - (y)(-1)$ or, $c = -z + y$ and exchanges the letters on the RHS gives $\boldsymbol{c = y - z}$

   Now all that we have done in this rather long-winded procedure is to multiply every term in the formula by (-1) or as you may remember, we have *changed the sign of every term* in order to eliminate the negative sign from the subject of our formula.

3. Now with the formula $x = yz$ we have just two terms and our subject z is attached to y by *multiplication*. So all we need do is *divide* it out. In other words apply the inverse arithmetic operation, then we get:

   $$\frac{x}{y} = \frac{yz}{y} \quad \text{or} \quad \frac{x}{y} = z$$

   and reversing the formula about the equals sign gives:

   $$\boldsymbol{z = \frac{x}{y}}$$

4. With the formula $y = \dfrac{a}{b}$, then b is attached to a by *division*, so we *multiply* it out to give:

   $$by = \frac{ab}{b} \quad \text{or} \quad by = a$$

   This leaves us with y attached to b by *multiplication* so to

   eliminate y we *divide* it out, then: $\dfrac{by}{y} = \dfrac{a}{y}$ or $\boldsymbol{b = \dfrac{a}{y}}$ as required.

In the above examples I have shown every step in full. We often leave out the intermediate steps, so, for example, if $p = \dfrac{q - m}{r}$ and we wish to make q the subject of the formula, then multiplying both sides by r gives $pr = q - m$ and adding m to both sides we get $pr + m = q$ and reversing the formula $q = pr + m$.

Transposition of formulae with common factors

What about transposing simple formula with *common factors*? You have already learnt to factorize, so now we can put that knowledge to good use.

Key point

When transposing a formula for a variable, you are making that variable the subject of the formula.

Key point

Always change the sign of a term, variable or number when you cross the equals (=) sign.

Key point

When transposing for a variable that appears more than once, always collect the terms containing the variable together, then factorize using a bracket.

Example 4.19

Transpose the following formulae to make c the subject:

1. $a = c + bc$
2. $2c = pq + cs$
3. $x = \dfrac{ab + c}{a + c}$.

1. All we need do here is to take out c as a common factor, then:

$$a = c(1 + b)$$

Now dividing through by the *whole* of the bracketed expression, we get:

$$\frac{a}{1 + b} = c$$

and reversing the formula we get:

$$c = \frac{a}{1 + b}$$

2. Transposition of this formula is essentially the same as in (1), except that we first need to collect all the terms involving the common factor onto one side of the formula so subtracting cs from both sides gives $2c - cs = pq$ and after taking out the common factor we get $c(2 - s) = pq$ and after division by the *whole of the bracketed expression* we get $c = \dfrac{pq}{(2 - s)}$ or

$c = \dfrac{pq}{2 - s}$ since there is no longer any need for the bracket.

3. Now on multiplication of both sides by $a + c$ we get $x(a + c) = ab + c$. Notice that we have placed $a + c$ in brackets. This is very important because x is multiplied by both a and c. When transferring complicated expressions from one side of the formula to the other, a convenient way of doing it is to *place the expression in a bracket*, then move it.

Now we can remove the brackets by multiplying out, having transferred the whole expression, so we get $ax + cx = ab + c$ and collecting the terms containing c onto one side gives $cx - c = ab - ax$ and taking out c as a common factor we get $c(x - 1) = ab - ax$ and again after dividing out the bracketed expression we get $c = \dfrac{ab - ax}{x - 1}$ or $c = \dfrac{a(b - x)}{x - 1}$.

Transposition of formulae involving powers and roots

You may remember from your study of fundamentals that when we write a number, say *25*, in index form we get $5^2 = 25$, where the *5* is the base and the *2* is the *index* or *power*. Look back at the work we did on indices, in particular, on *powers* and the *laws of indices*. We are going to use this knowledge to transpose formulae that involve terms with powers; they may be *positive*, *negative* or *fractional*, for example p^2, p^{-3} and $p^{\frac{1}{2}} = \sqrt{p}$, respectively.

If $x^2 = yz$ and we wish to make x the subject of the formula, all we need do is to take the *square root of both sides*, i.e. $\sqrt{x^2} = \sqrt{yz}$ or $x = \sqrt{yz}$. In index form this is the equivalent to $x^{(2)(\frac{1}{2})} = y^{(1)(\frac{1}{2})} z^{(1)(\frac{1}{2})}$ or $x^1 = y^{\frac{1}{2}} z^{\frac{1}{2}}$. Similarly if we are required to make x the subject of the formula $\sqrt{x} = yz$ then all that we need do is to square both sides, then $(\sqrt{x})^2 = (yz)^2$ or $x = y^2 z^2$. Suppose we wish to make p the subject in the formula $(\sqrt[3]{p})^2 = abc$. Then writing this formula in index form we have $p^{\frac{2}{3}} = a^1 b^1 c^1$ and to get p^1 we need to multiply both sides of the formula by the power $\frac{3}{2}$ so $p^{(\frac{2}{3})(\frac{3}{2})} = (a^1 b^1 c^1)^{\frac{3}{2}}$ or $p = (abc)^{\frac{3}{2}}$ or $p = (\sqrt{abc})^3$. What the above working shows is that if we wish to find the subject of a formula that has itself been raised to a power, we multiply it by its inverse power. It does not matter whether this power is greater than one (>1), or less than one (<1), in other words whether it is a power or a root, respectively.

Example 4.20

1. If $a = b\sqrt{c}$, make c the subject of the formula.

2. If $Z = \sqrt{R^2 + X^2}$, transpose the formula for X.

3. If $a^{\frac{3}{4}} + b^2 = \dfrac{c - d}{f}$, make a the subject of the formula.

1. Our subject c is under the square root sign, so our first operation must be to square both sides and release it!

    Squaring both sides: $a^2 = (b\sqrt{c})^2$ or $a^2 = b^2 (\sqrt{c})^2$
    then $a^2 = b^2 c$

    Dividing through by b^2:

    $$\frac{a^2}{b^2} = c \text{ and reversing } c = \frac{a^2}{b^2} \text{ so that:}$$

    $$c = \left(\frac{a}{b}\right)^2$$

2. Again we need to release X from underneath the square root sign.

Squaring both sides: $Z^2 = R^2 + X^2$

Subtracting R^2 from both sides: $Z^2 - R^2 = X^2$

and reversing $X^2 = Z^2 - R^2$

Then taking the square root of both sides we get:

$$X = \sqrt{Z^2 - R^2}$$

Note that we square root the *whole* of both sides!

3. Isolating the term involving a by subtracting b^2 from both sides we get:

$a^{\frac{3}{4}} = \left[\dfrac{c-d}{f}\right] - b^2$. Now, multiplying *all* of both sides by the inverse power, i.e. by $\left(\dfrac{4}{3}\right)$ we get

$$a^{(\frac{3}{4})(\frac{4}{3})} = \left[\left(\dfrac{c-d}{f}\right) - b^2\right]^{\frac{4}{3}} \text{ and so } a = \left[\left(\dfrac{c-d}{f}\right) - b^2\right]^{\frac{4}{3}}$$

Key point

When carrying out any transposition, remember that the object of the transposition is to isolate the term involving the subject, then obtain the subject by using multiplication or division.

Key point

Multiplying every term by (-1) is the same as changing the sign of every term.

Transposition of complex formulae and equations

So far we have been transposing relatively simple formulae, where the order in which we carried out the operations was reasonably obvious. With more complex formulae and equations, you may have doubts about the order of operations. If you are in any doubt, the following sequence should be followed.

1. Remove root signs, fractions and brackets (in an order which suits the particular problem).
2. Rearrange the formula for the subject, following the arithmetic operations.
3. Collect all terms on one side of the equation that contain the subject.
4. Take out the subject as a common factor, if necessary.
5. Divide through by the coefficient of the subject.
6. Take roots, as necessary.

Note that the *coefficient* is a decimal number multiplying a literal number in a formula. So, for example, in the simple formula $3b = cde$ the number *3* is the *coefficient of b* and on division by *3*, we get

$$b = \frac{cde}{3}$$

The above procedure is best illustrated by example.

Example 4.21

1. Given that $\dfrac{1}{f} = \dfrac{1}{u} + \dfrac{1}{v}$ make v the subject of the formula.

2. If $s = ut + \frac{1}{2}at^2$, transpose the formula for a.

3. If $\dfrac{D}{d} = \sqrt{\dfrac{f+p}{f-p}}$ transpose the formula for f.

1. Following the procedures we need first to clear fractions.

Remember you *cannot* just turn the fractions upside down! Only when there is a *single fraction* on each side of the equals are we allowed to invert them. I hope you can remember how to combine fractions! If you are unsure, look back now, and study the method we adopted for combining two or more algebraic fractions.

Then $\dfrac{1}{f} = \dfrac{1}{u} + \dfrac{1}{v}$ or $\dfrac{1}{f} = \dfrac{v+u}{uv}$ and clearing fractions by multiplying both sides by f and uv, we get $uv = f(v + u)$ and after multiplying out $uv = fv + fu$ and after gathering all terms containing the subject $uv - fv = fu$ then removing the subject as a common factor gives, $v(u - f) = fu$ and after division of both sides by $(u - f)$ we finally get, $v = \dfrac{fu}{u-f}$.

2. Following our procedure, there is really only one fraction which we can eliminate, it is $\frac{1}{2}$, if we multiply every term by the inverse of a $\frac{1}{2}$, that is $\frac{2}{1}$ we get $2s = 2ut + at^2$. Subtracting $2ut$ from both sides gives $2s - 2ut = at^2$ and dividing both sides by t^2, then $\dfrac{2s - 2ut}{t^2} = a$ and reversing the formula and pulling out the common factor gives $a = \dfrac{2(s - ut)}{t^2}$. Alternatively, remembering your laws of indices, we can bring up the t^2 term and write the formula for a as $a = 2t^{-2}(s - ut)$.

3. We again follow the procedure, first clearing roots, then fractions, in the following manner.

Squaring $\left(\dfrac{D}{d}\right)^2 = \dfrac{f+p}{f-p}$ or $\dfrac{D^2}{d^2} = \dfrac{f+p}{f-p}$ and multiplying both sides by the terms in the denominator, or *cross-multiplying* d^2 and $(f - p)$ gives $D^2(f - p) = d^2(f + p)$ and $D^2 f - D^2 p = d^2 f + d^2 p$, so collecting terms on one side containing the subject we get $D^2 f - d^2 f = d^2 p = D^2 p$. After pulling out common factors we have $f(D^2 - d^2) = (d^2 - D^2)p$ and dividing both sides by $(D^2 - d^2)$ yields the result $f = \dfrac{(d^2 - D^2)p}{D^2 - d^2}$.

Evaluation of formulae

We have spent some time transposing formulae and equations. We will now combine transposition with the substitution of numerical values into formulae and equations. In Example 4.21 we transposed the formula $s = ut + \frac{1}{2}at^2$ for a. This equation relates the *distance* s, with the *initial velocity u, acceleration a* and the *time t*. Suppose we want to find a numerical value for the acceleration a. We may first transpose the formula in terms of a, as we did earlier to give $a = \dfrac{2(s - ut)}{t^2}$. Then, if we are told that distance $s = 125$ m, initial velocity $u = 20$ m s^{-1} and time $t = 5$ s, we may substitute these directly into our rearranged formula for a to give

$$a = \frac{[2(125 - (20)(5)]}{5^2} = \frac{(250 - 200)}{25} = \frac{50}{25} = 2 \text{ m s}^{-2}.$$

In this procedure we have found a numerical value for one of the variables a, in other words we have *evaluated* the formula.

In this fairly simple formula, it was not strictly necessary to transpose the formula for a before substituting in the given numerical values. However, we need to be careful with this direct substitution method, especially if the formula is complex. It is easy to assign the incorrect numerical value to a variable or experience difficulties in simplifying numbers and letters for the desired unknown.

In the next example we combine the idea of substitution with that for solving a simple equation, where the power of the unknown is one.

If you are unsure what this means look back at your work on powers and exponents, where you will find numbers written in index form. As a brief reminder 5^2 is the number 5 raised to the power 2; in other words, five squared. If the literal number z is an unknown it is, in index form, z^1 or z raised to the power one. We normally ignore writing the power of a number when raised to the power one, *unless* we are simplifying expression where numbers are given in index form and we need to use the *laws of indices*, which you learnt earlier in the fundamentals section.

Example 4.22

If $a^2x + bc = ax$, find x, given $a = -3$, $b = -4$, $c = -1$.
In this case we will substitute the numerical values before we simplify the formula. Then:

$(-3)^2x + (-4)(-1) = (-3)x$

$9x + 4 = -3x$

$9x + 3x = -4$

$12x = -4$

$x = \dfrac{-4}{12}$ then $x = -\dfrac{1}{3}$

Notice the important use of brackets on the first line; this prevents us from making mistakes with signs!

In the next example, where we use the formula for centripetal force, we will solve for the unknown m, using both direct substitution and by transposing first and then substituting for the values.

Example 4.23

If $F = \dfrac{mV^2}{r}$ find m when $F = 2560$, $V = 20$ and $r = 5$. Then, by direct substitution:

$2560 = \dfrac{m(20)^2}{5}$ so $(2560)(5) = m(400)$

Test your knowledge 4.4

1. If $F = \dfrac{mv^2}{r}$, find F when $m = 40$, $v = 10$ and $r = 0.2$ and write your answer F in standard form.
2. Transpose the formulae $v = \pi r^2 h$ for r.
3. If the value of the resistance to balance a Wheatstone bridge is given by the formula
   $R_1 = \dfrac{R_2 R_3}{R_4}$. Find R_2, if $R_1 = 3$, $R_3 = 8$ and $R_4 = 6$.
4. If $Q = A_2 \sqrt{\dfrac{2gh}{1 - \left(\dfrac{A_2}{A_1}\right)^2}}$
   find Q, when $A_1 = 0.0201$, $A_2 = 0.005$, $g = 9.81$ and $h = 0.554$.
5. Make a the subject of the formula
   $S = \dfrac{n}{2}[2a + (n-1)d]$
6. Transpose the equation
   $\dfrac{x-a}{b} + \dfrac{x-b}{c} = 1$ for x.
7. If $X = \dfrac{1}{2\pi fC}$ calculate the value of C, when $X = 405.72$ and $f = 81.144$.
8. Simplify
   $\dfrac{1}{x-1} - \dfrac{1}{x-1} - \dfrac{3}{2(x^2-1)}$

Key point

A ratio can be presented as a *fraction* or using the *is to* (:) sign.

$400m = 12\,800$

$m = \dfrac{12\,800}{400}$ then $m = 32$

Alternatively, we can transpose the formula for m and then substitute for the given values.

$F = \dfrac{mV^2}{r}$ and $Fr = mV^2$ so $\dfrac{Fr}{V^2} = m$

then $m = \dfrac{Fr}{V^2}$ and $m = \dfrac{(2560)(5)}{(20)^2} = \dfrac{12\,800}{400} = 32$

giving the same result as before.

In our final example on substitution, we use a formula that relates electric charge Q, resistance R, inductance L and capacitance C.

Example 4.24

Find C if $Q = \dfrac{1}{R}\sqrt{\dfrac{L}{C}}$ where $Q = 10$, $R = 40\ \Omega$, $L = 0.1$.

$QR = \sqrt{\dfrac{L}{C}}$ and squaring both sides gives:

$(QR)^2 = \dfrac{L}{C}$ or $Q^2 R^2 = \dfrac{L}{C}$

$C(Q^2 R^2) = L$ then $C = \dfrac{L}{Q^2 R^2}$

Substituting for the given values, we get:

$C = \dfrac{0.1}{10^2 40^2} = 6.25 \times 10^{-7}$ farads.

Ratio and proportion

A *ratio* is a comparison between two similar quantities. We use ratios when determining the scale of things. For example, when reading a map we may say that the scale is 1 in 25 000 or 1 to 25 000. We can express ratios mathematically, either as fractions or in the form *1:25 000* read as *one to twenty-five thousand*.

Apart from maps, we as technicians are more likely to meet the idea of ratio when we need to read technical drawings or produce vector drawings to scale. For example, if we have a force of 100 N and we wish to represent its magnitude by a straight line of a specific length, then we may choose a scale, say, 1 cm = 10 N, so effectively we are using a scale with a ratio of 1:10. When dealing with ratios it is important to deal with the *same* quantities. If we need to work out the ratio between 20 pence and £2.00, then first we must put these quantities into the *same units*, i.e. 20 pence and 200 pence, so the ratio becomes 20:200 and in its simplest terms this is a ratio of 1:10, after division of both quantities by 20.

We may also express ratios as fractions, so in the case of 20 pence to 200 pence, this is 1:10 as before or $\dfrac{1}{10}$ as a fraction.

Example 4.25

Two lengths have a ratio of 13:7. If the second length is 91 metres, what is the first length?

The first length $= \dfrac{13}{7}$ of the second length $= \left(\dfrac{13}{7}\right)91 = 169$ metres.

Suppose now, that we wish to split a long length of electrical cable into *three parts* that are *proportional* to the amount of money contributed to the cost of the cable by three people. Then if the overall length of the cable is 240 metres and the individuals payed £30.00, £40.00 and £50.00, respectively, how much cable do they each receive?

This is a problem that involves *proportional parts*. The amount of money paid by each individual is in the ratio 3:4:5, giving a total of $3 + 4 + 5 = 12$ parts. Then the length of each part is $= \dfrac{240}{12}$ or 20 metres. So each individual receives, respectively, $20 \times 3 = 60$ m, $20 \times 4 = 80$ m, $20 \times 5 = 100$ m. A quick check will show that our calculations are correct, i.e. $60 + 80 + 100 = 240$ m, the total length of the original electrical cable.

Direct proportion

Two quantities are said to vary directly, or be in *direct proportion*, if they increase or decrease at the same rate. So, for example, we know that the fraction $\dfrac{6}{4}$ reduces to $\dfrac{2}{3}$ so we can write the proportion $\dfrac{6}{4} = \dfrac{3}{2}$; we read this as *6 is to 4 as 3 is to 2* or expressed mathematically 6:4::3:2, where the double colon (::) represents the word *as* in the proportion.

Now in this form, the *first* and *fourth* numbers in the proportion 6 and 2 in this case, are called the *extremes* and the *second* and *third* numbers, 4 and 3 in this case, are called the *means*. Now it is also true that from our proportion $\dfrac{6}{4} = \dfrac{3}{2}$ that $6 \times 2 = 4 \times 3$. So that we can say that in any *true* proportion, *the product of the means equals the product of the extremes*.

Example 4.26

A train travels 200 kilometres in 4 hours. How long will it take to complete a journey of 350 kilometres, assuming it travels at the same average velocity?

The key is to recognize the *proportion*. 200 km is proportional to 4 hours as 350 km is proportional to x hours. Then in symbols:

200 : 4 :: 350 : x and using our rule for *means* and *extremes*, we get

$200x = (4)(350)$ or $200x = 1400$ and $x = \dfrac{1400}{200}$ or $x = 7$ hours

---

The rule for the products of the means and extremes is very useful and should be remembered!

We can generalize the above rule, using algebra (literal numbers), then:

$\dfrac{x}{y} = \dfrac{a}{b}$, or $x : y :: a : b$, then $bx = ay$

In general we may also represent a proportion, by use of the proportionality sign, $\propto$. So, for example: $2a \propto 4a$, where $\propto$ is read as '*is proportional to*'.

Inverse proportion

If 30 men working on a production line produce 6000 components in 10 working days, we might reasonably assume that if we double the amount of men, we can produce the components in half the time. Similarly if we employ 20 men it would take longer to produce the same number of components. This situation is an example of *inverse proportion*. So in the above case, the number of men is reduced in the proportion of $\dfrac{20}{30} = \dfrac{2}{3}$ therefore it will take the *inverse proportion* of days to complete the same number of components, i.e. $\left(\dfrac{3}{2}\right)10$ or 15 days.

Example 4.27

Two gear wheels mesh together, as shown in Figure 4.1. One has 60 teeth, the other has 45 teeth. If the larger gear rotates at an angular velocity of 150 rpm, what is the angular velocity of the smaller gear wheel in rpm?

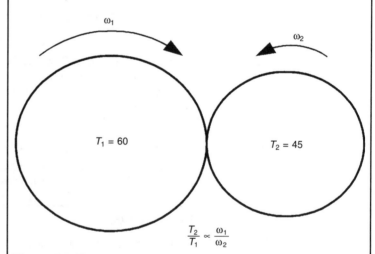

$$\frac{T_2}{T_1} \propto \frac{\omega_1}{\omega_2}$$

Figure 4.1 *Two gear wheels in mesh*

I hope you can see, from Figure 4.1, that the larger gear wheel will make fewer revolutions than the smaller gear wheel in a given time. Therefore, we are dealing with inverse proportion. The *ratio of teeth* of the smaller gear wheel compared to the larger gear wheel is $\dfrac{45}{60} = \dfrac{3}{4}$.

Therefore the ratio of angular velocities must be in the inverse proportion, $\dfrac{4}{3}$.

Then the velocity of the smaller gear wheel is $= \left(\dfrac{4}{3}\right)150 = 200$ rpm.

Constant of proportionality

We can write down the general expression for inverse proportion as $y \propto \dfrac{1}{x}$, where y is said to be inversely proportional to x.

Algebraically, using the proportion sign, *direct proportion* between any two quantities may be represented as $y \propto x$.

Now in order to *equate* the above expressions we need to introduce the *constant of proportionality* k. For example, if $2 \propto 4$, then $2 = 4k$, where $k = \dfrac{1}{2}$. We say that *k is the constant of proportionality*. It allows us to replace the proportionality sign ($\propto$) with the equal sign ($=$). In our simple example above, $k = \dfrac{2}{4}$, after transposition or $k = \dfrac{1}{2}$.

Now if in general $y \propto x$ then $y = kx$ or $\dfrac{y}{x} = k$, where k is the constant of proportionality. Similarly for *inverse proportion*, where $y \propto \dfrac{1}{x}$, then $y = \dfrac{k}{x}$ or $xy = k$.

> **Key point**
>
> When the constant of proportionality k is introduced, the proportion becomes an equality.

Example 4.28

The electrical resistance of a wire varies *inversely* as the *square* of its radius.

1. Write down an algebraic expression for this proportionality.
2. Given that the resistance is 0.05 ohms when the radius of the wire is 3 mm, find the resistance when the wire used has a radius of 4.5 mm.

1. It is not always the case that variables are proportional only to their first powers. In this case, the resistance of the wire varies inversely as the square of the radius. Now if R is the resistance and r the radius, then $R \propto \dfrac{1}{r^2}$ or $R = \dfrac{k}{r^2}$. This is the required algebraic expression.

2. When $R = 0.05$, $r = 3$, then $0.05 = \dfrac{k}{3^2}$ and $k = 0.45$.

   Therefore the final connecting equation is $R = \dfrac{0.45}{r^2}$ and when

   $r = 4.5$, $R = \dfrac{0.45}{4.5^2} = \dfrac{0.45}{20.25} = 0.022$ ohm.

The above example shows a typical engineering use for proportion. In the example that follows we can write down some familiar scientific relationships, using the rules for direct and inverse proportion.

Example 4.29

Write down the formulae to express the following:

1. The volume of a gas at constant temperature is inversely proportional to the pressure.

2. The electrical resistance of a wire varies directly as the length and inversely as the square of the radius.
3. The kinetic energy of a body is jointly proportional to its mass and the square of its velocity when the constant of proportionality $= \frac{1}{2}$.

1. This should be familiar to you as Boyle's law. If we use the symbol V for volume and p for pressure, then $V \propto \frac{1}{p}$ and introducing the constant of proportionality k, give the required relationship as:

$$V = \frac{k}{p} \quad \text{or} \quad pV = k \text{ (a constant)}$$

2. This is the same relationship that you met earlier, except the length l of the conductor is involved. So if we again use R for resistance and r for radius, then:

$$R \propto \frac{l}{r^2}$$

and again introducing the constant of proportionality we get:

$$R = \frac{kl}{r^2}$$

Note that in the above case the resistance R is a function of two variables, the length l and the radius r.

3. The kinetic energy KE, is also dependent on two variables, the mass m and the square of the velocity v^2, both variables being in direct proportion. So we may write down the relationship as KE $\propto mv^2$ and introducing the constant of proportionality, which in this case we are given as $\frac{1}{2}$, then the required relationship is

KE $= \frac{1}{2} mv^2$. You will be studying this relationship in your science.

Test your knowledge 4.5

1. A car travels 205 km on 20 litres of petrol. How much petrol is needed for a journey of 340 km?
2. Four men are required to produce a certain number of components in 30 hours. How many men would be required to produce the same number of components in 6 hours?
3. The cost of electroplating a square sheet of metal varies as the square of its length. The cost to electroplate a sheet of metal with sides of 12 cm is £15.00. How much will it cost to electroplate a square piece of metal with sides of 15 cm.
4. If $y - 3$ is directly proportional to x^2 and $y = 5$ when $x = 2$, find y when $x = 8$.
5. Write down the formula to express the height of a cone when it varies directly as its volume and inversely as the square of the radius.

You will be using the ideas of proportion in the next section where we consider the surface area and volume of regular solids.

Surface area and volume of regular solids

Before considering the surface area and volume of solids we will use some common formulae to find the area of a triangle, circle and parallelogram. The complete solution of triangles using trigonometric ratios and radian measure is left until we deal with these topics in the outcome on trigonometry. The formulae we are going to use are given, without proof, in Table 4.2.

Table 4.2

Shape	Area
Triangle	Half the base multiplied by the perpendicular height, or $A = \frac{1}{2}\,bh$.
Triangle	$A = \sqrt{s(s-a)(s-b)(s-c)}$ where a, b, c are the lengths of the sides and $s = \frac{1}{2}(a+b+c)$.
Parallelogram	A = base multiplied by the perpendicular height between the parallel sides. The base can be any side of the parallelogram.
Circle	$A = \pi r^2$ or $A = \frac{\pi d^2}{4}$ where r = radius and d = diameter of circle.
Trapezium	Half the sum of the parallel sides (a, b) multiplied by the vertical distance (h) between them, or $A = \left(\dfrac{a+b}{2}\right)h$.

Example 4.30

In the triangle *ABC* shown in Figure 4.2, side *AB* = 3 cm and side *AC* = 4 cm. Find the area of the triangle, using *both* of the formulae given in Table 4.2.

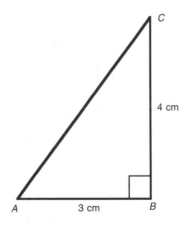

Figure 4.2 *Figure for Example 4.30*

Now we can see from the diagram that this is a *right-angled triangle*, therefore the area *A*, is found simply by using the formula

$A = \dfrac{1}{2}\,bh$, where the base can be taken as either side, containing the

rectangle. Then, $A = \dfrac{1}{2}\,(3)\,(4) = 6$ cm^2. Note that the other side, not used as the base, is at right angles to the base and is therefore the *perpendicular* height. If the triangle was not right angled, we would need to find the perpendicular height of all of the sides in order to find the area.

In our second formula, involving the sides of the triangle, we need to know side *BC*. Since this is a right-angled triangle we can find the third side (opposite the right angle) by using *Pythagoras' theorem (page 386)*. I am sure you are familiar with this theorem, it states that: *the sum of the square on the hypotenuse is equal to the sum of the squares of the other two sides.* In our case we have that $(BC)^2 = 3^2 + 4^2 = 9 + 16 = 25$, or $BC = \sqrt{25} = 5$.

We now have three sides and $s = \dfrac{1}{2}(a + b + c) = \dfrac{1}{2}(3 + 4 + 5) = 6$, therefore the area of the triangle

$$A = \sqrt{s(s-a)(s-b)(s-c)} = \sqrt{6(6-3)(6-4)(6-5)}$$

$$= \sqrt{6(3)(2)(1)} = \sqrt{36} = 6 \text{ cm}^2 \text{ as before.}$$

We will now demonstrate the use of the trapezium formula through another example.

Example 4.31

The cross-section of a metal plate is shown in Figure 4.3. Find its area correct to 4 significant figures (sf).

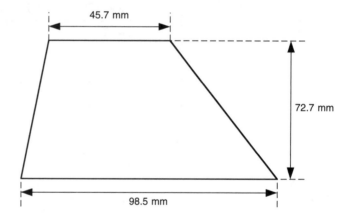

Figure 4.3 *Figure for Example 4.31*

Then using the area rule for a trapezium, where in this case the vertical height is 72.7 mm,

$$A = \left(\dfrac{a+b}{2}\right)h = \left(\dfrac{45.7 + 98.5}{2}\right)72.7 = (72.1)(72.7) = 5241.67$$

$$= 5242 \text{ mm}^2 \text{ (to 2 sf)}$$

The rule for the area of a circle I am sure you are familiar with, but we will use it to find the area of an *annulus*.

Example 4.32

Determine the area of the annulus shown in Figure 4.4, which has an inner radius of 5 cm and an outer radius of 8 cm.

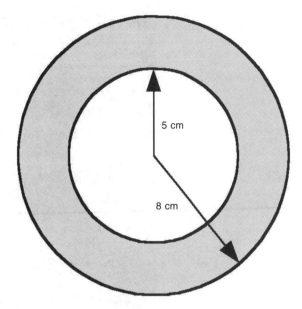

Figure 4.4 *Figure for Example 4.42 – the annulus*

The shaded area (similar to a doughnut in shape) is the area of the *annulus* we require. We know both the inner and outer radii, therefore we can treat this shape as the *difference* between the *outer* and *inner* *circles*. We know that the area of a circle $= \pi r^2$. Now our two circles have two different radii, where $R = 8$ cm and $r = 5$ cm. Then since the area of the *annulus A* is the difference between these two circles we may write:

$$A = \pi R^2 - \pi r^2 \quad \text{or} \quad A = \pi(R^2 - r^2)$$

then substituting the appropriate values of the radii gives,

$$A = \pi(8^2 - 5^2) = \pi(64 - 25) = 39\pi = 122.5 \text{ cm}^2$$

> **Key point**
>
> The circumference of a circle $= 2\pi r = \pi d$.

> **Key point**
>
> The area of a circle $= \pi r^2$
> $= \dfrac{\pi d^2}{4}$.

Note also, with respect to the circle, that its *circumference* $C = 2\pi r$ or $C = \pi d$, where again r = radius and d = diameter.

We are now in a position to tabulate some of the more common formulae that we will need to calculate the surface area and volume of regular solids (see Table 4.3).

Example 4.33

Find the volume and total surface area of a right cylinder, with a top and bottom, if the cylinder has a height of 12 cm and a base radius of 3 cm.

In this example it is simply a question of applying the appropriate formula. Then for the volume $V = \pi r^2 h = \pi(3)^2 \, 12 = 108\pi$ $= 339.29 \text{ cm}^3$.

Now the cylinder has a base and a top, therefore the surface area $S = 2\pi r(h + r)$.

Then, $S = 2\pi 3(12 + 3) = 90\pi = 282.74 \text{ cm}^2$.

We finish this short section on areas and volumes with one more example, leaving you to practise use of these formulae by completing the exercises in *Test your knowledge 4.6*.

Table 4.3

Solid	Volume	Surface area
Right circular cylinder *without* base and top	$V = \pi r^2 h$	$S = 2\pi rh$
Right circular cylinder *with* base and top	$V = \pi r^2 h$	$S = 2\pi rh + 2\pi r^2$ or $S = 2\pi r(h + r)$
Cone *without* base	$V = \frac{1}{3}\pi r^2 h$	$S = \pi rl$ where l = the slant height
Cone *with* base	$V = \frac{1}{3}\pi r^2 h$	$S = \pi rl + \pi r^2$ or $S = \pi r(l + r)$
Sphere	$\frac{4}{3}\pi r^3$	$S = 4\pi r^2$
Hollow pipe of uniform circular cross-section	$V = \pi(R^2 - r^2)l$	$S = 2\pi(R^2 - r^2)$ $+ 2\pi(R + r)$
Spherical shell	$V = \frac{4}{3}\pi(R^3 - r^3)$	$S = 4\pi(R^2 + r^2)$

Notes on table:

1. For the cylinder the height h is the vertical height. There are two formulae for the surface area of a cylinder dependent on whether or not it has a base and top. The area πr^2 is for the addition of the base *or* top, thus $2\pi r^2$ is for both.
2. The formulae for the surface area of the cone also take into consideration the cone with and without circular base. In the volume formula, the height h is again the vertical height from the base. While the surface area formulae use the slant height l.
3. The hollow pipe takes into account the surface area at the ends of the pipe when the cross-section is cut at right angles to its length. The volume is given by the cross-sectional area of the annulus, multiplied by the pipe length.
4. The surface area of the spherical shell includes both the inside and outside surface of the shell.

Test your knowledge 4.6

1. Find the volume of a circular cone of height 6 cm and base radius 5 cm.
2. Find the area of the curved surface of a cone (not including base) whose base radius is 3 cm and whose *vertical* height is 4 cm. (Hint: You need first to find the slant height.)
3. If the area of a circle is 78.54 mm^2, find its diameter to 2 significant figures.
4. A cylinder of base radius 5 cm has a volume of 1 litre (1000 cm^3). Find its height.
5. A pipe of thickness 5 mm has an external diameter of 120 mm. Find the volume of 2.4 m of pipe.

Example 4.34

Water flows through a circular pipe of internal radius 10 cm at 5 m s^{-1}. If the pipe is always three-quarters full, find the volume of water discharged in 30 minutes.

This problems requires us to find the volume of water in the pipe per unit time, in other words the volume of water in the pipe per second. Note that no length has been given.

The area of the circular cross-section $= \pi r^2 = \pi(10)^2 = 100\pi$, therefore the area of the cross-section of water

$$= \left(\frac{3}{4}\right)100\pi = 75\pi \text{ cm}^3 = (75\pi)10^{-4} \text{ m}^3.$$

Now since water flows at 5 m s^{-1}, the volume of water discharged

per second $= \dfrac{(5)(75\pi)10^{-4}}{1} = (375\pi)10^{-4}$ m^3 s^{-1}. The number of m^3 discharged in 30 minutes $= (30)(60)(375\pi)(10^{-4}) = 67.5\pi = 212$ m^3.

Solution of equations

In the following section we will be looking at the analytical and graphical solution of linear, quadratic and simultaneous equations. These equations appear a great deal in engineering problems and require a certain amount of algebraic rigour for their solution, as well as an understanding of graphical methods.

Solution of simple equations

Although you may not have realized, you have already solved simple equations analytically. However, before we start our study of the *graphical solution* of equations, here is an example which shows that in order to solve simple equations analytically, all we need do is apply the techniques you have learnt when transposing and manipulating formula. The important point about equations is that the *equality* sign must always be *present*!

Example 4.35

Solve the following equations:

1. $3x - 4 = 6 - 2x$
2. $8 + 4(x - 1) - 5(x - 3) = 2(5 - 2x)$

3. $\dfrac{1}{2x + 3} + \dfrac{1}{4x + 3} = 0$

1. For this equation, all we need do is to collect all terms involving the unknown x on to the left-hand side of the equation. Simply by using our rules for transposition of formula. Then:

    $3x + 2x - 4 = 6$ so $3x + 2x = 6 + 4$ or $5x = 10$ and so $x = 2$

2. In this equation we need first to multiply out the brackets, then collect all terms involving the unknown x onto one side of the equation and the numbers onto the other side, then divide out to obtain the solution.

    $8 + 4(x - 1) - 5(x - 3) = 2(5 - 2x)$

    $8 + 4x - 4 - 5x + 3 = 10 - 4x$

    So:

    $4x - 5x + 4x = 10 - 3 - 8$

    $-5x = -1$ and on division by -5

    $$x = \frac{-1}{-5} \text{ or } x = \frac{1}{5}$$

    Note the care taken with the signs! Also remember from your earlier work that a minus number divided by a minus number leaves us with a plus number. Alternatively, multiply top and bottom of the fraction $\dfrac{-1}{-5}$ by (-1), then from $(-)(-) = (+)$ we get

    $\dfrac{1}{5}$, as required.

3. To solve this equation we need to manipulate fractions, or apply the inverse arithmetic operation to every term! The simplification to obtain *x* using the rules for transposition is laid out *in full* below.

$$\frac{1}{2x+3} + \frac{1}{4x+3} = 0$$

$$\frac{1(2x+3)}{2x+3} + \frac{1(2x+3)}{4x+3} = 0(2x+3)$$

$$1 + \frac{2x+3}{4x+3} = 0$$

and

$$1(4x+3) + \frac{(2x+3)(4x+3)}{4x+3} = 0(4x+3)$$

$$(4x+3) + (2x+3) = 0$$

or

$$4x+3+2x+3 = 0$$

$$6x = -6$$

$$x = -1$$

We could have carried out the multiplication by the terms in the denominator in just one operation simply by multiplying every term by the product $(2x+3)(4x+3)$. Notice also that when multiplying any term by zero, the product is always zero.

Key point

For all linear equations the highest power of the unknown is 1 (one).

Graphical axes, scales and coordinates

To plot a graph, you know that we take two lines at right angles to each other (Figure 4.5). These lines being the *axes of reference*, where their intersection at the point zero is called the origin. When plotting a graph a suitable scale must be chosen, this scale need not be the same for both axes. In order to plot points on a graph, they

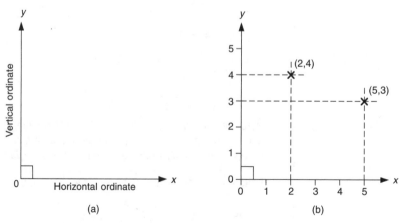

Figure 4.5 *Axes and coordinates of graphs*

are identified by their coordinates. The points (2, 4) and (5, 3) are shown in Figure 4.5(b). Note that the *x-ordinate* or *independent variable* is always quoted first. Also remember that when we use the expression, *plot s against t*, then all the values of the *dependent* variable *s* are plotted up the *vertical axis* and the other *independent variable* (in this case *t*) is plotted along the *horizontal* axis.

You met the concept of dependent and independent variables during your study on fundamentals. Just remember that the values of the dependent variable are determined by the values assigned to the independent variable. So, for example, in the simple equation $y = 3x + 2$, if $x = 2$ then $y = 8$ and if $x = -2$ then $y = -4$ and so on. So to plot a graph all we need do is:

1. Draw the two axis of reference at right angles to each other.
2. Select a suitable scale for the dependent and independent variable, or both.
3. Ensure that values of the dependent variable are plotted up the vertical axis.
4. Produce a table of values, as necessary, to aid your plot.

If the graph is either a straight line, or a smooth curve, then it is possible to use the graph to determine other values of the variables, apart from those given.

Key point

When plotting any variable *y* against *x*, the variable *y* is plotted on the vertical axis.

Example 4.36

Plot the graph of *y* against *x*, given the following coordinates:

x (m)	0	1	2	3	4	5	6	7	8	9	10
y (m)	2	5	8	11	14	17	20	23	26	29	32

And find the corresponding value of *y* when $x = 5.5$ and the value *x* when $y = 38$.

The graph is plotted in Figure 4.6; note that when we join the coordinate points, we get a straight line. The *x-axis* scale is *1 cm = 1 m* and the *y-axis* scale is *1 cm = 2 m*.

To find the value of *y* corresponding to $x = 5.5$, we find 5.5 on the horizontal axis and draw a vertical line up until it meets the graph at point P, then draw a horizontal line until it meets the vertical *y* ordinate and read out the value which is **18.5**.

Should we wish to find a value of *x* given *y*, we reverse this procedure. So to find the value of *x* corresponding to $y = 38$, we first find 38 on the *y*-axis and draw a horizontal line across to meet the line. However, in this case the line does not extend this far, using the tabulated values. It is therefore necessary to *extend or extrapolate* the line. In this particular case it is possible to do this, as shown; where reading vertically down we see that the intercept is at **x = 12**. This process involved extending the graph without data being available to verify the accuracy of our extended line. Great care must be taken when using this process to prevent excessive errors. In the case of a straight line graph, or linear graph, this is acceptable practice. This process is commonly known as graphical *extrapolation*.

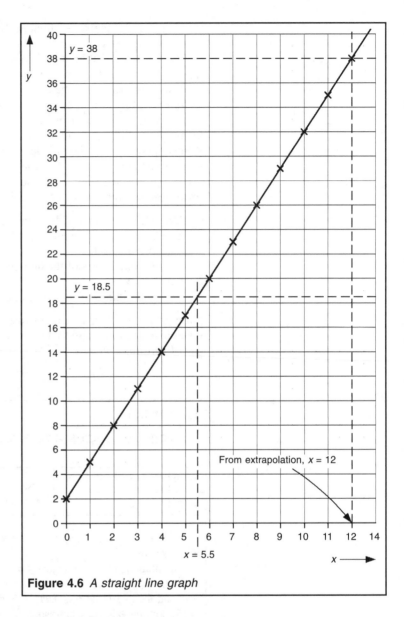

Figure 4.6 *A straight line graph*

Graphs of linear equations

In the above example all values of the coordinates are positive. This is not always the case and to accommodate negative numbers, we need to extend the axes to form a cross (Figure 4.7), where both positive and negative values can be plotted, on both axes.

Figure 4.7 not only shows the positive and negative axes, but also the plot of the equation $y = 2x - 4$. To determine the corresponding y ordinates shown, for values of x between -2 and 3, we use a table:

x	-2	-1	0	1	2	3
$2x$	-4	-2	0	2	4	6
-4	-4	-4	-4	-4	-4	-4
$y = 2x-4$	-8	-6	-4	-2	0	2

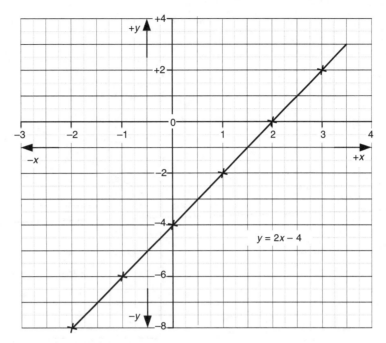

Figure 4.7 *Graph of y = 2x − 4*

So, for example, when $x = -2$, $y = 2(-2) -4 = -4-4 = -8$.

The scale used on the y-axis is 1 cm = 1 unit and on the x-axis 2 cm = 1 unit.

This equation, where the highest power of the variable x, y is 1.0, is known as an *equation of the first degree* or a *linear equation*. All *linear equations produce graphs that are always straight lines*.

Now every linear equation may be written in *standard form*, that is:

$y = mx + c$

So for our equation $y = 2x - 4$ which is in the standard form, $m = 2$ and $c = -4$. Also, every linear equation may be rearranged so that it is in standard form. For example:

$4y + 2 = 2x -6$ then rearranging for y

$4y = 2x - 6 - 2$ or $4y = 2x - 8$ and on divison by 4

$y = \dfrac{2}{4} x - \dfrac{8}{4}$ or $y = \dfrac{1}{2} x - 2$ where $m = \dfrac{1}{2}$ and $c = -2$

Determining m and c for the equation of a straight line

In Figure 4.8, point A is where the straight line cuts the y-axis and has coordinates $x = 0$ and $y = c$. Thus c in the equation $y = mx + c$ is the point where the line meets the y-axis, when the value of $x = 0$ or *the variable c = the y intercept when x = 0*.

Also from Figure 4.8, the value $\dfrac{BC}{AC}$ is called the *gradient* of the line. Now the length

$$BC = \left(\dfrac{BC}{AC} \right) AC = AC \times \text{gradient of the line}$$

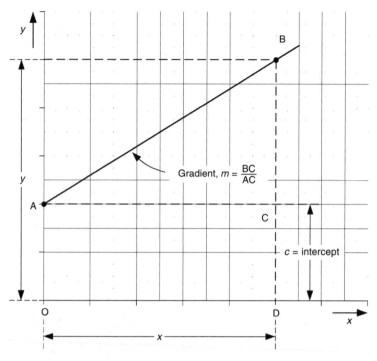

Figure 4.8 *Graph showing relationship for variables c and m*

$y = BC + CD = BC + AO$

$\quad = AC \times$ the gradient of the line $+ AO$

$\quad = x$ multiplied by the gradient of the line $+ c$

But $y = mx + c$. So it can be seen that $m =$ the gradient of the line and $c =$ the intercept on the y-axis.

Example 4.37

1. Find the law of the straight line illustrated in Figure 4.9.
2. If a straight line graph passes through the point (−1, 3) and has a gradient of 4, find the values of m and c and then write down the equation of the line.

1. Since the intercept c is at the origin, it can be read off the graph as −4. The value of m, the gradient of the line, is found by taking convenient values of x and y, then the gradient m from the graph

   $$= \frac{NP}{QP} = \frac{10\text{ cm}}{5\text{ cm}} = 5. \text{ So the equation of the line } y = mx + c \text{ is}$$

   $y = 5x + 4$.

2. We are given the gradient $m = 4$, therefore $y = 4x + c$ and this line passes through the point (−1, 3). So we know that $y = 3$ when $x = −1$ and substituting these values into the equation of the straight line gives $3 = 4\,(−1) + c$ and so $c = 7$. Then the equation of this line is $y = 4x + 7$.

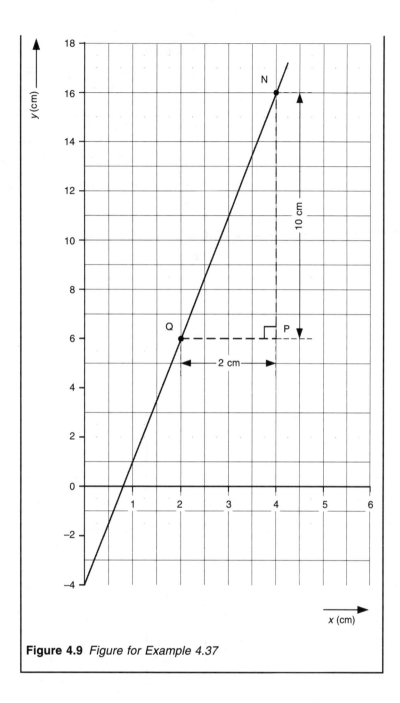

Figure 4.9 *Figure for Example 4.37*

Note that in the questions given in Example 4.37 the gradient (or slope) of the straight lines were both *positive*. Straight line graphs can also have a *negative gradient*, this will occur when the graph of the line slopes downwards to the right of the y-axis. Under these circumstances a negative value of y results, so that $m = \dfrac{-y}{x}$ and so the gradient m is negative.

We leave our study of linear equations and their straight line graphs with an example of the application of the law of a straight line, $y = mx + c$, to experimental data.

Example 4.38

During an experiment to verify Ohm's law, the following experimental results were obtained.

E (volts)	0	1.1	2.3	3.4	4.5	5.65	6.8	7.9	9.1
I (amperes)	0	0.25	0.5	0.75	1.0	1.25	1.5	1.75	2.0

Plot voltage against current and so determine the equation connecting E and I.

The resulting plot is shown in Figure 4.10.

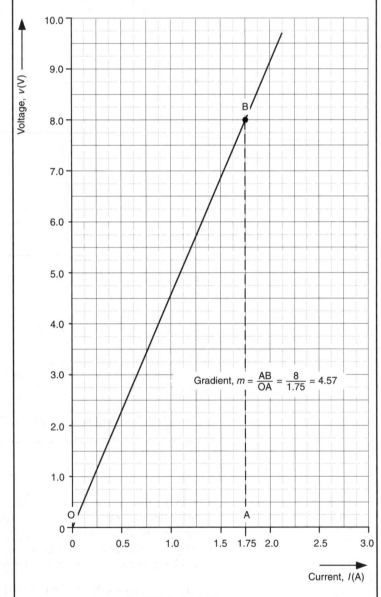

$$\text{Gradient, } m = \frac{AB}{OA} = \frac{8}{1.75} = 4.57$$

Figure 4.10 *Figure for Example 4.38*

From the plot, it can be seen that the experimental data produces a straight line. Therefore the equation connecting E and I is of the form $y = mx + c$. Since the graph goes *directly through the origin*, then the variable $c = 0$. Also from the graph taking suitable values of E and I, the gradient $m = 4.57$ correct to 3 significant figures. So the equation connecting E and I is $E = 4.57I$.

Quadratic equations

A quadratic equation is one in which the unknown variable is raised to the second power. For example, the equation $x^2 = 4$ is perhaps one of the simplest of quadratic equations. We can solve this equation by taking the square root of both sides, something which you are familiar with when transposing a formula. Then:

$$\sqrt{x^2} = \sqrt{4} \quad \text{or} \quad x = \pm 2.$$

Note that even for this simple equation there are two possible solutions, either $x = +2$ or $x = -2$, remembering your laws of signs! When we square a positive number we get a positive number $(+2)(+2) = +4$ or simply 4, also $(-2)(-2) = 4$, from the laws of signs.

In general a quadratic equation is of the type $ax^2 + bx + c = 0$, where the constants a, b, c can take *any* numerical value, positive or negative, decimal or fraction. Like linear equations, quadratic equations do not always appear in *standard form*, that is they are not always arranged in exactly the same order as their qualifying equation, $ax^2 + bx + c = 0$.

How is our simple equation $x^2 = 4$ related to its qualifying equation? Well, the coefficient of x^2 that is the number multiplying the x^2 term $a = 1$. What about the constant b? Well, there is no x term in our equation so $b = 0$. What about the constant c? Our equation is not in standard form, because the equation should be equated to zero. Then in standard form our equation becomes $x^2 - 4 = 0$, by simple transposition! So now we know that for our equation the constant term $c = -4$. A quadratic equation may contain only the square of the unknown variable, as in our simple equation, or it may contain the square and the first power of the variable, for example $x^2 - 2x + 1 = 0$. Also, the unknown variable may have *up to two* possible *real* solutions. The equations we deal with in this course will always have at least one real solution.

There are several ways in which quadratic equations may be *solved*, that is finding the values of the *unknown* variable. We shall concentrate on just three methods of solution: factorization, using the formula and solving by graphical methods.

Solution of quadratic equations by factorization

Take the equation $x^2 - 2x + 1 = 0$. If we ignore for the moment the fact that this is an equation and concentrate on the expression $x^2 - 2x + 1$, then you may remember how to find the *factors* of this expression! Look back now at your work on factors (page 337) to remind yourself.

I hope you were able to identify the factors of this expression as $(x - 1)(x - 1)$. Now all we need do is equate these factors to *zero* to solve our equation. Thus:

Key point

For all quadratic equations, the highest power of the independent variable is 2 (two).

$(x - 1)(x - 1) = 0$

then for the equation to *balance* either the first bracket $(x - 1) = 0$ or the second bracket (the same in this case) $(x - 1) = 0$. Thus solving this very simple linear equation gives $x = 1$, no matter which bracket is chosen. So, in this case, our equation only has *one solution* $x = 1$. Note that if any one of the bracketed expressions $(x - 1) = 0$, then the other bracket is multiplied by zero, i.e. $0(x - 1) = 0$. This is obviously true, because any quantity multiplied by zero is itself zero.

Example 4.39

Solve the equation $3x^2 - 5 = -2x - 4$ by factorization.

The first thing to note before we attempt a solution is that this equation *is not* in standard form. All we need do is transpose the equation to get it into standard form. You should by now be able to do the transposition with ease, so make sure you obtain:

$3x^2 + 2x - 1 = 0$

Now using the techniques for factorization that you learnt earlier, after trial and error you should find that:

$(3x - 1)(x + 1) = 0$ then either $3x - 1 = 0$ giving $x = \dfrac{1}{3}$

or $(x + 1) = 0$ giving $x = -1$

Note that in this case the equation has two different solutions, both can be checked for accuracy by substituting them into the *original* equation.

Then either: $3\left(\dfrac{1}{3}\right)^2 - 5 = -2\left(\dfrac{1}{3}\right) - 4$ or $\dfrac{3}{9} - 5 = -\dfrac{2}{3} - 4$, therefore

$-4\dfrac{2}{3} = -4\dfrac{2}{3}$ which is correct or: $3(-1)^2 - 5 = -4 - 2(-1)$ or $3 - 5 =$

$-4 + 2$, therefore $-2 = -2$, which is also correct. Note the need to manipulate fractions and be aware of the laws of signs, skills I hope you have acquired at this stage in your learning.

Solution of quadratic equations using formula

It is not always possible to solve quadratic equations by factorization. When we cannot factorize a quadratic expression, we may resort to use of the standard formula. Now we know that the standard form of the quadratic equation is:

$ax^2 + bx + c = 0$

and it can be shown that the solution of this equation is:

$$x = \frac{-b \pm \sqrt{b^2 - 4ac}}{2a}$$

Now this equation may look complicated but it is relatively simple to use. The coefficients a, b, c are the same coefficients as in the standard form of the quadratic. So in finding a solution for the variable x, all we need do is substitute the coefficients into the above formulae for the quadratic equation we are considering. All you need to remember is that *before* using the above formula, *always put the equation to be solved into standard form.* Also note that in the above formula, the whole of the numerator, including the −*b*, is divided by 2*a*.

Example 4.40

Solve the equation $5x(x + 1) - 2x(2x - 1) = 20$.

The above equation is not in standard form, in fact until we simplify it, we may not be aware that it is a quadratic equation. So simplifying, by multiplying out the brackets and collecting like terms, gives:

$$5x^2 + 5x - 4x^2 + 2x = 20 \quad \text{and so } x^2 + 7x - 20 = 0$$

This equation is now in standard form and may be solved using the formula. You may have attempted to try a solution by factorization first. If you cannot find the factors reasonably quickly, then you can always resort to the formula, unless told otherwise!

Then from, $x = \dfrac{-b \pm \sqrt{b^2 - 4ac}}{2a}$ we get

$$x = \frac{-7 \pm \sqrt{7^2 - (4)(1)(-20)}}{2(1)} \quad \text{and simplifying gives}$$

$$x = \frac{-7 \pm \sqrt{129}}{2} \quad \text{or } x = \frac{-7 \pm 11.358}{2} \quad \text{and so,}$$

$$x = \frac{-7 + 11.358}{2} \quad \text{or } x = \frac{-7 - 11.358}{2}$$

Giving, the values of the unknown x, correct to 3 significant figures as:

$$x = 2.18 \quad \text{or } x = -9.18.$$

We now consider our final method of solution of quadratic equations, using a graphical method.

Solution of quadratic equations using a graphical method

If we plot a quadratic function of the form $ax^2 + bx + c$ against x, the resulting curve is known as a *parabola*, and depending on the sign of the coefficient a, will determine which way up the curve sits (Figure 4.11).

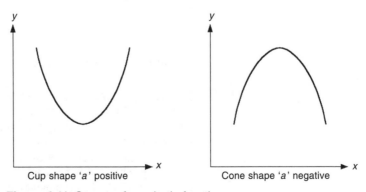

Figure 4.11 *Curves of quadratic functions*

The plotting of such curves requires a table of values to be set up in terms of the values of the independent and dependent variables. This procedure is best illustrated by example.

Example 4.41

Draw the graph of $y = x^2 - 3x + 2$, taking values of the independent variable x between 0 and 4.

x	0	1	2	3	4
x^2	0	1	4	9	16
$-3x$	0	-3	-6	-9	-12
2	2	2	2	2	2
y	2	0	0	2	6

Now from the table of values we can see that when $x = 1$ and $x = 2$, then $y = 0$. Under these circumstances it is advisable to consider a value of x between 1 and 2. Logically a value of 1.5 seems appropriate.

So, from the equation when $x = 1.5$, $y = 2.25 - 4.5 + 2 = -0.25$. The resulting plot is shown in Figure 4.12.

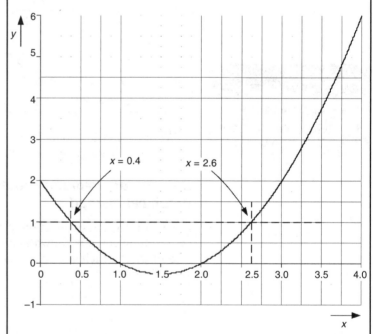

Figure 4.12 *Graph of the function $y = x^2 - 3x + 2$*

Now the points on the curve where it crosses the x-axis are $x = 1$ and $x = 2$. These are the points on the curve for which $y = 0$ or $x^2 - 3x + 2 = 0$. Therefore $x = 1$ and $x = 2$ are the solutions of the quadratic equation $x^2 - 3x + 2 = 0$.

Now from our graph, we can also solve any equation of the type $x^2 - 3x = k$, where k is a constant. If, for example, we wish to solve $x^2 - 3x + 1 = 0$, then comparing this equation with the equation of

Key point

The graphs of quadratic expressions and equations will always be parabolic in shape.

the plot, all we need do is add 1 to both sides to acquire the equation $y = x^2 - 3x + 2 = 1$. So that to solve this equation, we need the points on the curve, where $y = 1$. We then draw the line $y = 1$ and read off the corresponding values of x at these points. From the dashed line on the graph we obtain the solution of this modified equation as $x = 2.6$ or $x = 0.4$.

We finish our study of equations by considering simultaneous equations.

Simultaneous equations

Simultaneous equations involve more than one variable or unknown. We can solve a simple linear equation with one unknown using the laws of algebra you have already learnt. It is often required to represent an engineering problem that involves more than one unknown. For example, if an engineering problem involves the solution of an equation such as $3x = 2y = 12$, how do we go about solving it? Well, the answer is that a single equation with two unknowns is unsolvable, unless we know the value of one of the variables. However, if we have *two equations*, with *two unknowns*, it is possible to solve these equations *simultaneously*, that is at the same time. Three linear equations, with three variables, can also be solved simultaneously, in fact any number of linear equations with a corresponding number of unknowns (variables) can be solved simultaneously. However, when the number of variables is greater than three, it is better to solve the system of equations using a computer!

These *systems of equations* occur in many aspects of engineering; for example, when we model static and dynamic behaviour of solids and liquids. You will be pleased to know, that we will only be considering *two equations* simultaneously, involving *two unknowns*! Even so, the distribution of currents and voltages, for example in electrical networks, sometimes involves the solution of such equations with just two unknowns.

Analytical solution of simultaneous equations

Consider the pair of equations:

$$3x + 2y = 12 \tag{1}$$

$$4x - 3y = -1 \tag{2}$$

Now to solve these equations, all we need to do is use *elimination* and *substitution* techniques, working on both equations simultaneously.

Let us try to eliminate the variable x from both equations. This can be achieved by multiplying each equation by a constant. When we do this, we do not alter the nature of the equations. If we multiply equation (1) by the constant 4, and equation (2) by the constant 3, we get:

$$12x + 8y = 48$$

$$12x - 9y = -3$$

Test your knowledge 4.7

1. The values in the table below show how instantaneous current i varies with voltage v. Plot a graph of v against i and so find the value of v when i = 3.0.

v	15	25	35	50	70
i	1.1	2.0	2.5	3.2	3.9

2. Solve the following linear equations:
   (a) $5x - 1 = 4$
   (b) $3(x - 2) = 2(x - 1)$
   (c) $\dfrac{1}{p} + \dfrac{1}{p+1} = \dfrac{2}{p-1}$.

3. Solve the following simultaneous equations:

   (a) $\begin{aligned} 2x + 3y &= 8 \\ 2x - 3y &= 2 \end{aligned}$

   (b) $\begin{aligned} 5x + 4y &= 22 \\ 3x + 5y &= 21 \end{aligned}$

   (c) $\begin{aligned} \dfrac{a+b}{a-b} &= \dfrac{1}{2} \\ \dfrac{a+1}{b+1} &= 2 \end{aligned}$

Note that we have multiplied *every term* in the equations by the constant! Now, how does this help us to eliminate x? Well, if we now add both equations together we end up with the first term being $24x$, this is not very helpful. However, if we subtract equation (2) from equation (1), we get:

$$12x + 8y = 48$$
$$-(12x - 9y = -3)$$
$$0 + 17y = 51$$

From which we see that $y = 3$. Now having found one of the unknown variables, we can substitute its value into *either one of the original equations* in order to find the other unknown. Choosing equation (1), then from $3x + 2y = 12$ we get $3x + (2)(3) = 12$ or $3x = 6$ and therefore $x = 2$. So the required solution is $y = 3$ and $x = 2$.

When solving any equation, the solutions can always be checked by substituting their values into the original equation, so substituting the values into equation (2) gives $4(2) - (3)(3) = -1$, which is correct.

Graphical solution of two simultaneous equations

The method of solution is shown in the next example. For each of the linear equations, we plot their straight line graphs and where the plots intersect is the unique solution for both equations.

Example 4.42

Solve the following simultaneous equations graphically:

$$\frac{x}{2} + \frac{y}{3} = \frac{13}{6}; \quad \frac{2x}{7} - \frac{x}{4} = \frac{5}{14}$$

Now we first need to simplify these equations and rearrange them in terms of the independent variable y. I hope you can remember how to simplify fractions! Make sure that you are able to rearrange the equations and obtain:

$$2y = 13 - 3x$$

$$-7y = 10 - 8x$$

Now transposing in terms of y, we get:

$$y = \frac{13}{2} - \frac{3}{2}x$$

$$y = -\frac{10}{7} + \frac{8}{7}x$$

Now we can find the corresponding values of y for our chosen values of x. Using just four values of x, say 0, 1, 2 and 3, will enable us to plot the straight lines. Then:

x	0	1	2	3
$y = \dfrac{13}{2} - \dfrac{3}{2}x$	$\dfrac{13}{2}$	5	$\dfrac{7}{2}$	2
$y = -\dfrac{10}{7} + \dfrac{8}{7}x$	$-\dfrac{10}{7}$	$-\dfrac{2}{7}$	$\dfrac{6}{7}$	2

(d) $\dfrac{p}{2} + \dfrac{q}{3} = 2$

$2p + 3y = 13$

4. If $y = ax + b$, find the value of y when $x = 4$, given that $y = 4$ when $x = 1$ and that $y = 7$ when $x = 2$.

5. Solve graphically the following simultaneous equations:

(a) $\begin{aligned} 7x - 4y &= 37 \\ 6x + 3y &= 51 \end{aligned}$

(b) $\begin{aligned} \dfrac{x+1}{y+1} &= 2 \\ \dfrac{2x+1}{2y+1} &= \dfrac{1}{3} \end{aligned}$

6. Solve the following quadratic equations:

(a) $6x^2 + x - 2 = 0$
(b) $-2x^2 - 20x = 32$

(c) $f + \dfrac{1}{f} = 3$

(d) $\dfrac{1}{a+1} + \dfrac{1}{a+2} - \dfrac{2}{3} = 0$

7. Solve the equation

$\dfrac{3}{4}x^2 - x = \dfrac{5}{4}$ graphically.

8. Draw, using the same scale and axes, the graphs of $s = 2u + 3$ and $s = u^2 + u + 1$ from your graphs, solve the equation $u^2 - u = 2$.

From the plot shown in Figure 4.13, the intersection of the two straight lines yields the required result, that is $x = 3$ and $y = 2$.

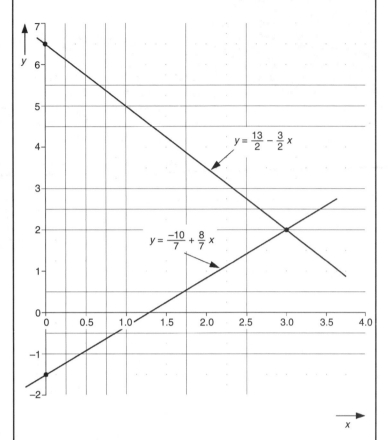

Figure 4.13 *Simultaneous graphs of the functions*

$y = \dfrac{13}{2} - \dfrac{3}{2}x$ and $y = \dfrac{-10}{7} + \dfrac{8}{7}x$

In this particular example, it would be easier to solve these equations using an algebraic method.

Logarithms and logarithmic functions

We have already studied the laws and use of indices, it is now time to briefly consider logarithms. In fact logarithms are really indices. The logarithm of a number is in fact its index. For example, $10^3 = 1000$, the left-hand side of this equation 10^3 is the number 1000 written in index form. The index 3 is in fact the *logarithm* of 1000. Check this by pressing your *log* button (which is the logarithm to the base ten) then key in the number 1000 and press the = button, you will obtain the number 3.

Manipulation of numbers, expressions and formulae that are in index form may be simplified by using logarithms. Another use for logarithms is being able to reduce the sometimes more difficult arithmetic operations of multiplication and division to those of addition and subtraction. This is often necessary when manipulating more complex algebraic expressions.

We start by considering the laws of logarithms in a similar

Key point

The power or index of a number, when that number is in index form, is also its logarithm, when taken to the base of the number.

manner to the way in which we dealt with the laws of indices earlier.

The laws of logarithms

The laws of logarithms are tabulated below, they are followed by simple examples of their use. In all these examples, we use *common logarithms*, that is logarithms to the *base 10*. Later we will look at one other type, the *Naperian logarithm*, or natural logarithm, where the *base* is the number *e*.

Number	Logarithmic law
1	If $a = b^c$, then $c = \log_b a$
2	$\log_a MN = \log_a M + \log_a N$
3	$\log_a \dfrac{M}{N} = \log_a M - \log_a N$
4	$\log_a (M^n) = n \log_a M$
5	$\log_b M = \dfrac{\log_a M}{\log_a b}$

Law 1

All these laws look complicated, but you have already used law 1 when you carried out the calculator exercise above. So again, we know that $1000 = 10^3$. Now if we wish to put this number into *linear* form (decimal form), then we may do this by *taking logarithms*.

Following *law 1*, where in this case $a = 1000$, $b = 10$ and $c = 3$, then $3 = \log_{10} 1000$. Now this fact you have already proved on your calculator! So you are probably wondering why we need to bother with logarithms? Well, in this case we are dealing with *common logarithms*, that is numbers in index form where the *base* of the logarithm is 10. We can also consider numbers in index form that are not to the base ten, as you will see later. We may also be faced with a problem where the index (power) is not known.

Suppose we are confronted with this problem: *find the value of x where* $750 = 10^x$. The answer is not quite so obvious, but it can easily be solved using our first law of logarithms. So, again following the law, that is taking logarithms to the appropriate base, we get $x = \log_{10} 750$ and *now* using our calculator, we get $x = 2.875$, correct to 4 significant figures.

Law 2

One pair of factors for the number 1000 is 10 and 100. Therefore according to the second law: $\log_a (10)(100) = \log_a 10 + \log_a 100$. If we choose logarithms to the base 10, then we already know that the $\log_{10} 1000 = 3$. Then using our calculator again, we see that $\log_{10} 10 = 1$ and $\log_{10} 100 = 2$. What this law enables us to do is to convert the *multiplication* of numbers in index form into that of *addition*. Compare this law with the first law of indices, you studied earlier! Remember also that we are at liberty to choose any base we wish, providing we are able to work in this base. Your calculator gives you logarithms to only two bases, 10 and *e*.

Law 3

This law allows us to convert the *division* of numbers in index form into that of *subtraction*. When dealing with the *transposition* of more complex formulae, these conversions can be particularly useful and help us with the transposition.

So using the law directly, for example:

$$\log 10 \frac{1000}{10} = \log_{10} 100 = \log_{10} 1000 - \log_{10} 10$$

or from your calculator $2 = 3 - 1$.

Law 4

This laws states that if we take the logarithm of a number in index form M^n this is equal to the logarithm of the base of the number $\log_a M$, multiplied by the index of the number $n \log_a M$. So, for example, $\log_{10} (100^2) = \log_{10} 10\ 000 = 2 \log_{10} 100$. This is easily confirmed on your calculator as $4 = (2)(2)$.

Law 5

This law is rather different from the others in that it enables us to *change the ba*se of a logarithm. This of course is very useful if we have to deal with logarithms or formulae involving logarithms that have a base not found on our calculator!

For example, suppose we wish to known the numerical value of $\log_2 64$, then using law 5, we have:

$$\log_2 64 = \frac{\log_{10} 64}{\log_{10} 2} = \frac{1.806179974}{0.301029995} = 6, \quad \text{interesting!}$$

If we *use law 1* in reverse then $\log_2 64$ is equivalent to the number $64 = 2^6$, which of course is now easily verified by your calculator! This example again demonstrates that given a number in *index form*, the *index* of that number is also its *logarithm*, providing the logarithm has the same base.

We will now consider, through example, one or two engineering uses for the laws of *common* and *natural* logarithms.

Example 4.43

An equation connecting the final velocity v of a machine with the

machine variables w, p and z is given by the formula $v = 20^{\left(\frac{w}{pz}\right)}$. Transpose the formula for w, and find its numerical value when $v = 15$, $p = 1.24$ and $z = 34.65$.

This formula may be treated as a number in *index form*. Therefore to find w as the subject of the formula, we need to apply the laws of logarithms. The first step in this type of problem is to *take logarithms of both sides*. The base of the logarithm chosen is not important provided we are able to find the numerical values of these logarithms, when required. Thus, we generally take logarithms to the base 10 or to the base e. As yet, we have not considered logarithms to the base e, so we will take *common logarithms* of both sides. However, if the number or expression is not to a base of logarithms we can manipulate, then we are at liberty to change this base using law 5!

So, $\log_{10} v = \log_{10} 20^{\left(\frac{w}{pz}\right)}$ at this stage, taking logarithms, seems to be of little help! However, if we now apply the appropriate logarithmic laws, we will be able to make w the subject of the formula.

Applying law 4 to the right-hand side of the expression we get

$$\log_{10} v = \left(\frac{w}{pz}\right) \log_{10} 20.$$

Then, finding the numerical value of $\log_{10} 20 = 1.30103$, we can now continue with the transposition:

$$\log_{10} v = \left(\frac{w}{pz}\right) 1.30103 \quad \text{or} \quad \frac{\log_{10} v}{1.30103} = \frac{w}{pz} \quad \text{and so} \quad w = \frac{(pz)(\log_{10} v)}{1.30103}$$

Having transposed the formula for w, we can substitute the appropriate values for the variables and find the numerical value of w. Then:

$$w = \frac{(1.24)(34.65)(\log_{10} 15)}{1.30103} = \frac{(1.24)(34.65)(1.17609)}{1.30103} = 38.84$$

Naperian logarithms and the exponential function

If you look at your calculator you will see the *ln* or *Naperian logarithm* button. The *inverse* of the Naperian logarithm function is e^x or exp x, the exponential function. This logarithm is sometime known as the *natural logarithm*, because it is often used to model naturally occurring phenomena, such as the way things grow or decay. In engineering, for example, the decay of charge from a capacitor may be modelled using the natural logarithm. It is therefore a very useful function, and both the natural logarithm and its inverse, the exponential function, are very important within engineering. We will now consider the transposition of a formula that involves the use of natural logarithms and the logarithmic laws.

Example 4.44

Transpose the formula, $b = \log_e t - a \log_e D$ to make t the subject.

First, note that the natural or Naperian logarithm may be expressed as $\log_e$ or ln, as on your calculator. *Do not mix-up* the expression $\log_e$, or its inverse e^x or exp x with the exponent function (EXP) on your calculator, which multiplies a number by powers of ten!

We first use the laws of logarithms, as follows:

$b = \log_e t - \log_e D^a$ from law 4

then:

$$b = \log_e \left(\frac{t}{D^a}\right) \quad \text{from law 3}$$

Now, for the first time we take the *inverse* of the natural logarithm or *antilogarithm*. Noting that *any* function multiplied by its inverse is 1 (one), then multiplying both sides of our equation by e, the *inverse* of ln ($\log_e$), we get:

$$e^b = \frac{t}{D^a}$$

(since e is the inverse or antilogarithm of $\log_e$ = ln or $(e)(\log_e) = 1$), then:

$$t = e^b D^a \quad \text{as required}$$

As mentioned before, the *exponential function e^x*, or exp x, and its inverse ln, natural logarithm, have many uses in engineering, because they can be used to model growth and decay. So the way solids expand, electrical resistance changes with temperature, substances cool, pressure changes with altitude or capacitors discharge can all be modelled by the *exponential* function.

Here are just two engineering examples of the use of the exponential function.

Example 4.45

If the pressure p at height h (in metres) above the ground is given by the relationship

$p = p_0 e^{\frac{h}{k}}$, where p_0 is the sea-level pressure of 101325 N m^{-2},

determine the value of the height h when the pressure at altitude p is 70129 N m^{-2} and $k = -8152$.

We need first to transpose the formula for h, this will involve taking *natural logarithms*, the inverse function of $e^{\frac{h}{k}}$. Before we do so we will first isolate the exponential term, then $\frac{p}{p_0} = e^{\frac{h}{k}}$ and taking

logarithms gives $\log_e \left(\frac{p}{p_0} \right) = \frac{h}{k}$, then $k \log_e \left(\frac{p}{p_0} \right) = h$, then

substituting the given values, $h = -8152 \log_e \left(\frac{70129}{101325} \right)$

$= (-8152) \log e\, (0.692) = (-8152)(-0.368) = 3000$ m, correct to 4 significant figures. Thus the altitude $h = 3000$ m.

The following example is concerned with the information contained in a radio communications message. It is not necessary to understand the background physics in order to solve the problem, as you will see.

Example 4.46

It can be shown that the information content of a message is given by $I = \log_2\left(\dfrac{1}{p}\right)$. Show using the laws of logarithms that the information content may be expressed as $I = -\log_2(p)$ and find the information content of the message if the chances of receiving the code (p) is $\dfrac{1}{16}$.

So we are being asked to show that $I = \log_2\left(\dfrac{1}{p}\right) = -\log_2(p)$;

the left-hand side of this expression may be written as $\log_2(p^{-1})$
I hope you remember the laws of indices! Now if we compare this expression with law 4 where $\log_a(M^n) = n\log_a M$ then in this case $M = p$ and $n = -1$, so $\log_2(p-1) = -1\log_2 p = -\log_2 p$, as required.
    Now to find the information content of the message, we need to substitute the given value of $p = \dfrac{1}{16}$ into the equation

$\log_2(p^{-1}) = \log 2\left(\dfrac{1}{p}\right) = \log_2(16)$. Now our problem is that we

cannot easily find the value of logarithms to the base 2. However, if

we use logarithmic law 5, then we get $\log_2 16 = \dfrac{\log_{10} 16}{\log_{10} 2} = 4$. Then

the information content of the message = 4.

I hope you were able to following the reasoning in the above two, quite difficult, examples. There is just one more application of the laws of logarithms that we need to cover. It is sometimes very useful when considering experimental data to determine if such data can be related to a particular law. If we can relate this data to the law of a straight line $y = mx + c$, then we can easily determine useful results. Unfortunately the data are not always related in this form. However, a lot of engineering data follow the general form $y = ax^n$, where as before x is the *independent variable*, y *is the dependent variable* and in this case a and n are *constants* for the particular experimental data being considered.
    We can use a technique involving *logarithms* to reduce equations of the form $y = ax^n$ to a linear form, following the law of the straight line $y = mx + c$. The technique is best illustrated by example.

Example 4.47

The pressure p and volume v of a gas at constant temperature are related by Boyle's law, which can be expressed as $p = cv^{-0.7}$, where c is a constant. Show that the experimental values given in the table follow this law and from an appropriate graph of the results, determine the value of the constant c.

Volume v (m^3)	1.5	2.0	2.5	3.0	3.5
Pressure p (10^5 N m^{-2})	7.5	6.2	5.26	4.63	4.16

The law is of the form $p = ax^n$. So taking common logarithms of both sides of the law $p = cv^{-0.7}$ we get $\log_{10} p = \log_{10}(cv^{-0.7})$ and applying law 2 and law 4 to the right-hand side of this equation gives $\log_{10} p = -0.7 \log_{10} v + \log_{10} c$; make sure you can see how to get this result. Then comparing this equation with the equation of a straight line $y = mx + c$, we see that $y = \log_{10} p$, $m = -0.7$, $x = \log_{10} v$ and $c = \log_{10} c$.

So we need to plot $\log_{10} p$ against $\log_{10} v$. A table of values and the resulting plot is shown below.

Volume v (m³)	1.5	2.0	2.5	3.0	3.5
$\log_{10} v$	0.176	0.301	0.398	0.447	0.544
Pressure p (10^5 N m⁻²)	7.5	6.2	5.26	4.63	4.16
$\log_{10} p$	0.875	0.792	0.721	0.666	0.619

Then from the plot it can be seen that the slope of the graph (Figure 4.14) is −0.7, and the y intercept at $\log_{10} v = 0$ is given as 1.0, or $\log_{10} c = 1.0$ and so $c = 10$. Therefore the plotted results do follow the law p $10v^{-0.7}$.

Test your knowledge 4.8

1. Transpose the formulae
   $q = rx^{\frac{s}{t}}$ for (t) and then find its value when $q = 30\pi$, $r = 3\pi$, $x = 7.5$ and $s = 16$.

2. The formula
   $P = T(1 - e^{-\mu\theta})v$
   relates the power (P), belt tension (T), angle of lap (θ), linear velocity (v) and coefficient of friction (μ) for a belt drive system. Transpose the formula for (μ) and find its value when $P = 2500$, $T = 1200$, $V = 3$ and $\theta = 2.94$.

3. In an experiment, values of current I and resistance R were measured, producing the results tabulated below.

R	0.1	0.3	0.5	0.7
I	0.00017	0.0015	0.0043	0.0083
R	0.9	1.1	1.3	
I	0.014	0.021	0.029	

Show that the law connecting I and R has the form $I = aR^b$, where a and b are constants and determine this law.

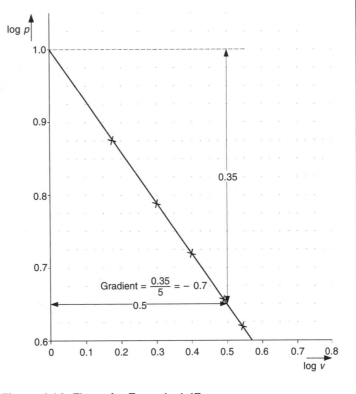

Figure 4.14 *Figure for Example 4.47*

This use of logarithms to manipulate experimental data, is very useful.

TRIGONOMETRY

In this, the second of the outcomes for the unit, we consider the solution of right-angled and other triangles, using basic trigonometric ratios. We then look again at the circle, in particular we study the concept of the radian and radian measure.

The second part of the outcome is concerned with the sine and cosine as oscillatory functions and the graphical methods needed to ascertain the time constant, frequency and amplitude of sinusoidal wave forms.

Trigonometric ratios – the triangle and circle

We start this section by looking at the solution of triangles, in other words, finding the missing angles and/or sides of triangles and the use to which we may put these trigonometric techniques.

Solution of triangles

You have already met Pythagoras' theorem. This theorem is particularly useful when dealing with vector quantities or phasors (rotating vectors) in electrical theory. I remind you again of this important theorem.

The theorem states that: *in a right-angled triangle, the square on the hypotenuse is equal to the sum of the squares on the other two sides.* Pictorially, the theorem is illustrated in Figure 4.15.

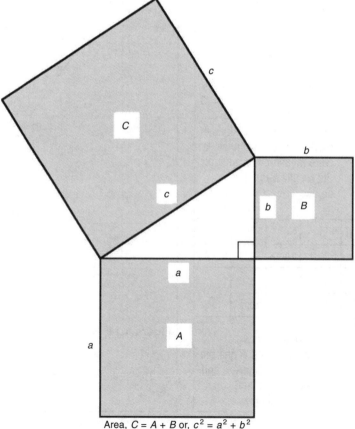

Area, $C = A + B$ or, $c^2 = a^2 + b^2$

Figure 4.15 *Illustration of Pythagoras' theorem*

For the triangle shown, $c^2 = a^2 + b^2$ or $a^2 = c^2 - b^2$ or $b^2 = c^2 - a^2$.

We can use this theorem to find the height of *equilateral* and *isosceles* triangles, where an equilateral triangle has all its sides of equal length and an isosceles triangle has two sides of equal length.

Example 4.48

Find the height of an isosceles triangle which has equal sides of length 13 cm and a base length of 10 cm. The triangle is illustrated in Figure 4.16.

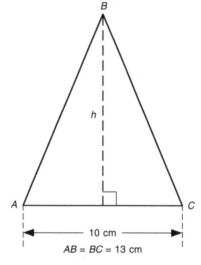

Figure 4.16 *Figure for Example 4.48 – the isosceles triangle*

Applying Pythagoras and noting that the hypotenuse is opposite the right angle, then: $h^2 = 13^2 - 5^2$ and so $h^2 = 169 - 25 = 144$, then square rooting both sides, $h = 12$ cm.

Note that the height h is always at right angles to the base and meets at tshe apex. A perpendicular can be constructed from any side of the triangle, providing it cuts the base at right angles and extends to the apex of the triangle.

Pythagoras has helped us to partially solve right-angled triangles. Another important technique is concerned with the ratio of the sides in similar triangles.

Similar triangles

Similar triangles have the same shape, one is simply an enlargement or reduction of the other. These triangles have two important properties, these are illustrated in Figure 4.17. From the figure you should note that their corresponding angles are equal and their corresponding sides are proportional. Thus in both triangles: taking $\hat{A}_1$ (*angle A_1*) as the reference, side a is the *opposite*, side b is the *adjacent* and side c is the *hypotenuse* and the *ratio* of these sides are equal, that is for both similar triangles:

$$\frac{a}{c} = \frac{3}{5} = \frac{6}{10}$$

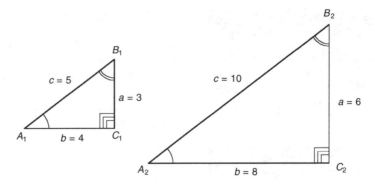

Figure 4.17 *Similar triangles (right angled in this case)*

$$\frac{b}{c} = \frac{4}{5} = \frac{8}{10}$$

$$\frac{a}{b} = \frac{3}{4} = \frac{6}{8}$$

Key point

Similar triangles do not have to be right angled, they may take any three internal angles, provided that corresponding angles are equal.

and in general the ratios of the corresponding sides are constant, that is the ratios: $\frac{a}{c}$, $\frac{b}{c}$ and $\frac{a}{b}$ are the same for *all similar right-angled triangles*.

Also note that for right-angled triangles to be similar, all corresponding angles must be the same. That is, for the triangles illustrated $A_1 = A_2$, $B_1 = B_2$ and of course they both have a right angle (90°).

Fundamental trigonometric ratios

I am sure that in your previous studies you have met the fundamental trigonometric ratios. However, for those with this gap in their knowledge and to serve as a reminder to others, they are repeated here.

For any right-angled triangle (Figure 4.18), then:

1. $\dfrac{\text{side opposite the angle}}{\text{hypotenuse}}$ is called the *sine* of the reference

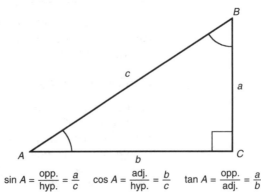

$$\sin A = \frac{\text{opp.}}{\text{hyp.}} = \frac{a}{c} \qquad \cos A = \frac{\text{adj.}}{\text{hyp.}} = \frac{b}{c} \qquad \tan A = \frac{\text{opp.}}{\text{adj.}} = \frac{a}{b}$$

Note: 1. The hypotenuse is opposite the right angle

2. $\tan A = \dfrac{\sin A}{\cos A}$

Figure 4.18 *The right-angled triangle*

angle and is often abbreviated to *sin*. Therefore:

$$\sin A = \frac{\text{opposite}}{\text{hypotenuse}} = \frac{a}{c}.$$

Note that from now on we will use only the capital letter to represent angles, dropping the ^ (hat) sign above the letter. Also note that the *sides* of the triangle are represented by *lower case* letters.

2. $\dfrac{\text{side adjacent to the angle}}{\text{hypotenuse}}$ is called the *cosine* of the reference angle and is often abbreviated to *cos*. Therefore:

$$\cos A = \frac{\text{adjacent}}{\text{hypotenuse}} = \frac{b}{c}.$$

3. $\dfrac{\text{side opposite the angle}}{\text{side adjacent to the angle}}$ is called the *tangent* of the angle and is often abbreviated to *tan*. Therefore: $\tan A = \dfrac{\text{opposite}}{\text{adjacent}} = \dfrac{a}{b}.$

These fundamental ratios are very important and should be remembered. One aid to memory is to use the acronym SOHCAHTOA (pronounced *sock-ca-tow-ah*). Where the letters mean: sine, opposite, hypotenuse; cosine, adjacent, hypotenuse; tangent, opposite, adjacent. There are other aids to memory, stick with what you know!

Note:
From Figure 4.18 the *tangent ratio*, with respect to $\angle A$, may be expressed as:

$$\tan A = \frac{\sin A}{\cos A} = \frac{\dfrac{a}{c}}{\dfrac{b}{c}} = \left(\frac{a}{c}\right)\left(\frac{c}{b}\right) = \frac{a}{b} = \tan A.$$

This relationship holds true for the tangent of *any* angle.

One final very important point concerning these ratios – remember that the sine, cosine and tangent function *must be followed by the angle they refer to*. Thus, for example, sin *A*, makes sense, but sin on its own is nonsense. Remember that the sin and the *A cannot* be separated!

Before we look at an example of the use of these ratios, we need to ensure that you can use your calculator to evaluate them.

Suppose, for example, that we wish to find the numerical value of sin 51°. That is find the value of the sine of 51°. The following is a typical sequence you need to follow:

(a) Ensure your calculator is selected in the degree mode (shown in the viewing window).
(b) Press the sine key.
(c) Key in 51.
(d) Press the = button.
(e) Read off the result as 0.77715 (correct to 5 decimal places).

Note, that with some calculators operations (b) and (c) may be reversed and on pressing the sine key the result is automatically shown in the viewing window.

Also note that fractions of a degree are best entered into your calculator in decimal form. If you are required to find the trigonometric ratio of an angle given in degrees and minutes, the following procedure may be used.

Find, using your calculator, the numerical value of cos 26° 45′ (45 minutes).

We first convert the 45 minutes into degrees, by dividing by 60, then $\frac{45'}{60} = 0.75°$. Now we follow the procedure as before:

(a) Ensure calculator is in degree mode.
(b) Press the cosine key.
(c) Key in 26.75.
(d) Press the = button.
(e) Read off the result as 0.89298 (correct to 5 decimal places).

Example 4.49

For the triangle shown in Figure 4.19 find the sine, cosine and tangent for ∠A (angle A) and ∠B.

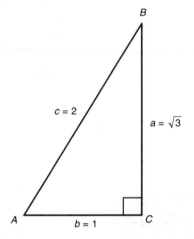

Figure 4.19 *Figure for Example 4.49*

Then for angle A, the opposite $= \sqrt{3}$, adjacent $= 1$ and the hypotenuse $= 2$. So the respective trigonometric ratios are $\sin A = \frac{\sqrt{3}}{2}$, $\cos A = \frac{1}{2}$ and $\tan A = \frac{\sqrt{3}}{1}$; in fractional form these are exact values for these ratios, so we will leave them in this form for the moment. Similarly for ∠B we get $\sin B = \frac{1}{2}$, $\cos B = \frac{\sqrt{3}}{2}$ and $\tan B = \frac{1}{\sqrt{3}}$. Note that this is a special *right-angled triangle* where

∠A = 60° and ∠B = 30° and the ratios of the sides are always as given above.

Inverse trigonometric ratios

Now if we wish to evaluate these ratios and *find the angle*, in degrees, that they represent, we need to consider the *inverse trigonometric ratio*. Thus, if we consider the trigonometric ratio

$\sin A = \dfrac{\sqrt{3}}{2}$, from the previous example, we need to find the *angle* represented by this ratio. In general terms, the inverse ratio is, in symbols, represented as $\sin^{-1} A$, or in words as *the angle A whose sign is*. So in our particular example $\sin A = 0.8660254$ and so the angle whose sign is 0.8660254, or $\sin^{-1} A = 0.8660254$, is the angle we are required to find! This inverse function of sine (in this case) is easily found on our calculator, as follows:

(a) Ensure calculator is in degree mode.
(b) Press the *shift* key.
(c) Press the sin key ($\sin^{-1}$ will appear in the display window).
(d) Key in desired value of trigonometric ratio (in our case 0.8660254).
(e) Press = button and read of angle in degrees (correct to 4 decimal places we get $A = 60°$).

Following this procedure for any of the trigonometric ratios we found in Example 4.49 will yield the following results:

Trigonometric ratio	Angle in degrees
$\sin A = 0.8660$	60
$\cos A = 0.5$	60
$\tan A = 1.732$	60
$\sin B = 0.5$	30
$\cos B = 0.866$	30
$\tan B = 0.577$	30

Solving triangles

So far we have used the trigonometric ratios sine, cosine and tangent to find angles, given the three sides of a triangle. We can in fact solve any right-angled triangle, given any side and two angles, using the trigonometric ratios, and where necessary combine these ratios with Pythagoras.

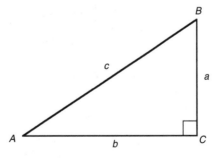

Figure 4.20 *Labelled right-angled triangle*

Consider the triangle shown in Figure 4.20, with sides a, b, c and angles $\angle A$, $\angle B$ and $\angle C$. Then from the figure:

$\sin A = \dfrac{a}{c}$,   therefore $a = c \sin A$

$\cos A = \dfrac{b}{c}$,   therefore $b = c \cos A$

$\tan A = \dfrac{a}{b}$,   therefore $a = b \tan A$

Also, from Pythagoras, $c^2 = a^2 + b^2$.

Example 4.50

In the right-angled triangle shown in Figure 4.21 find $\angle A$, $\angle B$ and side c.

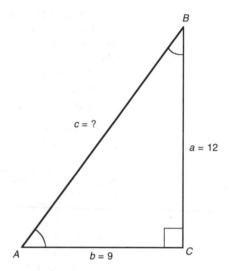

Figure 4.21 *Figure for Example 4.50*

We find angle A by noting that we are given the opposite side and the adjacent side, therefore I hope you can see that we should use the tangent ratio to find $\angle A$. Then:

$$\tan A = \frac{\text{opp.}}{\text{adj.}} = \frac{12}{9}$$

and using our calculator we find $A = 53.13°$.

To find side c, we could again use $\angle A$ with the sine ratio. Unfortunately, if we have made a mistake in our calculation of $\angle A$, we will create yet another error. So it is wise to use given information when we can. One way to do this is to find side c, using Pythagoras. Then:

$$c^2 = 12^2 + 9^2 = 144 + 81 = 225 \quad \text{and so side } c = 15$$

Now in this example, for the sake of completeness, we will check this result using:

$$\sin A = \frac{12}{h} \quad \text{or } h = \frac{12}{\sin 53.13°} = \frac{12}{0.79999} = \frac{12}{0.8} = 15$$

Note that the calculator value for $\sin 53.13°$ was 0.79999, rather than the exact value 0.8. This is because the original value for

$\tan A = \dfrac{12}{9} = 1.3333$. If you enter this recurring decimal value into your calculator, the resulting value of $\angle A$ is not quite exact and this non-exact value was keyed in again, when finding $\sin A$. Yet another reason to use Pythagoras and exact fractions!

Example 4.51

From two points *A* and *B*, on level ground and on opposite sides of a vertical pylon of height 26 m, the angles of elevation to the top of the pylon are 25° and 48°, respectively. Find the horizontal distance between the two points.

This type of problem requires us to draw a space diagram of the situation. The angle of elevation is that angle which starts at zero degrees along the horizontal and increases, as we look upwards. In our case the angle from the horizontal to the top of the pylon, viewed from both sides, as shown in Figure 4.22.

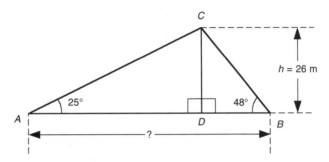

Figure 4.22 *Figure for Example 4.51*

It can be seen that the vertical pylon forms a common side between the two triangles *ACD* and *BCD*. We are required to find the distance between the two points *A* and *B*. This distance can be found by first finding the distance *AD* and then adding this distance to the side *DB*.

Then with respect to $\angle A = 25°$, we need to select the trigonometric ratio that involves the height of the pylon and the unknown adjacent side *AD*. I hope you can see that this will be the tangent ratio, where:

$$\tan A = \frac{CD}{AD}$$

and on substituting given values we get,

$$\tan 25° = \frac{26}{AD} \text{ so that } AD = \frac{26}{\tan 25} = \frac{26}{0.4663} = 55.76 \text{ m}$$

Similarly for $\angle B$ we have

$$\tan B = \frac{CD}{DB} \text{ or } \tan 48° = \frac{26}{DB}, \ DB = \frac{26}{\tan 48} = \frac{26}{1.1106}$$

and *DB* = 22.51 m.

Then the distance between the points *A* and *B* = *AD* + *DB* = 55.76 + 22.51 = 78.27 m.

In order to be able to solve some problems involving right-angled triangles it is necessary to understand the relationship between complementary angles. This really means that we need to memorize *two facts*.

The first is that there are 180° in a semicircle, which I am sure you know. So, for example, if one angle in a semicircle is 40°, its *complementary* angle is 180 − 40 = 140°.

The second fact is that there are also 180° in a triangle. So that if a right-angled triangle has two angles, say 90° and 55°, then the complementary angle is easily found as:

$180 - 90 - 55 = 35°$

Example 4.52

Calculate the length of *AB*, shown in Figure 4.23. Given that length *BC* = 200 mm.

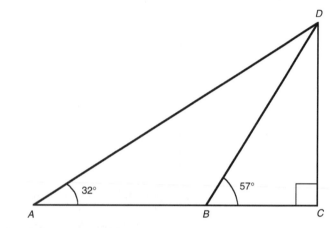

Figure 4.23 *Figure for Example 4.52*

To solve this problem, we need to remember the facts concerning complementary angles.

Then, $\angle ADC = 180 - 32 - 90 = 58°$ and with respect to the angle *CBD* = 57°, we can use the tangent ratio to find *CD*, then:

$\tan 57 = \dfrac{CD}{200}$ or $CD = 200 \tan 57$ and so $CD = (200)(1.5399)$

$CD = 308$ mm to the nearest mm

Now with respect to angle *ADC* = 58°, again using the tangent ratio, we get:

$\tan 58 = \dfrac{AC}{CD}$ or $AC = CD \tan 58 = (308)(1.6033) = 493$ mm, again to the nearest mm.

Then, required side $AB = AC - BC = 493 - 200 = 293$ mm.

In the above example we could go on to find the remaining sides for both triangles and so solve them completely.

Area of the right-angled triangle

In order to completely solve right-angled triangles, we also need to find their area. You will remember that we covered this topic earlier, when studying areas and volumes, in the outcome on algebra. The formula for the area of a *right-angled* triangle was given as:

area, *A* = half the base × the perpendicular height   or in symbols

$A = \dfrac{1}{2}bh$

Example 4.53

Find the area of the triangle shown in Figure 4.24.

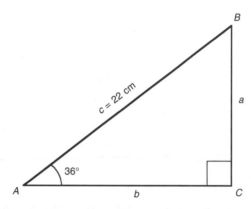

Figure 4.24 *Figure for Example 4.53*

In order to find the area, we first need to find the two sides a and b. Using the sine ratio and $\angle A$, then, $\sin 36 = \dfrac{a}{22}$ and $a = 22 \sin 36$ $= (22)(0.5878) = 12.93$ cm.

Now, we will use Pythagoras to find the remaining side b, then: $b^2 = c^2 - a^2 = 22^2 - 12.93^2 = 484 - 167.22 = 316.78$

therefore side $b = \sqrt{316.78} = 17.8$ cm.

Then the area of the triangle $A = \dfrac{1}{2} ba = \left(\dfrac{1}{2}\right)(17.8)(12.93) = 115$ cm^2.

Key point

The area of a right-angled triangle = half the base multiplied by the perpendicular height.

We will return to the solution of triangles, other than right-angled triangles, a little later. We first turn our attention to coordinates systems and the representation of angles $\geq 90°$.

Polar and rectangular coordinates

A point on a graph can be defined in several ways. The two most common ways use either rectangular or polar coordinates. You have already met rectangular (Cartesian) coordinates in your earlier graphical work.

Rectangular coordinates (Figure 4.25) use two *perpendicular* axes, normally labelled x and y. Where any point P is identified by its horizontal distance along the x-axis and its vertical distance up the y-axis. *Polar coordinates* give the distance r from the origin O and the angle θ of the line joining the origin and the point P with the x-axis.

Thus, for example, the point $(4, -3)$ is the rectangular or Cartesian coordinates for the point, that is four units to the right along the x-axis (Figure 4.26(a) and three units in the negative y direction, that is downwards.

The point $(25\angle128)$ gives the polar coordinates for the point P,

Key point

Rectangular coordinates are also known as Cartesian coordinates.

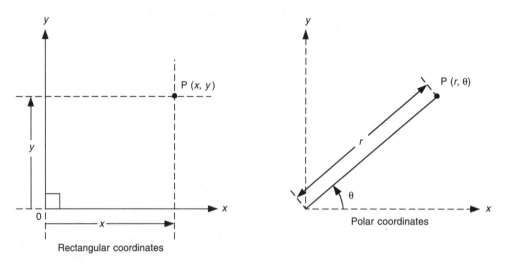

Figure 4.25 *Rectangular and polar coordinate systems*

that is 25 units in magnitude from the origin, at an angle of 128°, measured anticlockwise from the horizontal *x*-axis (Figure 4.26(b)).

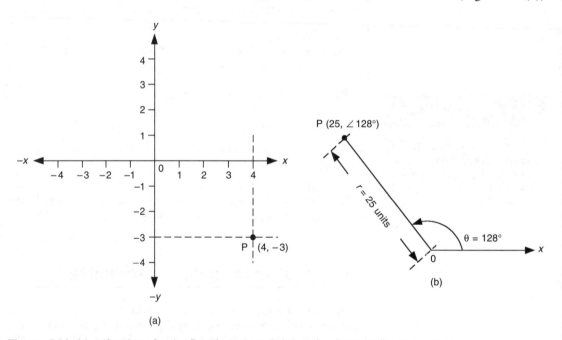

Figure 4.26 *Identification of point P, using rectangular and polar coordinate systems*

Converting rectangular and polar coordinates

A useful skill is to be able to convert rectangular to polar coordinates and vice versa. This is particularly helpful when dealing with sinusoidal functions and other oscillatory functions that you may meet in your later studies.

Consider Figure 4.27, which shows a set of rectangular and polar axes, combined.

Then to convert *rectangular to polar coordinates*, we use Pythagoras' theorem and the tangent ratio to give:

$$r = \sqrt{x^2 + y^2} \quad \text{and} \quad \tan \theta = \frac{y}{x}$$

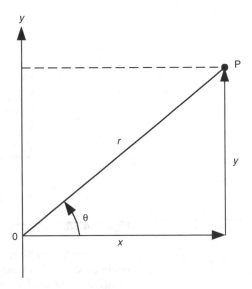

Figure 4.27 *Rectangular and polar coordinates*

To convert *polar to rectangular coordinates*, we use the sine and cosine ratios to give:

$\sin \theta = \dfrac{y}{r}$, therefore $y = r \sin \theta$ and $\cos \theta = \dfrac{x}{r}$, therefore $x = r \cos \theta$

Example 4.54

(a) Convert the rectangular coordinates (−5, −12) into polar coordinates.

(b) Convert the polar coordinates (150 ∠300) into rectangular coordinates.

(a) Using Pythagoras and the tangent ratio, we get:

$$r = \sqrt{(-5)^2 + (-12)^2} = \sqrt{25 + 144} = \sqrt{169} = 13$$

and $\tan \theta = \dfrac{-12}{-5} = 2.4$, therefore $\theta = 67.4°$

So the polar coordinates are **13 ∠67.4**.

(b) Using the sine and cosine ratios, to find y and x respectively, we get:

$y = r \sin \theta = 150 \sin 300 = (150)(-0.866) = -129.9$ and

$x = r \cos \theta = 150 \cos 300 = (150)(0.5) = 75$

So the rectangular coordinates are (**75, −129.9**).

Angles in any quadrant

If you key into your calculator cos 150, you get the value −0.866. This is the same numerically as cos 30 = 0.866, except that there has been a *sign* change. Whether any one trigonometric ratio is *positive* or *negative* depends on whether the projection is on the positive or negative part of the coordinate system. Figure 4.28

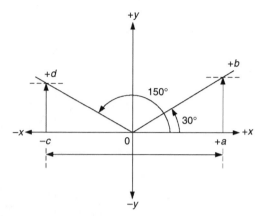

Figure 4.28 *Projection of the angles 30° and 150°*

shows the rectangular coordinate system on which two lines have been placed, at angles of 30° and 150°, respectively, from the positive horizontal *x-ordinate*.

Now if we consider the sine ratio for both angles, then we get:

$$\sin 30 = \frac{+ab}{+ob} \quad \text{and} \quad \sin 150 = \frac{+cd}{+od}$$

thus both these ratios are *positive* and therefore a positive value for sin 30 and cos 30 will result. In fact from your calculator sin 30 = sin 150 = 0.5.

Now, from the diagram we find that $\cos 30 = \frac{+oa}{+ob}$, which will again yield a positive value, in fact cos 30 = 0.866 but cos 150 $= \frac{-oc}{+od}$, a *negative* ratio that yields the negative value –0.866, which you found earlier.

If we continue to rotate our line in an *anticlockwise* direction, we will find that cos 240 = –0.5 and cos 300 = 0.5. Thus, dependent on which *quadrant* (quarter of a circle = 90°) the ratio is placed, depends whether or not the ratio is positive or negative. This is true for all three of the fundamental trigonometric ratios.

Figure 4.29(a) shows the *signs* for the sine, cosine and tangent

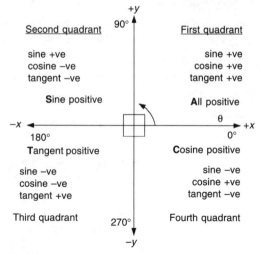

Figure 4.29 *Signs of angles of any quadrant*

functions. A way of remembering when the *sign* of these ratios is *positive*, is to use the word CAST, starting from the bottom right quadrant and reading anti-clockwise. Your calculator automatically shows the correct sign for any ratio of any angle, but it is worth knowing what to expect from your calculator.

Example 4.55

Find on your calculator the value of the following trigonometric ratios and verify that the *sign* is correct, by consulting Figure 4.29.

(a) sin 57
(b) cos 236
(c) tan 97
(d) sin 320
(e) cos 108
(f) tan 347
(g) sin 137
(h) cos 310
(i) tan 237

The values with their appropriate sign are tabulated below.

(a) 0.8387
(b) −0.5592
(c) −8.144
(d) −0.6428
(e) −0.3090
(f) −0.2309
(g) 0.6819
(h) 0.6428
(i) 1.5397

You can easily verify that all these values are in accord with Figure 4.29.

We finish solving triangles by considering triangles of any internal angles. This involves the use of the *sine* and *cosine rules*, which are given without proof.

General solution of triangles

We now extend our knowledge to the solution of triangles that are not right angled. In order to do this we need to be armed with just two additional formulae. These are tabulated below for reference.

Sine rule	$\dfrac{a}{\sin A} = \dfrac{b}{\sin B} = \dfrac{c}{\sin C}$
Cosine rule	$a^2 = b^2 + c^2 - 2bc \cos A$ $b^2 = a^2 + c^2 - 2ac \cos B$ $c^2 = a^2 + b^2 - 2ab \cos C$

The above rules can only be used in specific circumstances.

For the general triangle *ABC* shown in Figure 4.30, with sides *a*, *b*, *c* and angles $\angle A$, $\angle B$ $\angle C$, the *sine rule may only be used when either*:

- one side and any two angles are known or
- if two sides and an angle (not the angle between the sides) are known.

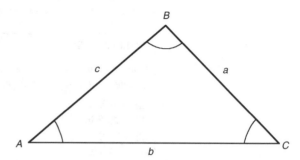

Figure 4.30 *The general triangle*

The cosine rule may only be used when either:

- three sides are known or
- two sides and the included angle are known.

Note 1:
When using the sine rule, the equality signs allow us to use any parts of the rule that may be of help. For example, if we have a triangle to solve, for which we know the angles $\angle A$ and $\angle C$ and side a, we would first use the rule with the terms $\dfrac{a}{\sin A} = \dfrac{c}{\sin C}$ to find side c.

Note 2:
When using the cosine rule, the version chosen will also depend on the information given. For example, if you are given sides a, b and the included angle C, then the formula $c^2 = a^2 + b^2 - 2ab \cos C$ would be selected to find the remaining side c.

Only *simple* examples of the use of this rule are given here. You will solve more difficult problems using these rules if you take the Further Mathematics for Technicians unit as part of your course profile.

Example 4.56

In a triangle ABC, $\angle A = 48°$, $\angle B = 59°$ and the side $a = 14.5$ cm. Find the unknown sides and angle.

The triangle ABC is shown in Figure 4.31.

When the triangle is sketched, it can be seen that we have two angles and one side, so we can use the *sine* rule. Remembering that the sum of the internal angles of a triangle $= 180°$, then $\angle C = 180 - 48 - 59 = 73°$. We will use the first two terms of the sine rule, $\dfrac{a}{\sin A} = \dfrac{b}{\sin B}$, to find the unknown side b. Then:

$$\frac{14.5}{\sin 48} = \frac{b}{\sin 59} \ \text{ or } \ b = \frac{(\sin 59)(14.5)}{\sin 48} = \frac{(0.8572)(14.5)}{0.7431} = 16.72 \text{ cm}$$

Similarly, to find side c we use $\dfrac{a}{\sin A} = \dfrac{c}{\sin C}$, which on substitution of the values gives,

$$\frac{14.5}{\sin 48} = \frac{c}{\sin 73} \quad \text{or} \quad c = \frac{(\sin 73)(14.5)}{\sin 48}$$

$$= \frac{(0.9563)(14.5)}{0.7431} = \frac{13.8664}{0.7431} = 18.66 \text{ cm}$$

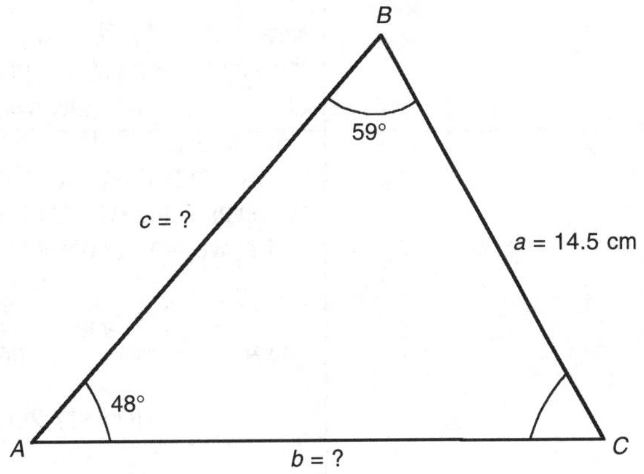

Figure 4.31 *Figure for Example 4.56*

When using the cosine rule, given three sides, it is necessary to transpose the formula to find the required angles. In the next example, we need to perform this transposition, which you should find relatively simple. If you have difficulties following the steps, you should refer back to the section on transposition of formula in the outcome on algebra.

Example 4.57

A flat steel plate is cut with sides of length 12 cm, 8 cm and 6 cm. Determine the three angles of the plate.

A diagram of the plate, suitably labelled, is shown in Figure 4.32, where side $a = 6$ cm, $b = 12$ cm and $c = 8$ cm.

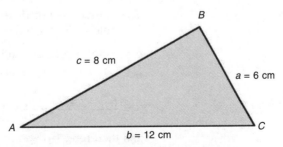

Figure 4.32 *Figure for Example 4.57 – steel plate*

Now, in this particular case we are free to choose any variant of the formula to find the corresponding angle. We will use:

$$b^2 = a^2 + c^2 - 2ac \cos B$$

Then transposing for $\cos B$,

$2ac \cos B = a^2 + c^2 - b^2$

and

$$\cos B = \frac{a^2 + c^2 - b^2}{2ac}$$

then

$$\cos B = \frac{6^2 + 8^2 - 12^2}{(2)(6)(8)} = \frac{36 + 64 - 144}{96} = \frac{-44}{96} = -0.4583$$

Now $\angle B = 117.28$, using a calculator. Note that cos B is negative, therefore $\angle B$ must lie outside the first quadrant, that is it must be greater than 90°. However, since it is the angle of a triangle it must also be less than 180°, thus $\angle B = 117.28°$ is its only possible value.

Now to find another angle, we could again use the cosine rule. However, since we now have an angle and two *non-included* sides, a and b, we are at liberty to use the simpler *sine* rule. Then:

$$\frac{a}{\sin A} = \frac{b}{\sin B} \quad \text{and so} \quad \frac{6}{\sin A} = \frac{12}{\sin 117.28}$$

or $\sin A = \dfrac{(6)(\sin 117.28)}{12} = \dfrac{(6)(0.8887)}{12} = 0.4444$

and from calculator $\angle A = 26.38°$.

Finally $\angle C = 180 - 117.28 - 26.38 = 36.34°$.

Area of any triangle

Now to complete our study of general triangles, we need to be able to calculate their area. This, of course, we have already done during our study of areas and volumes, in the previous outcome on algebra. Again, as we did for right-angled triangles, let us use one of the formula we learnt earlier to find the area of *any* triangle. The formula we will use is:

$$\sqrt{s(s-a)(s-b)(s-c)}$$

where a, b and c were the sides and $s = \dfrac{a+b+c}{2}$.

Then in the case of the triangle we have just been considering in Example 4.57, where:

$a = 6$ cm, $b = 12$ cm and $c = 8$ cm

then,

$$s = \frac{6 + 12 + 8}{2} = \frac{26}{2} = 13$$

and therefore,

the area $= \sqrt{13(13-6)(13-12)(13-8)} = \sqrt{(13)(7)(1)(5)}$

$$= \sqrt{455} = 21.33 \text{ cm}^2$$

Now the area of any triangle ABC can also be found using any of the following formulae.

area of any triangle $ABC = \frac{1}{2}\, ab \sin C$

or $= \frac{1}{2}\, ac \sin B$ or $= \frac{1}{2}\, bc \sin A$

These formulae are quoted here without proof and any variant may be used dependent on the information available. So again, for the triangle in Example 4.57, using the first of the formulae,

area of triangle $ABC = \frac{1}{2}\, ab \sin C = \frac{1}{2}(6)(12)(\sin 36.34)$

$$= (0.5)(72)(0.5926) = 21.33 \text{ cm}^2 \text{ as before.}$$

The circle

We have already been introduced to the way in which we find the circumference and area of a circle. Here we extend our knowledge of the circle, by identifying and defining certain elements of the circle, This is essentially about the geometry of the circle, which you will find useful when finding particular cross-sections, or when considering circular motion.

Elements and properties of the circle

The major elements of the circle are shown in Figure 4.33. You will be familiar with most, if not all, of these elements. However, for the sake of completeness, we will formally define them.

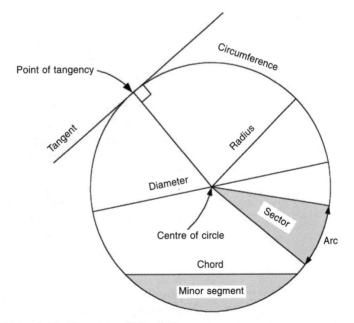

Figure 4.33 *Elements of the circle*

A point in a plane whose distance from a fixed point in that plane is constant lies on the *circumference* of a circle. The fixed point is called the centre of the circle and the constant distance is called the *radius*.

A circle may be marked out on the ground by placing a peg or spike at its centre. Then using a length of chord for the radius, we simply walk round with a pointer at the end of the cord and mark out the circumference of the circle.

A *chord* is a straight line which joins two points on the circumference of a circle. A *diameter* is a chord drawn through the centre of a circle.

A *tangent* is a line which just touches the circumference of a circle at one point (the point of tangency). This tangent line lies at right angles to a radius, drawn from the point of tangency.

A chord line cuts a circle into a *minor segment* and *major segment*. A *sector* of a circle is an area enclosed between two radii, and a length of the circumference, the *arc* length.

The radian and circular measure

Circular measure using degrees has been with us since the days of the Babylonians, when they divided a circle into 360 equal parts, corresponding to what they believed were the days in the year. An angle in degrees is a measure of rotation and an angle is formed when a line rotates with respect to a fixed line (Figure 4.34), when both lines have the same centre of rotation.

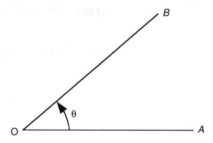

Figure 4.34 *The angle, as a measure of rotation*

The degree may be subdivided into minutes and seconds of arc, where the minute is $\frac{1}{60}$ of a degree (′) and a second is $\frac{1}{60}$ of a minute (″) or $\frac{1}{3600}$ of a degree of arc. We will restrict ourselves to angular measurement in degrees and decimal fractions of a degree, as you learnt earlier.

The *degree*, being an arbitrary form of circular measurement, has not always proved an appropriate unit for mathematical manipulation. Another less arbitrary unit of measure has been introduced, known as the *radian* (Figure 4.35); the advantage of this unit is its relationship with the *arc length* of a circle.

Note:
You can of course convert from degrees, minutes and seconds of arc to degrees in decimal form, using the button on your calculator marked as (° ′ ″). Simply enter the angle in decimal form, for example 83.37°, press the 'shift' function, then press the (° ′ ″) button, the answer is displayed as, 83° 22 minutes, 12 seconds.

A radian is defined as: *the angle subtended at the centre of a circle by an arc equal in length to the radius of the circle.*

Now we know that the circumference of a circle is given by

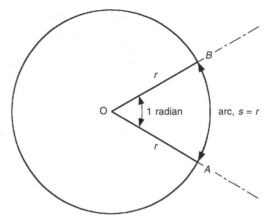

Figure 4.35 *The radian*

$C = 2\pi r$ where r is the radius. Therefore the circumference contains 2π radii. We have just been told that the arc length for 1 radian $s = r$. Therefore the whole circle must contain 2π radians, or approximately 6.28 radians. A circle contains 360°, so it follows that 2π rad = 360° or π rad = 180°. We can use this relationship to convert from degrees to radians and radians to degrees.

Example 4.58

(a) Express 60° in radians.

(b) Express $\dfrac{\pi}{4}$ rad in degrees.

(a) Since 180° = π rad then,

$$1° = \frac{\pi \text{ rad}}{180}$$

so,

$$60° = 60\left(\frac{\pi \text{ rad}}{180}\right)$$

$$60° = \frac{\pi \text{ rad}}{3} \quad \text{or } 1.047 \text{ rad (3 decimal places)}$$

Note that if we leave radians in terms of π we have an exact value to use for further mathematical manipulation. For this reason, it is more convenient to leave radians expressed in terms of π.

(b) We follow a similar argument, except we apply the reverse operations.

$$\pi \text{ rad} = 180°$$

then,

$$1 \text{ rad} = \frac{180°}{\pi}$$

so,

$$\frac{\pi}{4} \text{ rad} = \left(\frac{\pi}{4}\right)\frac{180°}{\pi}$$

and

$$\frac{\pi}{4} \text{ rad} = 30°$$

Key point

π rad = 180°, therefore

$1 \text{ rad} = \dfrac{180°}{\pi}$ degrees or

$\theta \text{ radians} = \dfrac{\theta\pi}{180}$ degrees.

Key point

π rad = 180°, therefore

$1° = \dfrac{\pi}{180}$ radians or

$\theta \text{ degrees} = \dfrac{\theta\pi}{180}$ radians.

To aid your understanding of the relationship between the degree and the radian, Figure 4.36 shows diagrammatically a comparison between some common angles using both forms of measure. Note that, in the figure, all angles in radian measure are shown in terms of π.

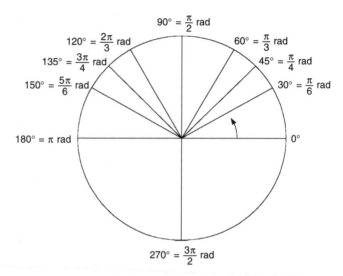

Figure 4.36 *Comparison of some common angles of measure in degrees and radians*

The area of a sector

It is often useful to be able to find the area of a sector when considering cross-sectional areas. To determine such areas, we first need to understand the relationship between the *arc length s* and the *angle* θ subtended at the centre of a circle by this arc length.

You have seen that the circumference of a circle subtends 2π rad. So if we consider the circumference to be an *arc of length* $2\pi r$, we can say that:

$$2\pi \text{ rad} = \frac{2\pi r}{r} \quad \text{where } r = \text{the radius}$$

or,

$$\text{the angle in radians} = \frac{\text{arc length }(s)}{\text{radius }(r)}$$

then,

$$\theta \text{ rad} = \frac{s}{r} \quad \text{or } s = r\theta$$

Always remember that when using this formula, the angle θ must be in *radians*.

The area of a sector is now fairly easy to find.

We know that the area of a circle = πr^2. So it follows that when dealing with a portion (sector) of a circle, like that shown in Figure 4.37, the ratio of the angle θ (in radians) of the sector to that of the angle for the whole circle in radians is $\frac{\theta}{2\pi}$, remembering that there are 2π radians in a circle (360°). Then the area of any portion of the circle such as the area of the sector = the area of the circle multiplied by the ratio of the angles, or in symbols:

Key point

When using the formulae for arc length or area of a sector of a circle, θ must always be in radians.

1. In a right angled triangle, the lengths of the shorter sides are 6 cm and 9 cm. Calculate the length of the hypotenuse.
2. All the sides of a triangle are 8 cm in length. What is the vertical height of the triangle?
3. In Figure 4.38, the angles of elevation of B and C from A are 32° and 62°, respectively. If $DC = 70$ m, calculate the length of BC.
4. A vertical radio mast has cable stays of length 64 m, extending from the top of the mast. If each wire makes an angle of 65° with the ground, find:

   (a) the distance each cable is from the base of the mast
   (b) the vertical height of the mast.

5. State the circumstances under which:

   (a) the sine rule may be used
   (b) the cosine rule may be used.

6. Use the sine rule to solve the triangle ABC where side $a = 37.2$ cm, side $b = 31.6$ cm and $\angle B = 37°$.
7. Use the cosine rule to solve the triangle ABC, where $a = 12$ cm, $b = 10$ cm and $c = 6$ cm. Also find the area of this triangle.
8. Define the radian.
9. Using the diagram of the right-angled triangle given in Figure 4.18 and Pythagoras' theorem, show that $\sin^2 \theta + \cos^2 \theta = 1$. Note that in this formula, $\sin^2 \theta$ and $\cos^2 \theta$ are short-hand ways of

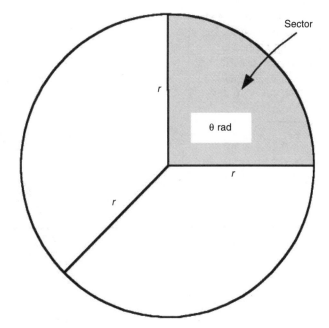

Figure 4.37 *Area of sector of circle*

$$\text{area of sector} = (\pi r^2)\left(\frac{\theta}{2\pi}\right) = \frac{r^2 \theta}{2} \quad (\theta \text{ in radians})$$

Example 4.59

(a) If the angle subtended at the centre of a circle by an arc length 4.5 cm is 120°, what is the radius of the circle?
(b) Find the angle of a sector of radius 20 cm and area 300 cm².

(a) We must first convert 120° into radians. This we can do very easily using the conversion factor we found earlier, then:

$$120° = \frac{120 \, \pi \, \text{rad}}{180} = \frac{2\pi}{3} \, \text{rad}, \quad \text{we will leave this angle in terms}$$

of π

Then from $s = r\theta$ we have $r = \frac{s}{\theta} = \frac{4.5}{2/3\pi} = 2.149$ cm (correct

to 3 decimal places).

(b) To find the angle of the sector we use the area of a sector

formula, that is, $A = \frac{1}{2} r^2 \theta$ or $\theta = \frac{2A}{r^2}$ and on substitution of

given values, we get

$$\theta = \frac{(2)(300)}{20^2} = \frac{600}{400} = 1.5 \, \text{rad}$$

If we wish to convert this angle to degrees, then

$$1.5 \, \text{rad} = (1.5)\frac{180°}{\pi} = 85.94° \text{ (correct to 2 decimal places)}.$$

representing $(\sin \theta)^2$ and $(\cos \theta)^2$, respectively. The short-hand method is nearly always used to represent trigonometric functions, raised to *any power*.

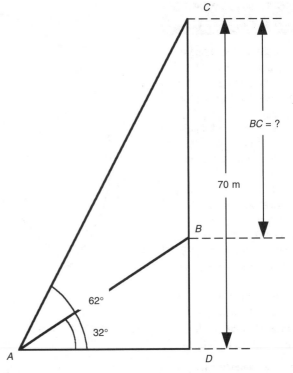

Figure 4.38 *Figure for Test your knowledge 4.9, question 3*

Trigonometric functions

In this short section, we will limit our study of trigonometric functions to the sine and cosine. In particular, we look at the nature of their graphs and the use to which these may be put. The graphs of these functions are very important, as the sine and cosine curves illustrate many kinds of oscillatory motion, which you are likely to meet in your future studies. The sine and cosine functions are used to model the oscillatory motion of currents, voltages, springs, vibration dampers, the rise and fall of the tides and many other forms of vibrating system, where the motion is *oscillatory*.

By oscillatory, we mean motion that vibrates back and forth about some mean value, during even *periods* of time. We start by plotting the sine and cosine curves, then consider their use for solving sine and cosine functions, in a similar manner to the graphs of algebraic equations we considered earlier.

Graphs of sine and cosine functions

The basic sine curve for $y = \sin x$ is a wave which lies between the values $+1$ and -1; it is therefore *bounded*. That is, the value of the dependent variable y reaches a *maximum* value of $+1$ and a *minimum* value of -1 (Figure 4.39). Also the curve is zero at multiples of $180°$ or at multiples of π rad.

The x-axis in Figure 4.39 is marked out in degrees and radians, which measure angular distance; the maximum and minimum values of y are also shown. Other things to know about this graph are the fact that it repeats itself every $360°$ or 2π rad. Also this curve reaches it first maximum value at $90°$ or $\frac{\pi}{2}$ rad; it reaches its second

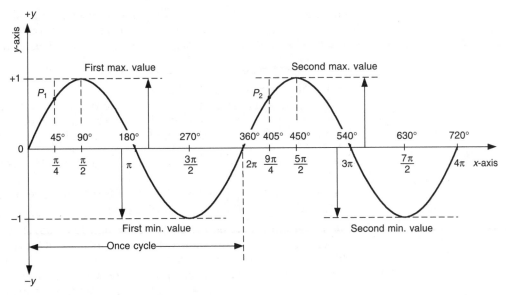

Figure 4.39 *Graph of y = sin x*

maximum 360° or 2π rad later, at 450° or $\frac{5\pi}{2}$ rad. Similarly, it reaches its first minimum value at 450° or $\frac{3\pi}{2}$ rad and again 360° or 2π rad later, at 630° or $\frac{7\pi}{2}$ rad. These *maximum* and *minimum* values are repeated periodically at 360° intervals. We therefore say that the sine wave has *periodic motion*, where any point on the wave, say p_1, repeats itself every 360° or 2π rad. These repetitions are known as *cycles*, as shown on Figure 4.39, where one complete cycle occurs every 360° or every 2π rad.

Now how do we plot values for sinusoidal functions? Look back at Figure 4.36 and note how we represented angular measure. In Figure 4.40, we can represent angular measure on the set of

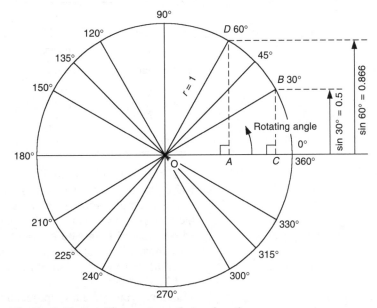

Figure 4.40 *Rotating angle and sine function*

rectangular coordinates, where the *angle* in degrees or radians is measured from the *positive x-axis*, and increases as it rotates in an *anticlockwise* direction, reaching a *positive maximum* value at 90° or $\frac{\pi}{2}$ rad. This maximum value is +1 when we make the radius of the circle $r = 1$, as in the diagram.

Now, the actual magnitude of this angle (its distance in the *y-direction*) is found using the *sine* function. For example, the height of the line *AB* in the triangle *OAB* can be found by noting that $\sin 30° = \frac{\text{opp.}}{\text{hyp.}} = \frac{AB}{1} = AB = 0.5$. Similarly, as the angle increases, say to 60° or $\frac{\pi}{3}$ rad, then $CD = \sin 60° = 0.866$. It reaches its first maximum value when $OE = \sin 90° = 1.0 = $ radius r; compare this value with the value on the curve of the sign function, shown in Figure 4.39! Now as the angle continues to increase, it moves into the second quadrant, where the magnitude of the rotating angle gradually reduces until it reaches 180° or π rad, when its value becomes zero, once more. As we move into the third quadrant, the magnitude of the rotating angle (vector) once again starts to increase, but in a negative sense, until it reaches its maximum value at 270° or $\frac{3\pi}{2}$ rad where $\sin 270° = -1$. Finally in the fourth quadrant, it reduces from the negative maximum (minimum) value until it once again reaches zero. The behaviour of this point is plotted as the curve shown in Figure 4.39, where the curve is produced by connecting the magnitude of this point for many values of the angle between 0 and 360°, after which the pattern repeats itself every 360°.

A table of values for the magnitude of the rotating angle is given below. Check that these values match the plot of the sine curve shown in Figure 4.39.

x = angle θ	$y = \sin θ$	x = angle θ	$y = \sin θ$
0	0		
30° or $\frac{\pi}{6}$ rad	0.5	210°	−0.5
45° or $\frac{\pi}{4}$ rad	0.7071	225°	−0.7071
90° or $\frac{\pi}{2}$ rad	1.0	270°	−1.0
120° or $\frac{2\pi}{3}$ rad	0.8660	300°	−0.866
135° or $\frac{3\pi}{4}$ rad	0.7071	315°	−0.7071
150° or $\frac{5\pi}{6}$ rad	0.5	330°	−0.5
180° or π rad	0	360°	0

The above table is similar to that you would need to produce when plotting *any sine function* graphically.

For example, suppose you were required to plot the curve for the function $y = 2 \sin θ$. What happens to the values of y in the above table? I hope you can see that every value of y is *doubled*. That means the first *maximum* value for this function will be $y = 2 \sin 90° = (2)(1) = 2$; similarly for all other angles, the y values will be doubled.

I hope you can now appreciate that if $y = 3 \sin θ$, then the magnitude of the y values will all be *trebled*. Then in general: *the magnitude of the plotted y values is dependent on the value of the constant a, when y = a* sin θ. The magnitude of the *y* values are referred to as their *amplitude*. Then the *maximum amplitude a* will occur when sin θ *is a maximum*, that is when sin θ = 1.0. This we know from the table above, to occur first at θ = 90° and then to occur every 360° or 2π rad, later. The *minimum* value of the *amplitude* will first occur when sin θ = –1.0, this again can be seen to occur first when θ = 270° and repeat itself every 360° thereafter.

What do you think happens if we plot the graph of *y* = sin 2θ?

Well if $\theta = \frac{\pi}{4}$ rad, then $y = \sin(2)\left(\frac{\pi}{4}\right) = \sin\frac{\pi}{2} = 1.0$. If we compare this with the plotted values above, then the function *y* = sin 2θ has reached its first maximum, *twice as fast* as the function *y* = sin θ. The effect of this is to increase the number of oscillations (cycles) in a given angular distance. This is illustrated in Figure 4.41.

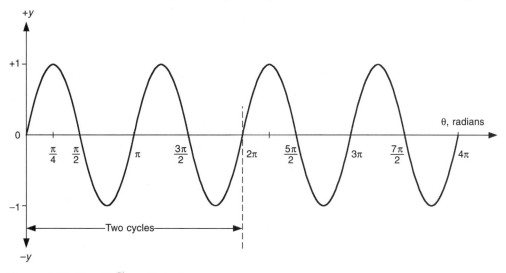

Figure 4.41 *Graph of y = 2 sin θ*

You should check a few of the plotted values to verify your understanding.

The cosine function

So far we have concentrated our efforts on the sine function. This is because the cosine function is very similar to the sine function, except that it reaches its first maximum and minimum values at different angles to that of the sine function. In all other respects it is identical.

Consider again Figure 4.40. In the case of the *cosine function, we start* our rotating angle in the *vertical position*, that is along the line *OE*. This means that what was 90° for the sine function *is now 0° for the cosine function*. This is illustrated in Figure 4.42.

Now, the cosine of the angle 30° is given by the *height of the y-ordinate*, in a similar manner to the sine function, then *y* = cos 30° = 0.866. Similarly the cosine of 90° is again the height of the *y-ordinate*, which can be seen to be zero, i.e. cos 90° = 0, which can easily be checked on your calculator. The net result is that all the cosine function values for the given angle are 90° *in advance* of the

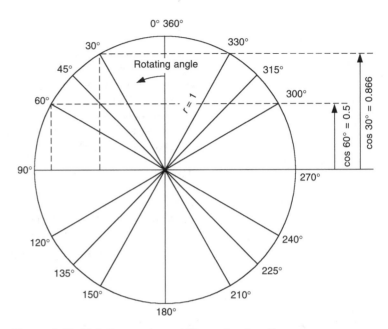

Figure 4.42 *Rotating angle and the cosine function*

sine function. So, for example, the cosine function starts with its maximum at 0°, which is 90° in advance of the first maximum for the sine function. A plot of the cosine function $y = \cos\theta$ for angles between 0 and 4π rad is shown in Figure 4.43.

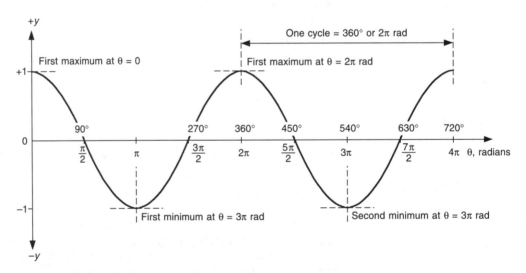

Figure 4.43 *Graph of y = cos θ*

Key point

The cosine function exhibits oscillatory motion identical to that of the sine function except that the periodic motion of the cosine function is advanced by $\frac{\pi}{2}$ radians.

It can be seen from Figure 4.43 that apart from the 90° *advance*, the cosine function follows an identical pattern to that of the sine function. Therefore we can write that:

$$y = \cos\theta = \left(\sin\theta + \frac{\pi}{2}\right)$$

We finish this short section with a couple of examples of the use of graphical plots of these functions and how they can be used to find solutions to simple trigonometric equations.

Example 4.60

Draw the graph of the function $y = 2 \sin \theta + 3 \cos \theta$ for values of θ between 0 and 90°. From the graph find:

(a) the maximum amplitude of the function
(b) a value of θ which satisfies the equation $2 \sin \theta + 3 \cos \theta = 3.5$.

(a) Our first task is to set up a table of values and find the corresponding values for θ and y. We will use an interval of 10°.

θ	0	10	20	30	40	50	60	70	80	90
$2 \sin \theta$	0	0.35	0.68	1.0	1.29	1.53	1.73	1.88	1.97	2.0
$3 \cos \theta$	3	2.95	2.82	2.60	2.30	1.94	1.50	1.03	0.52	0
$y = 2 \sin \theta + 3 \cos \theta$	3	3.3	3.5	3.6	3.59	3.47	3.23	2.91	2.49	2.0

The table shows only two decimal place accuracy, but when undertaking graphical work, it is difficult to plot values with any greater accuracy. Note also that we seem to have a maximum value for y when $\theta = 30°$. It is worth plotting a couple of intermediate values either side of $\theta = 30°$ to see if there is an even higher value of y. I have chosen $\theta = 27°$ and $\theta = 33°$. Then when $\theta = 27°$, $y = 3.58$ and when $\theta = 33°$, $y = 3.61$, the latter value is very slightly higher, so may be used as the maximum.

The plot is shown in Figure 4.44 where it can be seen that within the accuracy of the plot, the maximum value of the amplitude for the function is $y = 3.6$.

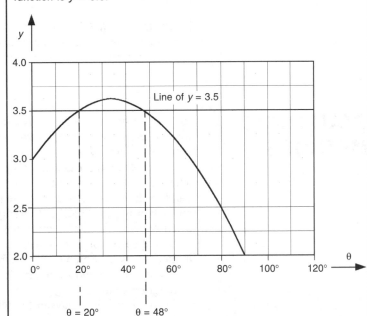

Figure 4.44 *Graph of $y = 2 \sin \theta + 3 \cos \theta$*

(b) Now the appropriate values for the solution of the equation: $2 \sin \theta - 3 \cos \theta = 3.5$ are read off from the graph, where the line $y = 3.5$ intersects with the curve $y = 2 \sin \theta + 3 \cos \theta$. The solutions are that when $y = 3.5$, $\theta = 2.0°$ and $\theta = 48°$.

Test your knowledge 4.10

1. Find the amplitude of the following functions, when
   $$\theta = \frac{\pi}{6}, \frac{\pi}{4} \text{ and } \frac{\pi}{2}.$$
   (a) $y = 2 \sin \theta$
   (b) $y = \cos 2\theta$
   (c) $y = \frac{1}{2} \sin 2\theta$.

2. Plot the graph of the function $y = \sqrt{1 - \cos^2 \theta}$ between $\theta = 0°$ and $\theta = 360°$, using steps of 30° and comment on its shape.

3. Draw the graph of the function $y = 4 \cos \theta - 1$ for values of θ from 0° to 90°. From your graph:
   (a) read of the values of y, when $\theta = 24°$, 30° and 72°
   (b) find the value of θ that satisfies the equation $4 \cos \theta - 1 = 2.6$.

4. Find by a graphical method the maximum amplitude of the function $5 \sin \theta + 12 \cos \theta$.

5. Draw the graph of $y = \tan \theta$ for values of θ from 0° to 180° using intervals of 10° and comment on the nature of the graph as θ approaches 90°.

Example 4.61

For the following trigonometric functions, find the first maximum amplitude and the angular distance it occurs from $\theta = 0°$. Comment on the general form of each function.

1. $y = 4.2 \cos \theta$
2. $y = 3 \sin 2\theta$

3. $y = \sin \left(\theta - \dfrac{\pi}{2} \right)$

1.   The maximum amplitude for all the functions is given when the amplitude a is multiplied by 1.0 in each case.
     We know that for $\cos \theta$ this first occurs when $\theta = 0$, so the maximum amplitude is 4.2 at an angular distance of $0°$ from the reference angle.
     The graph will follow exactly the form of the graph $y = \cos \theta$, except that every value will be amplified by a factor of 4.2.
2.   In this case the maximum amplitude is 3, and it first occurs when $2\theta = 90°$, that is at $+45°$ to the reference angle.
     This graph will complete each cycle, in half the angular distance, when compared to $y = \sin \theta$.
3.   This function has a maximum amplitude of $a = 1.0$, which first occurs when $\theta - \dfrac{\pi}{2} = \dfrac{\pi}{2}$ rad, therefore $\theta = \dfrac{\pi}{2} + \dfrac{\pi}{2} = \pi$ rad.
     That is the first maximum occurs $180°$ after the reference angle. When compared to the function $y = \sin \theta$, each value is found to be lagging by $\dfrac{\pi}{2}$.

If you are finding it difficult to envisage what is happening, sketch these functions on the same axes and make comparisons.

STATISTICAL METHODS

Your view of statistics has probably been formed from what you read in the papers, or what you see on the television. Results of surveys to show which political party is going to win the election, why men grow moustaches, if smoking damages your health, the average cost of housing by area, and all sorts of other interesting data! Well, statistics is used to analyse the results of such surveys and when used correctly, it attempts to eliminate the bias which often appears when collecting data on controversial issues.

Statistics is concerned with collecting, sorting and analysing numerical facts, which originate from several observations. These facts are collated and summarized, then presented as tables, charts or diagrams, etc.

In this brief introduction to statistics, we look at two specific areas. First, we consider the collection and presentation of data in their various forms. Then we look at how we measure such data, concentrating on finding *average values* and seeing how these average values may vary.

If you study statistics beyond this course, you will be introduced to the methods used to make predictions based on numerical data and the probability that your predictions are correct. At this stage in your learning, however, we will be considering only the areas of data manipulation and measurement of central tendency (averages), mentioned above.

Data manipulation

In almost all scientific, engineering and business journals, newspapers and government reports, statistical information is presented in the form of charts, tables and diagrams, as mentioned above. We now look at a small selection of these presentation methods, including the necessary manipulation of the data used to produce them.

Charts

Suppose, as the result of a survey, we are presented with the following statistical data.

Major category of employment	Number employed
Private business	750
Public business	900
Agriculture	200
Engineering	300
Transport	425
Manufacture	325
Leisure industry	700
Education	775
Health	500
Other	125

Now ignoring for the moment the accuracy of this data, let us look at typical ways of presenting this information in the form of charts, in particular the *bar chart* and *pie-chart*.

The bar chart

In its simplest form, the bar chart may be used to represent data by drawing individual bars (Figure 4.45), using the figures from the raw data (the data in the table).

Now the scale for the vertical axis, the number employed, is easily decided by considering the highest and lowest values in the table, 900 and 125, respectively. Therefore we use a scale from 0 to 1000 employees. Along the horizontal axis, we represent each category by a bar of even width. We could just as easily have chosen to represent the data using column widths instead of column heights.

Now the simple bar chart in Figure 4.45 tells us very little that we could not have determined from the table. So another type of bar chart that enables us to make comparisons, the *proportionate* bar chart, may be used.

In this type of chart, we use *one bar*, with the same width throughout its height, with horizontal sections marked off in proportion to the whole. In our example, each section would represent the number of people employed in each category, compared with the total number of people surveyed.

Number employed

Figure 4.45 *Bar chart representing number employed by category*

In order to draw a proportionate bar chart for our employment survey, we first need to total the number of people who took part in the survey, this total comes to 5000. Now, even with this type of chart we may represent the data either in proportion by height or in proportion by percentage. If we were to chose height, then we need to set our vertical scale at some convenient height, say 10 cm. Then we would need to carry out ten simple calculations to determine the height of each individual column.

For example, given that the height of the total 10 cm represents 5000 people, then the height of the column for those employed in

private business $= \left(\dfrac{750}{5000} \right) 10 = 1.5$ cm; this type of calculation is

then repeated for each category of employment. The resulting bar chart is shown in Figure 4.46.

Example 4.62

Draw a proportionate bar chart for the employment survey shown in the table using the percentage method.

For this method all that is required is to find the appropriate percentage of the total (5000) for each category of employment. Then, choosing a suitable height of column to represent 100%, mark on the appropriate percentage for each of the ten employment categories. To save space, only the first five categories of employment have been calculated in full.

1. Private business $= \left(\dfrac{750}{5000} \right) \times 100 = 15\%.$

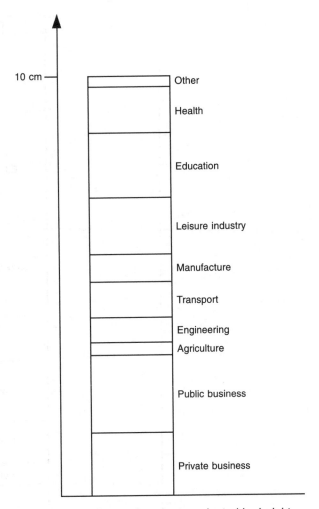

Figure 4.46 *A proportionate bar chart graduated by height*

2. Public business $= \left(\dfrac{900}{5000} \right) \times 100 = 18\%$.

3. Agriculture $= \left(\dfrac{200}{5000} \right) \times 100\% = 4\%$.

4. Engineering $= \left(\dfrac{300}{5000} \right) \times 100\% = 6\%$.

5. Transport $= \left(\dfrac{425}{5000} \right) \times 100 = 8.5\%$.

Similarly: manufacture = 6.5%, leisure industry = 14%, education = 15.5%, health = 10% and other category = 2.5%.
 Figure 4.47 shows the completed bar chart.

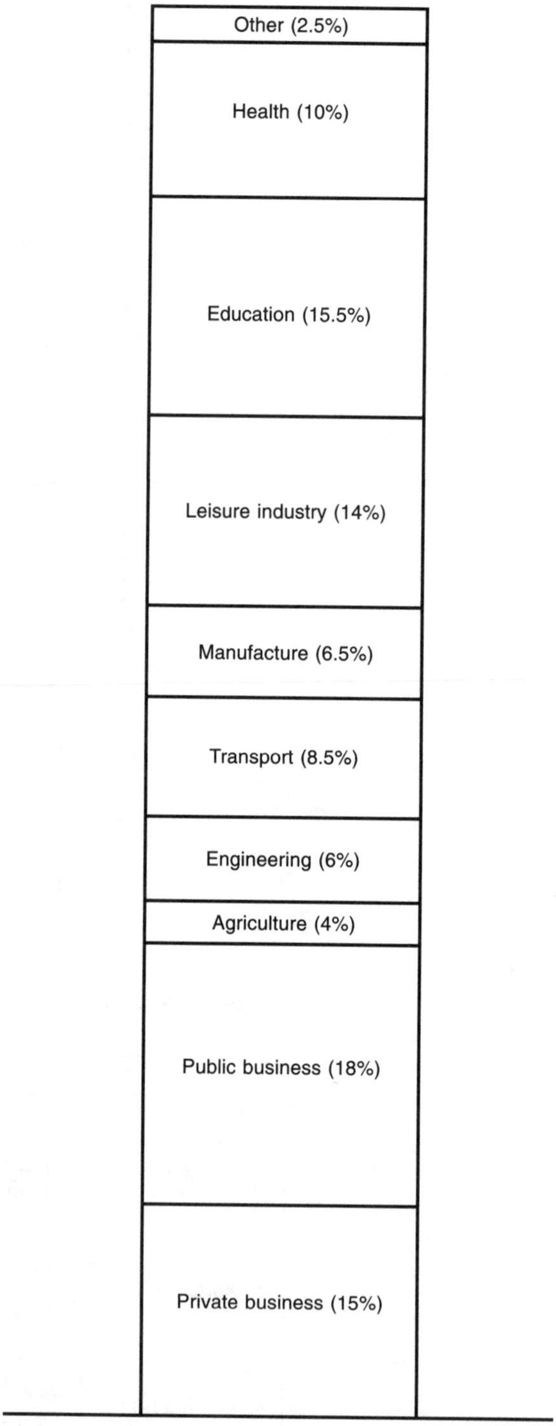

Figure 4.47 *Proportionate percentage bar chart for employment by industry*

Other categories of bar chart include *horizontal bar charts*, where for instance Figure 4.45 is turned through 90° in a clockwise direction. One last type may be used to depict data given in chronological (time) order. Thus, for example, the horizontal *x*-axis is used to represent hours, days, years, etc., while the vertical axis shows the variation of the data with time.

Example 4.63

Represent the following data on a chronological bar chart.

Year	Number employed in general engineering (thousands)
1995	800
1996	785
1997	690
1998	670
1999	590

Since we have not been asked to represent the data on any *specific bar chart* we will use the simplest, involving only the raw data. Then the only concern is the *scale* we should use for the vertical axis. To present a *true* representation, the scale should start from zero and extend to, say, 800 (Figure 4.48(a)). If we wish to emphasize a *trend*, that is the way the variable is rising or falling with time, we could use a very much exaggerated scale (Figure 4.48(b)). This immediately emphasizes the *downward trend* since 1995. Note that this data is *fictitious* (made up) and used here merely for emphasis!

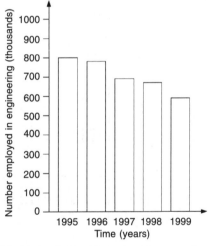

Figure 4.48 *(a) Chronological bar chart in correct proportion*

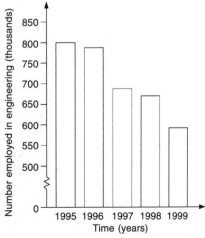

Figure 4.48 *(b) Chronological bar chart with graduated scale*

Pie-chart

In this type of chart the data is presented as a proportion of the total, using the angle or area of sectors. The method used to draw a pie chart is best illustrated by example.

Example 4.64

Represent the data given in Example 4.63 on a pie-chart.

Then, remembering that there are 360° in a circle and that the total number employed in general engineering (according to our figures) was 800 + 785 + 690 + 670 + 590 = 3535 (thousand), then we manipulate the data as follows:

Year	Number employed in general engineering (thousands)	Sector angle (to nearest half degree)
1995	800	$\left(\dfrac{800}{3535}\right) \times 360 = 81.5°$
1996	785	$\left(\dfrac{785}{3535}\right) \times 360 = 80°$
1997	690	$\left(\dfrac{690}{3535}\right) \times 360 = 70.5°$
1998	670	$\left(\dfrac{670}{3535}\right) \times 360 = 68°$
1999	590	$\left(\dfrac{590}{3535}\right) \times 360 = 60°$
Total	3535	$= 360°$

The resulting pie-chart is shown in Figure 4.49.

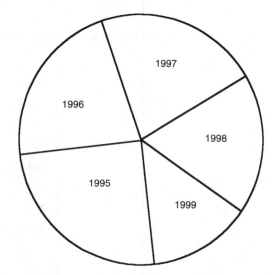

Figure 4.49 *Figure for Example 4.64 – pie-chart for numbers employed in general engineering for the period 1995–1999*

Key point

Charts and graphs offer an effective visual stimulus for the presentation of statistical data.

Other methods of visual presentation include pictograms and ideographs. These are diagrams in pictorial form, used to present information to those who have a limited interest in the subject matter or who do not wish to deal with data presented in numerical form. They have little or no practical use when interpreting engineering or other scientific data and apart from acknowledging their existence, we will not be pursuing them further.

Frequency distributions

One of the most common and most important ways of organizing and presenting raw data is through use of *frequency distributions*.

Consider the data given below that show the time in hours that it took 50 individual workers to complete a specific assembly line task.

Data for assembly line task

1.1	1.0	0.6	1.1	0.9	1.1	0.8	0.9	1.2	0.7
1.0	1.5	0.9	1.4	1.0	0.9	1.1	1.0	1.0	1.1
0.8	0.9	1.2	0.7	0.6	1.2	0.9	0.8	0.7	1.0
1.0	1.2	1.0	1.0	1.1	1.4	0.7	1.1	0.9	0.9
0.8	1.1	1.0	1.0	1.3	0.5	0.8	1.3	1.3	0.8

From the data you should be able to see that the shortest time for completion of the task was 0.5 hours, the longest time was 1.5 hours. The *frequency* of appearance of these values is *once*. On the other hand, the number of times the job took 1 hour appears 11 times, or it has a *frequency of 11*. Trying to sort out the data in this ad-hoc manner is time consuming and may lead to mistakes. To assist with the task we use a *tally chart*. This chart simply shows how many times the *event* of completing the task in a specific time takes place. To record the *frequency of events* we use the number 1 in a tally chart and when the frequency of the event reaches 5, we score through the existing four 1s to show a frequency of 5. The following example illustrates the procedure.

Example 4.65

Use a tally chart to determine the frequency of events for the data given above on the assembly line task.

Time (hours)	Tally	Frequency
0.5	1	1
0.6	11	2
0.7	1111	4
0.8	1111 1	6
0.9	1111 111	8
1.0	1111 1111 1	11
1.1	1111 111	8
1.2	1111	4
1.3	111	3
1.4	11	2
1.5	1	1
		Total 50

> We now have a full numerical representation of the *frequency of events*. So, for example, 8 people completed the assembly task in 1.1 hours or the time 1.1 hours has a frequency of 8. We will be using the above information later on, when we consider measures of central tendency.

The times in hours given in the above data are simply numbers. When data appears in a form where it can be *individually counted* we say that it is *discrete* data. It goes up or down in *countable* steps. Thus the numbers 1.2, 3.4, 8.6, 9, 11.1, 13.0 are said to be *discrete*. If, however, data are obtained by measurement, for example the heights of a group of people, then we say that this data is *continuous*. When dealing with continuous data, we tend to quote its limits, that is the limit of accuracy with which we take the measurements. So, for example, a person may be 174 ± 0.5 cm in height. When dealing numerically with continuous data or a large amount of discrete data, it is often useful to *group* this data into *classes or categories*. We can then find out the numbers (frequency) of items within each group.

Table 4.4 shows the height of 200 adults grouped into 10 classes.

Table 4.4

Height (cm)	Frequency
150–154	4
155–159	9
160–164	15
165–169	21
170–174	32
175–179	45
180–184	41
185–189	22
190–194	9
195–199	2
	Total 200

Key point

The grouping of frequency distributions is a means for clearer presentation of the facts.

Key point

The frequencies of a distribution may be added consecutively to produce a graph known as a cumulative frequency distribution or ogive.

The main advantage of grouping is that it produces a clear overall picture of the frequency distribution. In the table the first class interval is 150–154. The end number 150 is known as the *lower limit* of the class interval; the number 154 is the *upper limit*. The heights have been measured to the nearest centimetre. That means within ±0.5 cm. Therefore, in effect, the first class interval includes all heights between the range 149.5 and 154.5 cm these numbers are known as the lower and upper class *boundaries*, respectively. The *class width* is always taken as the *difference between the lower and upper class boundaries*, not the upper and lower limits of the class interval.

The histogram

The histogram is a special diagram that is used to represent a frequency distribution, such as that for grouped heights, shown in

Table 4.4. It consists of a set of rectangles, whose areas represent the frequencies of the various classes. Often when producing these diagrams, the class width is kept the same, so that the varying frequencies are represented by the height of each rectangle. When drawing histograms for grouped data, the *midpoints* of the rectangles represent the midpoints of the class intervals. So for our data they will be 152, 157, 162, 167, etc.

An adaptation of the histogram, known as the *frequency polygon*, may also be used to represent a frequency distribution.

Example 4.66

Represent the above data showing the frequency of the height of groups of adults on a histogram and draw in the frequency polygon for this distribution.

All that is required to produce the histogram is to plot frequency against the height intervals, where the intervals are drawn as class widths.

Then, as can been seen from Figure 4.50, the area of each part of the histogram is the *product of frequency × class width*. The frequency polygon is drawn so that it connects the *midpoint* of the class widths.

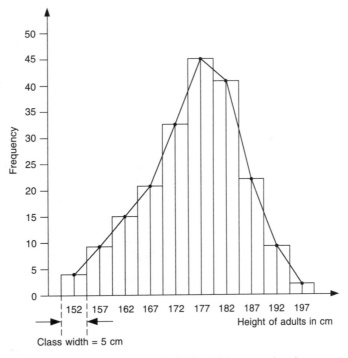

Figure 4.50 *Figure for Example 4.66 – histogram showing frequency distribution*

Statistical measurement

When considering statistical data it is often convenient to have one or two values which represent the data as a whole. Average values are often used. For example, we might talk about the average height of females in the United Kingdom being 170 centimetres, or that the average shoe size of British males is size 9. In statistics we may represent these average values using the mean, median or mode of the data we are considering.

If we again consider the hypothetical data on the height of females. We may also wish to know how their individual heights vary or deviate from their average value. Thus, we need to consider measures of *dispersion*, in particular, *mean deviation, standard deviation and variance* for the data concerned. These *statistical averages* and the way they *vary* are considered next.

The arithmetic mean

The arithmetic mean or simply the *mean* is probably the average with which you are most familiar. For example, to find the arithmetic mean of the numbers 8, 7, 9, 10, 5, 6, 12, 9, 6, 8, all we need do is to add them all up and divide by how every many there are, or more formally:

$$\text{arithmetic mean} = \frac{\text{arithmetic total of all the individual values}}{\text{number of values}}$$

$$= \frac{\sum n}{n}$$

where the Greek symbol $\sum$ = the sum of the individual values $x_1 + x_2 + x_3 + x_4 + \ldots x_n$ and n = the number of these values in the data.

So, for the *mean* of our *ten* numbers, we have:

$$\text{mean} = \frac{\sum n}{n} = \frac{8 + 7 + 9 + 10 + 5 + 6 + 12 + 9 + 6 + 8}{10}$$

$$= \frac{80}{10} = 8$$

Now, no matter how long or complex the data we are dealing with, *provided* that we are only dealing with individual values (discreet data), the above method will always produce the arithmetic mean. The mean of all the *x values* is given the symbol $\bar{x}$, pronounced, *x bar*.

Example 4.67

The height of 11 females was measured as follows: 165.6 cm, 171.5 cm, 159.4 cm, 163 cm, 167.5 cm, 181.4 cm, 172.5, 179.6 cm, 162.3 cm, 168.2 cm, 157.3 cm. Find the mean height of these females. Then, for $n = 12$:

$$\bar{x} = \frac{\begin{array}{c}165.6 + 171.5 + 159.4 + 163 + 167.5 + 181.4 + 172.5 \\ + 179.6 + 162.3 + 168.2 + 157.3\end{array}}{12}$$

$$\bar{x} = \frac{1848.3}{11} = 168.03 \text{ cm}$$

Mean for grouped data

What if we are required to find the mean for *grouped data*? Look back at Table 4.4 showing the height of 200 adults, grouped into *ten* classes. In this case, the *frequency* of the heights needs to be taken into account.

We select the *class midpoint x*, as being the average of that class and then multiply this value by the frequency (*f*) of the class, so that a value for that particular class is obtained (*fx*). Then, by adding up all class values in the frequency distribution, the total value for the distribution is obtained ($\Sigma\,fx$). This total is then divided by the *sum of the frequencies* (Σf) in order to determine the mean. So, for grouped data:

$$\bar{x} = \frac{f_1x_1 + f_2x_2 + f_3x_3 + \ldots + f_nx_n}{f_1 + f_2 + f_3 + \ldots + f_n} = \frac{\Sigma\,(f \times \text{midpoint})}{\Sigma f}$$

This rather complicated looking procedure is best illustrated by example.

Example 4.68

Determine the mean value for the heights of the 200 adults using the data in Table 4.4.

The values for each individual class are best found by producing a table, using the class *midpoints* and *frequencies*. Remembering that the *class midpoint* is found by dividing the *sum of the upper and lower class boundaries by 2*. So, for example, the mean value for the first class interval is $\frac{149.5 + 154.5}{2} = 152$. The completed table is shown below.

Midpoint (*x*) of height (cm)	Frequency (*f*)	fx
152	4	608
157	9	1413
162	15	2430
167	21	3507
172	32	5504
177	45	7965
182	41	7462
187	22	4114
192	9	1728
197	2	394
Total	$\Sigma f = 200$	$\Sigma fx = 35\,125$

I hope you can see how each of the values was obtained. When dealing with relatively large numbers be careful with your arithmetic, especially when you are keying in variables into your calculator!

Now that we have the required total the mean value of the distribution can be found.

mean value $\bar{x} = \dfrac{\Sigma fx}{\Sigma f} = \dfrac{35\,125}{200} = 175.625 \pm 0.5$ cm

Notice that our mean value of heights has the same margin of error as the original measurements. The value of the mean cannot be any more accurate than the measured data from which it was found!

Median

When some values within a set of data vary quite widely, the arithmetic mean gives a rather poor representative average of such data. Under these circumstances another more useful measure of the average is the *median*.

For example, the mean value of the numbers 3, 2, 6, 5, 4, 93, 7 is 20, which is not representative of any of the numbers given. To find the median value of the same set of numbers, we simply place them in *rank order*, that is 2, 3, 4, 5, 6, 7, 93. Then we select the middle (median) value. Since there are seven numbers (items) we choose the fourth item along, the number 5, as our *median value*.

If the number of items in the set of values is *even*, then we add together the value of the *two middle terms* and divide by 2.

Example 4.69

Find the mean *and* median value for the set of numbers 9, 7, 8, 7, 12, 70, 68, 6, 5, 8. The arithmetic mean is found as:

$$\text{mean } \bar{x} = \frac{9 + 7 + 8 + 7 + 12 + 70 + 68 + 6 + 5 + 8}{10} = \frac{200}{10} = 20$$

This value is not really representative of any of the numbers in the set.

To find the *median* value, we first put the numbers in *rank order*, that is:

5, 6, 7, 7, 8, 8, 9, 12, 68, 70

Then from the ten numbers, the two middle values, the 5th and 6th values along, are 8 and 8. So the median value $= \dfrac{8 + 8}{2} = 8$.

Mode

Yet another measure of central tendency for data containing extreme values is the *mode*. Now the *mode* of a set of values containing discreet data is the value that occurs most often. So for the set of values 4, 4, 4, 5, 5, 5, 5, 6, 6, 6, 7, 7, 7, the mode or *modal value is 5*, as this value occurs four times. Now it is possible for a set of data to have more than one mode, for example the data used in Example 4.69 above has two modes, 7 and 8, both of these numbers occurring twice and both occurring more than any of the others. A set of data may not have a modal value at all, for example the numbers 2, 3, 4, 5, 6, 7, 8 all occur once and there is no mode.

A set of data that has one mode is called *unimodal*; data with two modes are *bimodal* and data with more than two modes are known as *multimodal*.

When considering *frequency distributions* for grouped data, the *modal class* is that group which occurs most frequently. If we wish to find the actual *modal value* of a frequency distribution, we need to draw a histogram.

Example 4.70

Find the modal class and modal value for the frequency distribution on the height of adults given in Table 4.4.

Referring back to the table, it is easy to see that the class of heights which occurs most frequently is 175–179 cm, which occurs 45 times.

Now, to find the modal value we need to produce a histogram for the data. We did this for Example 4.66. This histogram is shown again in Figure 4.51 with the modal shown.

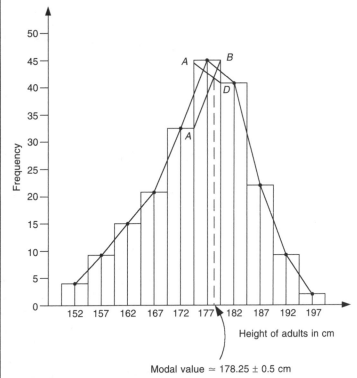

Modal value ≈ 178.25 ± 0.5 cm

Figure 4.51 *Histogram showing frequency distribution and modal value for height of adults*

From Figure 4.51 it can be seen that the modal value = 178.25 ± 0.5 cm.

This value is obtained from the intersection of the two construction lines, *AB* and *CD*. The line *AB* is drawn diagonally from the highest value of the preceding class, up to the top right-hand corner of the modal class. The line *CD* is drawn from the top left-hand corner of the modal group to the lowest value of the next class, immediately above the modal group. Then, as can be seen, the *modal value* is read off where the projection line meets the *x*-axis.

Key point

The mean, median and mode are statistical averages, or measures of central tendency for a statistical distribution.

Mean deviation

We talked earlier of the need not only to consider statistical averages, which give us some idea of the position of a distribution, but also the need to consider how the data are *dispersed* or *spread* about this average value. Figure 4.52 illustrates this idea, showing how the data taken from two distributions is dispersed about the same mean value.

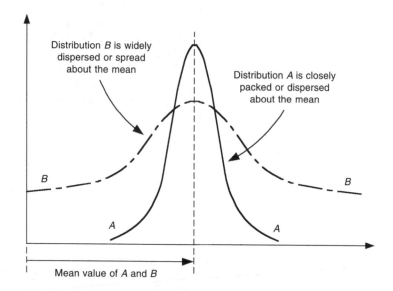

Figure 4.52 *Deviation from the mean value for a distribution*

A measure of dispersion which is often used is the *mean deviation*. To determine the deviation from the statistical average (mean, median or mode), we proceed in the following way.

We first find the statistical average for the distribution, the mean, median or mode ($\bar{x}$). We then find the difference between this average value and each of the individual values in the distribution. We then add up all these differences and divide by the number of individual values in the distribution. This all sounds rather complicated, but the mean deviation may be calculated quite easily using the formula:

$$\text{Mean deviation} = \frac{\Sigma |x - \bar{x}|}{n}$$

where x = a data value in the distribution, $\bar{x}$ = the statistical average, mean, median or mode, as before, and n = the number of individual items in the distribution as before. The $| \, |$ brackets tell us to use the *positive* value of the result contained within the brackets. So, for example, if $x = 12$ and $\bar{x} = 16$, then $|x - \bar{x}| = |12 - 16| = |-4| = +4$; we use the *positive* value, even though in this case the result was negative.

For *frequency distributions*, using grouped data, we find the deviation from the mean, using a similar formula to that we used to find the arithmetic mean. Where the only addition is to multiply the individual differences from the mean by their frequency, then, for a frequency distribution:

$$\text{mean deviation} = \frac{\Sigma f |x - \bar{x}|}{\Sigma f}$$

Example 4.71

Calculate the mean deviation from the arithmetic mean for the data shown in the table.

Length of rivet (mm)	9.8	9.9	9.95	10.0	10.05	10.1	10.2
Frequency	3	18	36	62	56	20	5

The easiest way to tackle this problem is to set up a *table of values* in a similar manner to the table we produced for Example 4.68. The headings for such a table being taken from the above formula for finding the mean deviation for a frequency distribution.

Table of values

Rivet length (x)	f	fx	$\lvert x - \bar{x} \rvert$	$f \lvert x - \bar{x} \rvert$
9.8	3	29.4	0.202	0.606
9.9	18	178.2	0.108	1.944
9.95	36	358.2	0.058	2.088
10	62	620	0.008	0.496
10.05	56	562.8	0.03	1.68
10.1	20	202	0.092	1.84
10.2	5	51	0.192	0.92
Total	$\Sigma f = 200$	$\Sigma fx = 2001.6$		$\Sigma f(\lvert x - \bar{x} \rvert = 9.574$

The arithmetic mean $\bar{x} = \dfrac{\Sigma fx}{\Sigma f} = \dfrac{2001.6}{200} = 10.008.$

$\bar{x} = 10.008$ was required to complete the last two columns in the table. Then the *mean deviation* from the mean of the rivet lengths is:

$$\frac{\Sigma f \lvert x - \bar{x} \rvert}{\Sigma f} = \frac{9.574}{200} = 0.04787 \text{ mm} \approx 0.05 \text{ mm}$$

This small average deviation from the arithmetic mean for rivet length is what we would expect in this case. The deviation being due to very small manufacturing errors. This is, therefore, an example of a frequency distribution tightly packed around the average for the distribution.

Key point

The mean deviation is a measure of the way a distribution deviates from its average value.

Standard deviation

The most important method in determining how a distribution is *dispersed* or *spread* around its average value is known as *standard deviation*. To find this measure of dispersion requires just one or two additional steps from those we used to find the mean deviation.

These additional mathematical steps involve further manipulation of the $\lvert x - \bar{x} \rvert$ or $f \lvert x - \bar{x} \rvert$ values we needed to find when calculating the mean deviation for discreet or grouped data. The additional steps require us to first *square* these differences, then find *their* mean and finally take their *square root* to reverse the squaring process. This strange way of manipulating these differences is known as the *root mean square deviation or standard deviation*, which is identified using the Greek symbol sigma (σ).

Thus, for frequency distributions with grouped data, we can represent these *three further processes* mathematically, as follows:

1. Square the differences and multiply by their frequency $= f|x - \bar{x}|^2$.

2. Sum all of these values and find their mean $= \dfrac{\Sigma f|x - \bar{x}|^2}{\Sigma f}$;

   this is a similar step to the way in which we found the mean deviation. The value of the deviation found at this stage is known as the *variance*.

3. Now take the square root of these mean squares to reverse the squaring process $= \sqrt{\dfrac{\Sigma f|x - \bar{x}|^2}{\Sigma f}}$.

Then the *standard deviation*, $\sigma = \sqrt{\dfrac{\Sigma f(x - \bar{x})^2}{\Sigma f}}$.

The | | brackets have been replaced by ordinary brackets in this final version of the formula. This is because, when we square any quantity, whether positive or negative, the result is always positive, by the law of signs! It is therefore no longer necessary to use the special brackets.

This particular value of deviation is more representative than the mean deviation value we found before, because it takes account of data that may have large differences between items, in a similar way to the use of the mode and median when finding average values.

When considering discreet ungrouped date, we apply the same steps as above to the differences $|x - \bar{x}|$ and obtain $\sqrt{\dfrac{\Sigma(x - \bar{x})^2}{n}}$, therefore for *ungrouped* data:

standard deviation $\sigma = \sqrt{\dfrac{\Sigma(x - \bar{x})^2}{n}}$

Note that once again we have removed the special brackets for the same reason as given above for grouped data.

Key point

The standard deviation as a measure of deviation from the statistical average takes into account data with extreme values, that is data that it statistically skewed.

Example 4.72

For the set of numbers 8, 12, 11, 9, 16, 14, 12, 13, 10, 9, find the arithmetic mean and the standard deviation.

Like most of the examples concerning central tendency and deviation measure, we will solve this problem by setting up a table of values. We will also need to find the arithmetic mean before we are able to complete the table, where in this case for non-grouped data $n = 10$.

Then, $\bar{x} = \dfrac{\Sigma x}{n} = \dfrac{8 + 12 + 11 + 9 + 16 + 14 + 12 + 13 + 10 + 9}{10}$

$\qquad = \dfrac{114}{10} = 11.4$

Table of values

X	$(x - \bar{x})$	$(x - \bar{x})^2$
8	−3.4	11.56
12	0.6	0.36
11	−0.4	0.16
9	−2.4	5.76
16	4.6	21.16
14	2.6	6.76
12	0.6	0.36
13	1.6	2.56
10	−1.4	1.96
9	−2.4	5.76
$\Sigma x = 114$		$\Sigma (x - \bar{x})^2 = 56.4$

Then from the table of values:

standard deviation $\sigma = \sqrt{\dfrac{\Sigma (x - \bar{x})^2}{n}} = \sqrt{\dfrac{56.4}{10}} = \sqrt{5.64} = 2.375$

Another measure of dispersion is the *variance* which is simply the value of the standard deviation before taking the square root, so in this example:

variance $= \dfrac{\Sigma (x - \bar{x})^2}{n} = \dfrac{56.4}{10} = 5.64.$

So when finding the standard deviation, you can also find the variance.

Finally, make sure you can obtain the values given in the table!

We finish our short study of standard deviation with one more example for grouped data.

Example 4.73

Calculate the standard deviation for the data on rivets given in Example 4.71. For convenience the data from Example 4.71 are reproduced here.

Length of rivet (mm)	9.8	9.9	9.95	10.0	10.05	10.1	10.2
Frequency	3	18	36	62	56	20	5

Now in Example 4.71 we calculated the arithmetic mean and *mean deviation*. Using a table of values, we obtained:

Rivet length (x)	F	fx	$\|x - \bar{x}\|$	$f\|x - \bar{x}\|$
9.8	3	29.4	0.202	0.606
9.9	18	178.2	0.108	1.944
9.95	36	358.2	0.058	2.088
10	62	620	0.008	0.496
10.05	56	562.8	0.03	1.68
10.1	20	202	0.092	1.84
10.2	5	51	0.192	0.92
Total	$\Sigma f = 200$	$\Sigma fx = 2001.6$		$\Sigma f\|x - \bar{x}\| = 9.574$

The arithmetic mean we found as $\bar{x} = \dfrac{\Sigma fx}{\Sigma f} = \dfrac{2001.6}{200} = 10.008$.

So, having found the mean, all we need do now to find the *standard deviation* is modify the table by adding in the extra steps. We then obtain:

Rivet length (x)	f	fx	$(x - \bar{x})$	$(x - \bar{x})^2$	$f(x - \bar{x})^2$
9.8	3	29.4	−0.202	0.040804	0.122412
9.9	18	178.2	−0.108	0.011664	0.209952
9.95	36	358.2	−0.058	0.003364	0.121104
10	62	620	−0.008	0.000064	0.003968
10.05	56	562.8	0.03	0.0009	0.0504
10.1	20	202	0.092	0.008464	0.16928
10.2	5	51	0.192	0.036864	0.18432
Total	200	2001.6			0.861436

Then from the table $\Sigma f = 200$, $\Sigma f(x - \bar{x})^2 = 0.861436$ and

standard deviation $\sigma = \sqrt{\dfrac{\Sigma f(x - \bar{x})^2}{\Sigma f}} = \sqrt{\dfrac{0.861436}{200}} = 0.066$ mm

This value is slightly more accurate than the value we found in Example 4.71 for the mean deviation = 0.05 mm, but as you can see, there is also a lot more arithmetic manipulation! Again, you should make sure that you are able to obtain the additional values shown in the table.

CALCULUS

Introduction

Meeting the calculus for the first time is often a rather daunting business. In order to appreciate the power of this branch of mathematics we must first attempt to define it. So, what is the calculus and what is its function?

Imagine driving a car or riding a motor cycle starting from rest, over a measured distance, say one kilometre. If your time for the run was 25 seconds, then we can find your average speed over the measured kilometre from the fact that speed = distance/time. Then using consistent units, your average speed would be 1000 m/25 s or 40 m s^{-1}. This is fine but suppose you were testing the vehicle and we needed to know its *acceleration* after you had driven 500 m. In order to find this we would need to determine how the vehicle speed *was changing* at this exact point because *the rate at which your vehicle speed changes is its acceleration*. To find such things as rate of change of speed, we can use *the calculus*.

The calculus is split into two major areas: the *differential calculus* and the *integral calculus*.

The *differential calculus* is a branch of mathematics concerned with finding how things *change with respect to variables such as time, distance or speed*, especially when these changes are *continually* varying. In engineering, we are interested in the study of motion and the way this motion in machines, mechanisms and vehicles varies with time. The way in which pressure, density and temperature change with height or time. Also, how electrical quantities vary with time, such as electrical charge, alternating current, electrical power, etc. All these areas may be investigated using the *differential calculus*.

The *integral calculus* has two primary functions. It can be used to find the length of arcs, surface areas or volumes enclosed by a surface. Its second function is that of *anti-differentiation*, for example if we use the differential calculus to find the rate of change of distance of our motor bike, with respect to time. In other words we have found its instantaneous speed. We can then use the *inverse process*, the integral calculus to determine the original distance covered by the motor bike from its instantaneous speed.

The mathematical process we use when applying the differential calculus is known as *differentiation* and when using the integral calculus, the mathematical process we apply is known as *integration*.

Before we can apply the calculus to meaningful engineering problems, we need first to understand the notation and ideas that underpin these applications. Thus at this level we spend the majority of our time looking at the basic arithmetic of the calculus that will enable us to *differentiate* and *integrate* a very small number of mathematical functions. If you study further mathematics, you will gain sufficient knowledge to be able to apply the calculus to realistic engineering problems.

We start our study with some introductory terminology and notation, which you will need in order to carry out *calculus arithmetic*.

Key point

The differential calculus is concerned with rates of change.

Key point

The integral calculus is anti-differentiation and is concerned with summing things.

Functions

When studying any new topic, you are going to be introduced to a range of new terms and definitions, unfortunately the calculus is no exception! We have mentioned functions throughout your study of mathematics. It is now time to investigate the concept of the *function* in a little more detail, before we consider differentiation and integration of such functions.

A *function is a many–one mapping or a one–one mapping*. An example of a one–one mapping (a function) is a car which has a unique licence plate number. All cars have a licence number, but for each vehicle that number is unique, so we say that the licence plate is a function of the vehicle. There are many people who may have an intelligence quotient (IQ) of 120, this is an example of a many–one mapping or function. Many people will map to an IQ of 120. What about mathematical functions?

Consider the function $y = x^2 + x - 6$; this is a mathematical function because for any one value given to the *independent variable* x, we get a corresponding value for the *dependent variable y*. We say that y *is a function of x*. For example, when $x = 2$ then $y = (2)^2 + (2) - 6 = 0$. When dealing with mathematical functions we often represent them using $f(x)$ as the dependent variable, instead of y, i.e. $f(x) = x^2 + x - 6$, where the letter inside the brackets represents the independent variable. So, for example, $f(t) = t^2 + t - 6$ is the

Key point

A function is a many–one or one–one mapping.

function f with respect to the independent variable t, which may, for example, represent time.

Now when we assign a value to the *independent variable*, this value is placed inside the bracket and the expression is then evaluated for the chosen value. So if $t = 3$, then we write $f(3) = (3)^2 + (3) - 6 = 6$; similarly, $f(-3) = (-3)^2 + (-3) - 6 = 9 - 3 - 6 = 0$. In fact any value of the independent variable may be substituted in this manner.

Example 4.74

If the distance travelled in metres by a slow moving earth vehicle is given by the function $f(t) = \dfrac{t^2 + t}{2} + 50$, find the distance travelled by this vehicle at $t = 0$, $t = 2.4$ and $t = 5.35$ seconds.

So this is a function that relates distance $f(t)$ and time (t) in seconds. Therefore to find the dependent variable $f(t)$, all we need do is substitute the time variable t into the function. Then for

$$t = 0 \text{ seconds}, \ f(0) = \frac{(0)^2 + (0)}{2} + 50 = 50 \text{ m}$$

and, similarly, when

$$t = 2.4 \text{ seconds}, \ f(2.4) = \frac{(2.4)^2 + 2.4}{2} + 50 = 54.08 \text{ m}$$

and, similarly, when

$$t = 5.35 \text{ seconds}, \ f(5.35) = \frac{(5.35)^2 + 5.35}{2} = 50 = 66.99 \text{ m}$$

We can extend this idea a little further by considering *how the distance changes with time* for the function $f(t) = \dfrac{t^2 + t}{2} + 50$. We will show graphically how the distance $f(t)$, for this quadratic function, varies with time t, between $t = 0$ and $t = 10$ seconds.

Example 4.75

1. Draw the graph for the function $f(t) = \dfrac{t^2 + t}{2} + 50$, relating the distance $f(t)$ in metres to the time t in seconds, between $t = 0$ and $t = 10$ seconds, using intervals of 1.0 second.
2. From your graph find:

    (a) the distance at time $t = 6.5$ seconds
    (b) the time it takes to reach a distance of 90 m.

3. What does the *slope* of the graph indicate?

1. You have drawn graphs of quadratic functions when you studied your algebra. We will set up a table of values in the normal manner and then use these values to plot the graph (Figure 4.53).

t	0	1	2	3	4	5	6	7	8	9	10
t^2	0	1	4	9	16	25	36	49	64	81	100
$+t$	0	2	6	12	20	35	42	56	72	90	110
$\div 2$	0	1	3	6	10	15	21	28	36	45	55
$+50$	50	50	50	50	50	50	50	50	50	50	50
$f(t)$	50	51	53	56	60	65	71	78	86	95	105

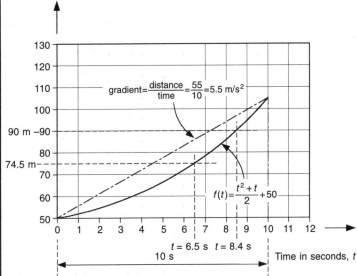

Figure 4.53 *Determining the slope (gradient) of the function*

$$f(x) = \frac{t^2 + t}{2} + 50$$

2. Note that the graph is parabolic in shape, which is to be expected for a quadratic function.

   Then from the graph the distance at time 6.5 s is approximately 74.5 m.

   The time it takes to reach 90 m approximately = 8.4 seconds.

3. Unfortunately, the *gradient or slope* of the graph *varies* (it is curved), but an *approximation to the gradient* can be found using a straight line which joins the points (0,50) and (10,105), as shown. Then from the graph it can be seen that:

   the gradient $= \dfrac{\text{distance}}{\text{time}} = \dfrac{55}{10} = 5.5 \text{ m s}^{-1}$

   which of course is *speed*.

   In effect, what we have found is the average speed over the 10 seconds.

The differential calculus

The gradient of a curve and graphical differentiation

Now, suppose we wish to find the speed of the vehicle identified in Example 4.75 over a slightly shorter period of time, say between 1 and 9 seconds. Then we know that the speed is given by the *slope or gradient* of the graph at these points (Figure 4.54). This process is continued for time periods 3 to 8 seconds and finally 3 to 4 seconds, the resultant speeds can be seen to be 5.5 m s^{-1}, 6.2 m s^{-1} and 5 m s^{-1}, respectively.

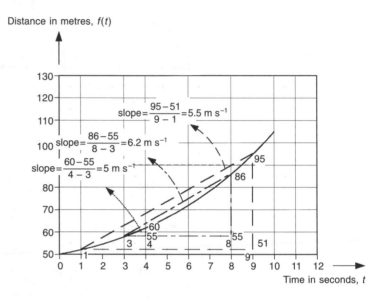

Figure 4.54 *Determining the gradient to a tangent at a point*

We could continue this process, taking smaller and smaller time periods, so that eventually we would be able to find the *gradient or slope of a point* on the graph, in other words we could find the *gradient at an instant in time*. Now you know from your study of the circle (page 403) that a *tangent line touches a circle (or curve) at just one point. Therefore finding the gradient of the slope of a curve at a point is equivalent to finding the gradient of a tangent line at that same point* (Figure 4.55).

In the case of our vehicle (Figure 4.54) it can be seen that the *gradient is in fact the speed*, so if we were able to find the *gradient of the tangent* at any instant in time, we would be finding the *instantaneous* speed.

Now this process of trying to find the gradient at a point (the tangent) is long and tedious. However, it can be achieved very easily using the *differential calculus*, that is by *differentiating* the function.

Thus, in the case of our speed example, by finding the slope at a point (the slope of its tangent), we have in effect *graphically differentiated* the function, or found the way that *distance f(x) changes at any instant in time t*, the instantaneous speed!

This may all sound rather complicated, but by *applying certain rules*, we will be able to carry out the *differentiation* process and so find out how functions change at any instant in time. However, there are a few things to learn before we get there!

Key point

To find the gradient of the tangent at a point of a function, we differentiate the function.

Key point

Finding the slope of a curve at a point is graphical differentiation.

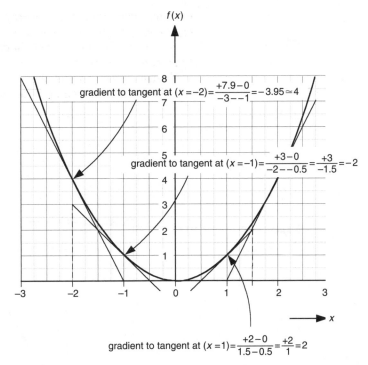

Figure 4.55 *Finding the gradient of a curve at a point (the derived function)*

Example 4.76

1. Draw the graph of the function $f(x) = x^2$ for values of x from $x = -3$ to $x = 3$.
2. Find the slope of the tangent lines drawn at $x = -1$, $x = 1$ and $x = -2$ and comment on your results.

1. The graph of the function $f(x) = x^2$ is shown in Figure 4.55. It can be seen that it is symmetrical about zero and is parabolic in shape.
2. From the graph it can be seen that the gradient to the tangent lines at the points −1,1 and −2 are −2,2 and −4, respectively. There seems to be a pattern in that at $x = -1$, the corresponding gradient $= -2$. So the gradient is twice as large as the independent variable x. This is also true for the gradients at $x = 1$ and $x = -2$, which are again twice as large. This pattern is no coincidence, as you are about to see!

For the function $f(x) = x^2$ we have just shown that on three occasions using three different independent variables, the gradient of the slope of the tangent line is twice the value of the independent variable or more formally:

gradient of the tangent at $f(x) = 2x$

The process of *finding the gradient of the tangent at a point* is known as *graphical differentiation*. What we have done is found the *differential coefficient* of the function $f(x) = x^2$. In other words we have found an algebraic expression of how this function varies, as we increase or decrease the value of the independent variable.

In functional notation, the process of *finding the differential coefficient* of a function $f(x)$ or *finding the slope to the tangent at a point*, or *finding the derived function*, is given the special symbol $f'(x)$, read as '*f* prime'.

We can generalize the above procedure for finding the derived function. Consider again part of our function $y = x^2$ (Figure 4.56).

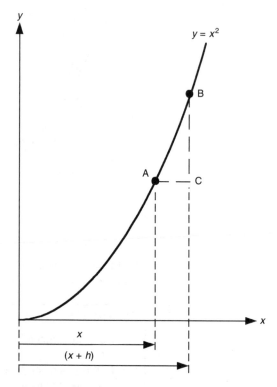

Figure 4.56 *Graph of the function $y = x^2 - 9$ showing turning point (TP)*

Suppose that A is the point on our curve $y = x^2$, with ordinate x and that B is another point on the curve with ordinate $(x + h)$. The y-ordinate of A is x^2 and the y-ordinate of B is $(x + h)^2$. Then:

$$BC = (x + h)^2 - x^2 = 2hx + h^2 \quad \text{and} \quad AC = (x + h) - x = h$$

Then the gradient of $AB = \dfrac{2hx + h^2}{h} = 2x + h$ (if h is not zero). Now as h gets smaller and smaller (tends to 0), the gradient tends to (approaches) $2x$.

Therefore, as we found graphically, *the gradient of the tangent is $2x$*, or *the derived function is $2x$*.

There are also other ways of representing the *differential coefficient or derived function*, which we are about to consider.

Example 4.77

Find the derived function (gradient of the curve) for $y = 2x^2 - 2x - 6$ at the point (x, y).

We use the same procedure as before. Identifying another point on the curve, say $(x + h, y + k)$. Then find the slope of the line joining the two points and bring the two points closer and closer together, until they coincide and become the tangent to the slope of the curve, in other words, the derived function. We proceed as follows.

For the two points on the *x*-ordinate the *y*-ordinates are:

$$y + k = 2(x + h)^2 - 2(x + h) - 6 \tag{1}$$

$$y = 2x^2 - 2x - 6 \tag{2}$$

Then expanding equation (1) and simplifying, *using your algebra*, we get:

$$y + k = 2(x^2 + hx + hx + h^2) - 2x - 2h - 6$$

$$y + k = 2(x^2 + 2hx + h^2) - 2x - 2h - 6$$

$$y + k = 2x^2 + 4hx + 2h^2 - 2x - 2h - 6$$

$$\text{or: } y + k = 2x^2 - 2x - 6 + 4hx + 2h^2 - 2h \tag{3}$$

Now subtracting equation (2) from equation (3) gives $k = 4hx + 2h^2 - 2h$, therefore on division by h, the gradient of the chord

$$\frac{k}{h} = 4x + 2h - 2 \text{ and } \frac{k}{h} \text{ tends to } (4x - 2) \text{ as } h \text{ tends to } 0.$$

Thus, $4x - 2$ *is the derived function of* $y = 2x^2 - 2x - 6$, *which is also the gradient of the function at the point (x, y).* So, for example, at the point (3,–3) the gradient is [4(3) – 2] = 10.

You will be pleased to know that we do not need to repeat this rather complicated method of finding the derived function (the tangent to a point of the slope). As you will soon see, *all derived functions* of simple algebraic expressions (polynomials) can be found using a simple rule!

Before we look at this rule, we need to consider the different ways in which the derived function can be expressed.

Notation for the derivative

There are several ways in which we can represent and describe the *differential coefficient or derived function*. Below are listed some of the most common methods used to describe the derived function that you will find in text books and literature dealing with the *differential calculus*.

These differing terms for finding the *derived function* include:

- find the derived function of . . .
- find the derivative of . . .
- find the differential coefficient for . . .
- differentiate . . .
- find the rate of change of . . .
- find the tangent to the function . . .
- find the gradient of a function at a point . . . .

This differing terminology is often confusing to beginners. It is further complicated by the fact that *different symbols* are used for the *differentiation process* (finding the derived function), based on the convention chosen.

We have been dealing with functional notation, where for a function $f(x)$ *the first derivative or first derived function* is given as $f'(x)$. If we were to carry out the *differentiation process again on the first derived function*, then we say we have found the *second* derivative, $f''(x)$ and so on.

We have used functional notation merely to introduce the idea of the *mathematical function*. We look next at the more common *Leibniz notation*, which we will use from now on throughout the remainder of this book.

In *Leibniz notation*, the mathematical function is represented conventionally as $y(x)$ and its derived function or differential coefficient is represented as $\dfrac{dy}{dx}$. This expression for the derived function can be thought of as finding the slope to the tangent of a point of a particular function, where we take a smaller and smaller bit of $x(dx)$ and divide it into a smaller and smaller bit of $y(dy)$, until we get the slope of a point $\dfrac{dy}{dx}$.

So, in Leibniz notation, for the function $y = x^2$ the differential coefficient is represented as $\dfrac{dy}{dx} = 2x$, which we found earlier.

The second derivative in Leibniz notation is represented as $\dfrac{d^2 y}{dx^2}$; the third derivative is $\dfrac{d^3 y}{dx^3}$ and so on.

One other complication arises with all notation, in that *the notation differs according to the variable being used*! For example, if our mathematical function is $s(t)$, then in Leibniz notation its first derivative would be $\dfrac{ds}{dt}$, we are differentiating the variable s with respect to t. In the same way as with $\dfrac{dy}{dx}$ we are differentiating the variable y with respect to x. So in general, $\dfrac{dy}{dx}, \dfrac{ds}{dt}, \dfrac{du}{dv}$, each represents the first derivative of the functions y, s and u, respectively.

One final type of notation which is often used in mechanics is *dot* notation. Where, for example, $\dot{v}$, $\ddot{s}$, etc., means that the function is differentiated once ($\dot{v}$) or twice ($\ddot{s}$) and so on. This notation will not be used in this book, but you may meet it in your further studies.

So much for all the hard theory, we are now going to *use one or two rules*, to carry out the differentiation process which, once mastered, is really quite simple!

Key point

In Leibniz notation $\dfrac{dy}{dx}$ means that we find the first derivative of the function y with respect to x.

Key point

In functional notation $f'(x)$ and $f''(x)$ are the first and second derivatives of the function f, respectively.

Differentiation

As you will be aware by now, the word *differentiate* is one of many ways of saying that we wish to *find the derived function*. Again, going back to the simple function $y = x^2$, when we *differentiated* this function, we found that its *derived function* was $\dfrac{dy}{dx} = 2x$. In a similar manner when we carried out the differentiation process on the function $y = 2x^2 - 2x - 6$ we obtained $\dfrac{dy}{dx} = 4x - 2$. If we were to carry out the rather complex process we used earlier on the following functions, $y = 3x^2$, $y = x^3$ and $y = x^3 + 3x^2 - 2$, we would obtain $\dfrac{dy}{dx} = 6x$, $\dfrac{dy}{dx} = 3x^2$ and $\dfrac{dy}{dx} = 6x^2 + 6x$, respectively. I wonder if you can see a pattern in these results? They are grouped below for your convenience; can you spot a pattern?

$$y = x^2, \frac{dy}{dx} = 2x$$

$$y = 3x^2, \frac{dy}{dx} = 6x$$

$$y = 2x^2 - 2x - 6, \frac{dy}{dx} = 4x - 2$$

$$y = x^3, \frac{dy}{dx} = 3x^2$$

$$y = x^3 + 3x^2 - 2, \frac{dy}{dx} = 3x^2 + 6x$$

I hope you spotted that we seem to multiply by the index (power) of the unknown, then we subtract one (1) from the index of the unknown. So, for example, with the function $y = 3x^2$, the index is 2 and $(2)(3) = 6$. Also the original index (power) of x was 2 and on subtracting 1 from this index we get $x^{(2-1)} = x^1 = x$, so we finally get $\dfrac{dy}{dx} = 6x$.

This technique can be applied to *any unknown raised to a power*. We can write this rule in general terms:

If $y = x^n$ then $\dfrac{dy}{dx} = nx^{n-1}$

Or, in words: *To find the differential coefficient of the function $y = x^n$ we first multiply the unknown variable by the index and then subtract 1 from the index to form the new index.*

Again, with this rather wordy explanation, you will appreciate the ease with which we can express this rule using a formula!

You may be wondering why the constant (number) in the above functions seems to just disappear. If you remember how we performed the differentiation process graphically, by finding the slope at a point on the function, then for a constant such as $y = -6$, its graph is simply a straight horizontal line cutting the y-axis at -6, therefore, its slope is *zero*, thus its *derived function is zero*. This is true for any constant term no matter what its value.

If the function we are considering has more than one term, for example $y = x^3 + 3x^2 - 2$, then we simply *apply the rule in sequence to each and every term*.

Example 4.78

Differentiate the following functions with respect to the variable:

1. $y = 3x^3 - 6x^2 - 3x + 8$

2. $y = \dfrac{3}{x} - x^3 + 6x^{-3}$

3. $s = 3t^3 - \dfrac{16}{t^2} + 6t^{-1}.$

1. In this example we can simply apply the rule $\dfrac{dy}{dx} = nx^{n-1}$ to each term in succession, so:

$$\frac{dy}{dx} = (3)(3)x^{3-1} - (2)(6)x^{2-1} - (1)(3)x^{1-1} + 0$$

and remembering that any number raised to the power zero is one, i.e. $x^0 = 1$, then:

$$\frac{dy}{dx} = 9x^{3-1} - 12x^{2-1} - 3x^{1-1} + 0$$

$$\frac{dy}{dx} = 9x^2 - 12x^1 - 3x^0$$

$$\frac{dy}{dx} = 9x^2 - 12x - 3$$

2. In this example, we need to simplify *before* we use the rule. The simplification involves clearing fractions. Then remembering that $x = x^1$ and that *from your laws of indices* when you bring a number in index form over the fraction line we *change its sign*, $y = \dfrac{3}{x} - x^3 + 6x^{-3}$ becomes:

$$y = \frac{3}{x^1} - x^3 + 6x^{-3}$$

or,

$$y = 3x^{-1} - x^3 + 6x^{-3}$$

and applying the rule,

$$\frac{dy}{dx} = (-1)(3)x^{-1-1} - (3)x^{3-1} - (3)(6)x^{-3-1}$$

$$\frac{dy}{dx} = -3x^{-2} - 3x^2 - 18x^{-4}$$

Notice how we have dealt with *negative* indices. The rule can also be used when *fractional* indices are involved.

3. The only change with this example is that it concerns different variables. In this case we are asked to differentiate the function s with respect to the variable t.

So proceeding as before and *simplifying first*, we get:

$$s = 3t^3 - 16t^{-2} + 6t^{-1}$$

and then differentiating

Key point

To find the first derivative of functions of the type $y = ax^n$, we use the rule that $\dfrac{dy}{dx} = nax^{n-1}$.

$$\frac{ds}{dt} = (3)(3)t^{3-1} - (-2)(16)t^{-2-1} + (-1)(6)t^{-1-1}$$

$$\frac{ds}{dt} = 9t^2 + 32t^{-3} - 6t^{-2}$$

Note that you must take care with your signs!

The second derivative

In the above example we found, in all cases, the *first derivative*. If we wish to find the *second derivative of a function*, all we need do is differentiate again. So in Example 4.78 question 1, for the function $y = 3x^2 - 6x^2 - 3x + 8$, then:

$$\frac{dy}{dx} = 9x^2 - 12x - 3$$

and differentiating this function again, we get,

$$\frac{d^2y}{dx^2} = (2)(9)x^{2-1} + (1)(-12)x^{1-1} = 18x - 12x^0 = 18x - 12$$

Notice the *Leibniz* terminology for the *second differential* in the above example.

Similarly for the function $s = 3t^3 - 16t^{-2} + 6t^{-1}$, then:

$$\frac{ds}{dt} = 9t^2 + 32t^{-3} - 6t^{-2}$$

and on differentiating this function again, we get,

$$\frac{d^2s}{dt^2} = (2)(9)t^{2-1} + (-3)(32)t^{-3-1} + (-6)(-2)t^{-2-1}$$

$$= 18t - 96t^{-4} + 12t^{-3}$$

Notice once again, the care needed with signs in this example.

Key point

$\frac{d^2y}{dx^2}$, $f''(x)$ and $\ddot{x}$ are all ways of expressing the second derivative.

Rate of change

One application of the differential calculus is to find instantaneous rates of change. The example given at the beginning of this section concerned our ability to find how the speed of a motor vehicle changed at a particular point in time. *In order to find the rate of change of any function*, we simply *differentiate that function* (find its gradient) *at the particular point concerned*.

So, for example, given that $y = 4x^2$ let us find its rate of change at the points $x = 2$ and $x = -4$. Then all we need do is differentiate the function and then substitute in the desired points.

Then, $\frac{dy}{dx} = (2)(4)x = 8x$ and when $x = 2$, $\frac{dy}{dx} = (8)(2) = 16$, thus the slope of the function at $x = 2$ is 16 and this tells us how the function is changing at this point. Similarly, when $x = -4$, $\frac{dy}{dx} = 8x = (-4)(8) = -32$. In this case the negative sign indicates a negative slope, so the function is changing in the opposite sense, compared with what was happening when $x = 2$.

Example 4.79

The distance (s) covered by a missile is given by $s = 4.905t^2 + 10t$. Determine its rate of change of distance with respect to time (its speed) after (a) 4 seconds and (b) 12 seconds have elapsed.

This is a simple rate of change problem, hidden in this rather wordy question! To find *rate of change* of distance with respect to time, we need to find the differential coefficient of the function. Then applying the rule:

$$\frac{ds}{dt} = (2)(4.905)t^{2-1} + 10t^{1-1} = 9.81t + 10$$

Now substituting for the desired times:

when $t = 4$, $\quad \dfrac{ds}{dt} = (9.81)(4) + 10 = 49.24$

and when $t = 12$, $\quad \dfrac{ds}{dt} = (9.81)(12) + 10 = 127.72$

Since $\dfrac{ds}{dt} = v$ (speed), then what the above results tell us is that after 4 seconds the missile has reached a speed of 49.24 m s^{-1} and after 12 seconds the missile reaches a speed of 127.72 m s^{-1}.

Thus, for very little effort, the differential calculus has enabled us to find instantaneous rates of change, which TP of practical use!

Key point

The rate of change of distance with respect to time is velocity.

Turning points

Another useful application of the differential calculus is in finding the turning point(s) of a function. We have already seen differentiation being used to determine rates of change; *turning points* enable us to tell when these rates of change are at a minimum value or maximum value.

Consider Figure 4.57, which shows the graph of the function $y = x^2 - 9$.

If we consider the slope of the function as it approaches the turning point (TP) from the left, the slope is negative. The slope of the graph as it moves away to the right of the turning point is positive. At *some point* TP, the slope went from a *negative value to a positive value*, in other words at the TP the slope (*gradient*) is *equal to zero*. Now we know that the gradient of the function $y = x^2 - 9$ is found by differentiating the function. Therefore, when $\dfrac{dy}{dx}$ of $y = x^2 - 9$ is zero, there must be a TP, because at this point the slope of the function is a *horizontal straight* line and its *slope is zero*.

So applying the rule $\dfrac{dy}{dx} = 2x$, and for a TP $\dfrac{dy}{dx} = 2x = 0$, which implies that $x = 0$.

Now if $x = 0$, $y = (0)^2 - 9 = -9$.

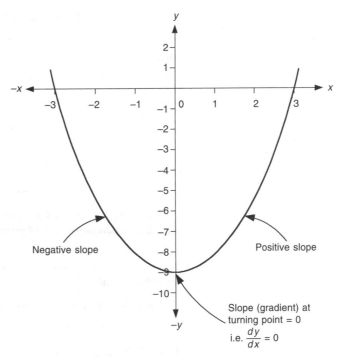

Figure 4.57 *Family of curves for y = x² + c*

So this function turns at the point (0,9), as can be seen in Figure 4.57.

You should note from Figure 4.56 that at the TP the function has a *minimum* value. You are not required to find maximum or minimum values in this unit. However, the technique used in finding TPs is the first stage in trying to establish whether such points indicate a maximum or minimum value for a particular function. This technique will thus prove useful if you intend to take your mathematical studies further.

Differentiation of elementary trigonometric and exponential functions

We have so far concentrated our attention on functions of the type $ax^n \pm ax^{n-1} \pm ax^{n-2} \pm \ldots \pm ax^3 \pm ax^2 \pm ax \pm a$, this general class of functions are known as *polynomials*.

There are, however, other mathematical functions that you have already met. These include *trigonometric functions* such as the sine and cosine. In addition you have met the *exponential function* e^x and its mathematical inverse the *Naperian logarithm* ln x.

Finding the differential coefficient of these functions can be achieved by graphically differentiating them in a similar manner to the way we originally found the derived function for $y = x^2$. If we were to carry out this exercise, we would be able to establish patterns and subsequent rules, as we did for polynomial functions.

Rather than going through this tedious process, you will be pleased to note that these rules have been listed below (without proof) for your convenience!

Table of some standard derivatives

Rule Number	y	$\dfrac{dy}{dx}$
1	x^n	nx^{n-1}
2	ax^n	nax^{n-1}
3	$\sin ax$	$a \cos ax$
4	$\cos ax$	$-a \sin ax$
5	e^{ax}	ae^{ax}
6	$\ln ax$	$\dfrac{\dfrac{dy}{dx}(ax)}{ax}$

Now one or two of the above rules may look a little complex, but in practice they are all fairly straightforward to use. The easiest way to illustrate their use is through the examples that follow.

Example 4.80

Differentiate the following with respect to the variable:

(a) $y = \sin 3x$ (b) $u = \cos 2\theta$ (c) $y = 5 \sin 2\theta - 3 \cos \theta$.

(a) In this example we may follow rule 3 in the table directly, noting that $a = 3$, then $\dfrac{dy}{dx} = a \cos ax = 3 \cos 3x$.

(b) The same approach is needed to solve this little problem, but noting that when we differentiate the cosine function, it has a *sign* change. Also we are differentiating the function u with respect to θ. Then the differential coefficient, using rule 4, is given as:

$$\frac{du}{d\theta} = -2 \sin 2\theta$$

(c) With this final problem, we simply use rule 3 for differentiating sine, followed by rule 4 for differentiating cosine. Noting that the numbers 5 and −3 are not the constant a given in the formulae in the table. We simply multiply these numbers by a when carrying out the differentiation process.
So,

$$\frac{dy}{d\theta} = (2)(5) \cos 2\theta \, (-3)(-1) \sin \theta = 10 \cos 2\theta + 3 \sin \theta$$

Note the effect of the sign change when differentiating the cosine function!

Key point

The sign of the differential of the cosine function is always negative.

Example 4.81

1. Find the differential coefficients of the following function, $y = e^{-2x}$.

2. Find $\dfrac{d}{dx}(6 \log_e 3x)$.

3. Differentiate $v = \dfrac{e^{3\theta}}{2} - \pi \ln 4\theta$.

The above functions involve the use of rule 5 and rule 6.

1. This is a direct application of rule 5 for the exponential function, where $a = -2$. Remember we are differentiating the function y with respect to the variable x; the base e is simply a constant (a number). As mentioned before the value of $e \simeq 2.71828$. It is a number like π; it has a limitless number of decimal places! Then,

$$\frac{dy}{dx} = (-2)\, e^{-2x} = -2e^{-2x}$$

2. This is yet another way of being asked to differentiate a function. What it is really saying is find $\dfrac{dy}{dx}$ of the function $y = 6 \log_e 3x$.

    Remember that when dealing with the Naperian log function, $\log_e f(x) = \ln f(x)$, both methods of representing the Naperian log function are in common use; so all we need do is apply rule 6, where the constant $a = 3$ in this case.

$$\frac{d}{dx}(6 \log_e 3x) = (6)\frac{\dfrac{dy}{dx}(3x)}{3x} = \frac{(6)(3)}{3x} = \frac{18}{3x} = \frac{6}{x}$$

    Note, when finding this differential we also had to apply rule 1 to the top part of the fraction. Providing you follow rule 6 exactly, laying out all your working, you should not make mistakes.

3. For this example, we need to apply rule 5 to the exponential function and then rule 6 to the Naperian log function, noting that $-\pi$ is a constant and does not play any part in the differentiation. We simply multiply the differential by it at the end of the process. Therefore:

$$\frac{dv}{d\theta} = \frac{3e^{3\theta}}{2} + (-\pi)\frac{\dfrac{dy}{d\theta}(4\theta)}{4\theta} \quad \text{and} \quad \frac{dy}{d\theta} = 1.5e^{3\theta} + (-\pi)\frac{4}{4\theta}$$

$$\frac{dv}{d\theta} = 1.5e^{3\theta} - \frac{4\pi}{4\theta} \quad \text{and} \quad \frac{dv}{d\theta} = 1.5e^{3\theta} - \frac{\pi}{\theta}$$

    this may look rather complicated but all we have done is followed rule 6 as before.

Now being able to find the differential coefficient of the functions in the above examples is all very well, but what use is it all?

Well, as was the case with the general rule for differentiating polynomial functions, we can also apply these rules to solving simple *rate of change* problems. In our final example for the *differential calculus*, we apply rule 5 and rule 6 to rate of change of current in an electrical circuit and rate of discharge from an electrical capacitor. This is not as difficult as it sounds!

Key point

We always differentiate when finding rates of change.

Test your knowledge 4.13

1. When differentiating polynomial functions of the form $y = ax^n$ write down the expression for finding $\dfrac{dy}{dx}$.

2. For the function $f(x) = 16x^2 - 3x^3 - 12$ find $f(3)$ and $f(-2)$.

3. Differentiate the following functions with respect to the variables given:

    (a) $y = 6x^2 - 3x - 2$

    (b) $s = 3t^2 - 6t^{-1} + \dfrac{t^{-3}}{12}$

    (c) $p = \dfrac{r^3 - r^2}{r^{-1}} + 12r - 6$

    (d) $y = 3x^{\frac{9}{2}} - 5x^{\frac{3}{2}} + \sqrt{x}$.

4. Plot the graph of the function $y = \sin 2\theta$ between $\theta = 0$ and $\theta = 2\pi$ rad, using the techniques you learnt in

---

Example 4.82

1. An alternating voltage is given by the function $v = \sin 2\theta$, where θ is the angular distance travelled and v is the instantaneous voltage at that angular distance (in radians). Determine the way the voltage is changing with respect to distance at $\theta = 2$ and $\theta = 4$ rad.

your trigonometry, making sure that (θ) is in *radians*. Then find the value of the slope at the point where θ = 2 rad. Compare your result with the answer to Example 4.82 question 1.

5. If $y = x^2 - 2x + 1$, find the coordinates (x, y) at the point where the gradient is 6.

$\left(\text{Hint: The gradient is } \dfrac{dy}{dx}.\right)$

6. Determine the rate of change of the function

$$y = \frac{x^4}{2} - 3x^3 + x^2 - 3,$$

at the point where $x = -2$.

7. What is the rate of change of the function $y = 4e^x$ when $x = 2.32$?

8. Differentiate the functions: (a) $y = \ln x$; (b) $y = 3 \ln x$; (c) $y = \ln 3x$, and comment on your answers.

9. An alternating current is given by the function $i = \cos 3\theta$, find the rate of change of current when θ = 1 rad.

10. Find the rate at which a capacitor is discharging at $t = 3$ ms and $t = 3.8$ ms, when the amount of charge on the capacitor is given by the function $Q = 2.6 \log_e t$.

2. Suppose the charge in a capacitor discharges according to the function $Q = \ln 3t$, where Q = charge (in coulombs) and t = time in milliseconds (ms). Determine the *rate of discharge* at $t = 4$ ms.

1. All we are being asked is to find the rate of change of the voltage after a particular angular distance has been covered by the alternating (sinusoidal) function. This means we need to find the differential coefficient (the rate of change function) and then simply substitute in the appropriate values.

So, $\dfrac{dv}{d\theta} = 2 \cos 2\theta$, which is the rate of change of voltage with respect to distance.

Then, at θ = 2 rad, remembering it is *radian* measure!, we get:

$$\frac{dv}{d\theta} = 2\cos(2)(2) = 2\cos 4 = (2)(-0.653) = -1.3073$$

and the voltage is changing negatively. This value is the slope of the graph of $v = \sin 2\theta$ at the point θ = 2 rad.

Similarly at θ = 4 rad, then

$$\frac{dv}{d\theta} = 2\cos(2)(4) = 2\cos 8 = (2)(-0.1455) = -0.291$$

to 3 decimal places

Again a negative slope, but with a shallower gradient.

2. The rate of discharge in this case means the rate of change of charge with respect to time. So it is a *rate of change problem* involving the differential coefficient of the function. Then following rule 6 and also using rule 1, we have:

$$\frac{dQ}{dt} = \frac{\dfrac{dQ}{dt}(3t)}{3t} = \frac{3}{3t} = \frac{1}{t}$$

then when $t = 4$ ms or 4×10^{-3} seconds,

$$\frac{dQ}{dt} = \frac{1}{t} = \frac{1}{4 \times 10^{-3}} = 250 \text{ C/s (coulombs per second)}.$$

If you were to put in higher values of time you will find that the rate of discharge decreases.

The integral calculus

In this short section we are going to look at the *integral calculus*, which we mentioned earlier. It has something to do with finding areas and is also the inverse process of finding the derived function. The integral calculus is all about *summing things*, that is finding the whole thing from its parts, as you will see shortly.

We start by considering *integration* (the arithmetic of the integral calculus) as the *inverse of differentiation*.

Integration as the inverse of differentiation

We know that for the function $y = x^2$ the derived function $\dfrac{dy}{dx} = 2x$. So reversing the process involves finding the function whose derived function is $2x$. One answer will be x^2, but is this the only possibility? The answer is *no*, because $2x$ is also the derived function of $y = 2x + 5$, $y = x^2 - 20.51$, $y = x^2 + 0.345$, etc. In fact $2x$ is the derived function of $y = x^2 + c$, where c is any constant. So when we are finding the inverse of the derived function, in other words, when we are integrating, we must always allow for the possibility of a constant being present by putting in this *arbitary constant c*, which is known as the *constant of integration*. Then in general terms, the inverse of the derived function $2x$ is $x^2 + c$.

Thus, whenever we wish to find the inverse of any derived function, that is whenever we *integrate the derived function*, we must include the *constant of integration c*.

When carrying out the *anti-differentiation process* or *integration*, we can only find a *particular value* for this constant c when we are given some additional information about the original function. For example, if we are told that for $y = x^2 + c$, $y = 2$, when $x = 2$, then by substituting these values into the original function, we find that $2 = 2^2 + c$, from which we find that $c = -2$, so the particular function becomes $y = x^2 - 2$. This is now one of a *whole family* of functions $y = x^2 + c$, illustrated graphically in Figure 4.58.

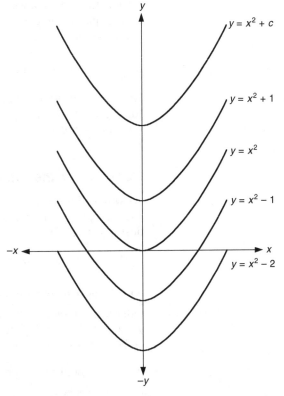

Figure 4.58 *Family of curves* $y = x^2 + c$

Tabulated below are a few familiar *polynomial functions*, on which has been carried out this inverse differentiation or *integration* process. When we integrate a derived function, the expression we obtain is often known as the *prime function* (F). See if you can spot a pattern for the derivation of these prime functions.

Derived function	Prime function (F)
$\dfrac{dy}{dx} = 1$	$y = x + c$
$\dfrac{dy}{dx} = x$	$y = \dfrac{x^2}{2} + c$
$\dfrac{dy}{dx} = x^2$	$y = \dfrac{x^3}{3} + c$
$\dfrac{dy}{dx} = x^3$	$y = \dfrac{x^4}{4} + c$

Apart from the mandatory constant of integration, I hope you can see that the power or index of x increases by 1 over that of the derived function. Then we divide the prime function by this new power or index. Then, in general:

If $\dfrac{dy}{dx} = x^n$, then the prime function is $y = \dfrac{x^{n+1}}{n+1} + c$

This rule is valid for all values of n *except* $n = -1$.

If $n = -1$, then in finding the prime function we would be trying to divide by $n + 1 = -1 + 1 = 0$ and as you are well aware from your earlier study of the laws of arithmetic, division by 0 is not allowed! In this particular case we adopt a special rule, which is given below, without proof.

If $\dfrac{dy}{dx} = x^{-1} = \dfrac{1}{x}$, then the prime function is $y = \ln |x|$

Notice that we have to take the *modulus or positive value of* x when finding corresponding values of y. This is because the *ln* (log_e) *function is not valid for all numbers* ≤ 0 (less than or equal to zero).

Notation for the integral

As with differentiation, when we carry out integration, we need to use the appropriate mathematical notation in order to convey our desire to integrate.

If y is a function of x, then $\displaystyle\int y\, dx$ represents the integral of y with respect to the variable x. The integral sign $\displaystyle\int$ is the Greek letter S, and indicates that when carrying out the integration process, we are really carrying out a *summing* process.

Note that in the same way that the d cannot be separated from the y in dy, neither can the $\displaystyle\int$ be separated from dx if the integration is with respect to x.

So, for example, if we wish to find the prime function F, that is if we wish to integrate the function x^2, then this is represented as $\displaystyle\int x^2\, dx$ and using the general rule, we see that $\displaystyle\int x^2\, dx = \dfrac{x^{n+1}}{n+1} + c = \dfrac{x^{2+1}}{2+1} + c = \dfrac{x^3}{3} + c$, which is in agreement with the prime function or the *integral* shown in the table.

Integration

We have seen from above how to integrate elementary polynomial functions using the basic rule. In Example 4.83 we use the rule successively to integrate the general polynomial expression with respect to the variable concerned.

Example 4.83

Integrate the following functions with respect to the variables given:

1. $y = 3x^3 + 2x^2 - 6$.
2. $s = 5t^{-3} + t^4 - 2t^2$.

3. $p = r^{-1} + \dfrac{r^4}{2}$.

1. What we are being asked to do is find the prime function $F(y)$, or using the conventional notation we have just learnt, then we must find:

$$F(y) = \int 3x^3 + 2x^2 - 6\,dx$$

In this case, all we need do is successively apply the basic rule, i.e.

$$\int 3x^3 + 2x^2 - 6\,dx = (3)\frac{x^{3+1}}{3+1} + (2)\frac{x^{2+1}}{2+1} + (-6)x^{0+1} + c$$

$$= \frac{3x^4}{4} + \frac{2x^3}{3} - 6x + c$$

2. In this question, we again apply the basic rule but in terms of the different variables, so:

$$\int 5t^{-3} + t^4 - 2t^2\,dt = (5)\frac{t^{-3+1}}{-3+1} + \frac{t^{4+1}}{4+1} + (-2)\frac{t^{2+1}}{2+1} + c$$

$$= \frac{5t^{-2}}{-2} = \frac{t^5}{5} - \frac{2t^3}{3} + c$$

3. With this question I hope you spotted immediately that for part of the function r^{-1}, we cannot apply the general rule but must apply the special case where $n = -1$. So for the integration, we proceed as follows:

$$\int r^{-1} + \frac{r^4}{4}\,dr = \ln|r| + \left(\frac{1}{4}\right)\frac{r^{4+1}}{4+1} + c = \ln|r| + \frac{r^5}{20} + c$$

Notice also that dividing by 4 is the same as multiplying by a $\frac{1}{4}$ and that we multiply tops by tops and bottoms by bottoms to obtain the final values.

Key point

When finding indefinite integrals, we must always include the constant of integration.

Some common integrals

We have seen now how to integrate polynomial expressions. We can also apply the inverse differentiation process to the *sinusoidal*, *exponential and Naperian logarithm functions*. The table below shows the prime functions (the *integral*) for the basic functions we dealt with during our study of the differential calculus.

Table of some standard integrals

Rule number	Function (y)	Prime function $\left(\int y\,dx\right)$
1	$x^n (n \neq -1)$	$\dfrac{x^{n+1}}{n+1} + c$
2	$x^{-1} = \dfrac{1}{x}$	$\ln \lvert x \rvert$
3	$\sin ax$	$-\dfrac{1}{a} \cos ax$
4	$\cos ax$	$\dfrac{1}{a} \sin ax$
5	e^{ax}	$\dfrac{1}{a} e^{ax}$
6	$\ln x$	$x \ln x - x$

If you compare the integrals of the sine and cosine functions, you should be able to recognize that the integral is the inverse of the differential. This is also clearly apparent for the exponential function. The only 'strange' integral that seems to have little in common with its inverse is that for the Naperian logarithm function. The mathematical verification of this integral is beyond the level for this unit. However, you will learn the techniques of the calculus necessary for its proof if you study the further mathematics unit.

We will demonstrate the use of these *standard integrals* through the examples that follow.

Example 4.84

1. Find $\displaystyle\int (\sin 3x + 3 \cos 2x)\,dx$.

2. Integrate the function $s = e^{4t} - 6e^{2t} + 2\ dt$.

3. Find $\displaystyle\int 6 \log_e t\,dt$.

1. This integral involves using rules 3 and 4 sequentially. The integral may be written as:

$$\int \sin 3x + \int 3 \cos 2x = -\frac{1}{3} \cos 3x + (3)\frac{1}{2} \sin 2x + c$$

$$= -\frac{1}{3} \cos 3x + \frac{3}{2} \sin 2x + c$$

Any integral involving expressions separated by $\pm$ may be integrated separately. Note also that the constant multiplying the function (3) in this case, *does not play any part in the integration*, it just becomes a multiple of the result.

2. This is just a direct integral involving the successive use of rule 5 and the use of rule 1 for the last term. Then:

$$\int e^{4t} - 6e^{2t} + 2 \; dt = \frac{1}{4}e^{4t} - (6)\left(\frac{1}{2}\right)e^{2t} + 2t + c$$

$$= \frac{1}{4}e^{4t} - 3e^{2t} + 2t + c$$

3. This integral demonstrates the direct use of rule 6, where the constant is taken behind the integral sign until the process is complete and then brought back in as the multiplier of the integral. Remembering also that $\log_e t = \ln t$. Then:

$$\int 6 \log_e t \; dt = 6 \int \log_e t \; dt = 6(t \log_e t - t) + c$$

or $\qquad = 6(t \ln t - t) + c$

Simple applications of the integral

In the differential calculus we considered *rates of change*, one particular application involved determining the *rate of change of distance with respect to time*. In other words, differentiating the function involving distance to find the derived function which gave the *velocity*. Look back at Example 4.79 if you cannot remember this procedure. If we carry out the inverse operation, that is we *integrate the velocity function* we will get back to the *distance function*. Taking this idea one step further, if we differentiate the velocity function, we will find the rate of change of velocity with respect to time, in other words we will find the acceleration function ($m\,s^{-2}$). So again, if we integrate the acceleration function, we get back to the velocity function.

Example 4.85

The acceleration of a missile moving vertically upwards is given by $a = 4t + 4$. Find the formulae for both the velocity and the distance of the missile, given that $s = 2$ and $v = 10$ when $t = 0$.

In this application it is important to recognize that *acceleration is rate of change of velocity*, or $\dfrac{dv}{dt} = 4t + 4$. This of course is a *derived function*, therefore in order to find v, we need to carry out anti-differentiation, that is *integration*. When we do this we find the prime function $F(x)$ by integrating both sides of the derived function as follows:

$$\int \frac{dv}{dt} = \int 4t + 4 \; dt$$

and so $F(x) = v = \dfrac{4t^2}{2} + 4t + c = 2t^2 + 4t + c$

We now have the original equation for the velocity $v = 2t^2 + 4t + c$. We can now use the given information to find the *particular equation* for the velocity. We know that when the velocity $= 10$, the time $t = 0$.

Therefore, substituting into our velocity equation gives 10 = (2)(0) + (4)(0) + *c*, or 10 = *c*. So our particular equation for velocity is:

$$v = 2t^2 + 4t + 10$$

We are also asked to find the formula for the *distance*. Again, recognizing that *velocity is the rate of change of distance with respect to time* we may write the velocity equation in its derived form as:

$$\frac{ds}{dt} = 2t^2 + 4t + 10$$

then integrating as before to get back to distance, we get:

$$\int \frac{ds}{dt} = \int 2t^2 + 4t + 10 \, dt$$

or $F(x) = s = \dfrac{2t^3}{3} + \dfrac{4t^2}{2} + 10t + c = \dfrac{2t^3}{3} + 2t^2 + 10t + c$

We now have the original equation for distance

$$s = \frac{2t^3}{3} + 2t^2 + 10t + c$$

Again, using the given information that *s* = 2 and *v* = 10, when *t* = 0, the particular equation for distance can be found. On substitution of time and distance into our distance equation we get that 2 = 0 + 0 + 0 + *c*, or *c* = 2. So our particular equation for distance is:

$$s = \frac{2t^3}{3} + 2t^2 + 10t + 2$$

Key point

If we integrate the acceleration function we obtain the velocity function. If we integrate the velocity function we obtain the distance function.

Area under a curve

The above example illustrates the power of the integral calculus in being able to find velocity from acceleration and distance from velocity. We know from:

$$\text{velocity (speed in a given direction)} = \frac{\text{distance}}{\text{time}}$$

that distance = velocity × time

So if we set velocity against time on a velocity time graph, the area under the graph (velocity × time) will be equal to the distance. Therefore if we know the rule governing the motion, we could in our case find any distance covered within a particular time period, by integrating the velocity time curve over this period.

Consider Figure 4.59, which shows a velocity time graph, where the motion is governed by the relationship $v = -t^2 + 3t$ or $\frac{ds}{dt} = -t^2 + 3t$. Then to find the *distance* equation for the motion, all we need do is *integrate* the velocity equation, as in Example 4.85.

The important point to note is that *when we integrate* and *find the distance equation*, this is the same as *finding the area under the graph* because the area under the graph = velocity × time = distance.

From the graph it can be seen that when time *t* = 0, the velocity *v* = 0, also that when *t* = 3, *v* = 0. So the area of interest is contained between these two time *limits*.

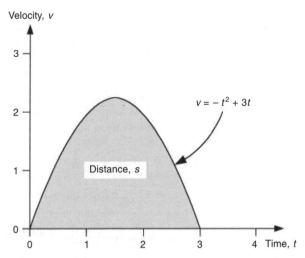

Figure 4.59 *Area under the graph for the function y = 2x² + 3*

Now integrating our velocity equation in the normal manner gives:

$$\int v = \frac{ds}{dt} = \int -t^2 + 3t \, dt = s = \frac{-t^3}{3} + \frac{3t^2}{2} + c$$

This *distance equation is equivalent to the area under the graph, between time t = 0 and t = 3.*

At time $t = 0$ the distance travelled $s = 0$ from the graph. The *constant of integration c* can be found by substituting these values of time and distance into our distance equation:

$$s = \frac{-t^3}{3} + \frac{3t^2}{2} + c$$

so $0 = 0 + 0 + c$, therefore $c = 0$ and our particular distance equation is $s = \frac{-t^3}{3} + \frac{3t^2}{2}$. Now between our limits of time $t = 0$ to $t = 3$ seconds, the *area under the graph indicates the distance travelled.* So at time $t = 0$, the distance travelled $s = 0$. At time $t = 3$, the area under the graph is found by substituting time $t = 3$ into our distance equation, then:

$$s = \frac{-t^3}{3} + \frac{3t^2}{2} = \frac{-(3)^3}{3} + \frac{(3)(3^2)}{2} = \frac{-27}{3} + \frac{27}{2}$$
$$= -9 + 13.5 = 4.5$$

Thus, in the above example, the area under the graph = 4.5 = the distance travelled.

The definite integral

When we *integrate between limits*, such as the time limits given for the motion discussed above, we say that we are finding the *definite integral*. All the integration that we have been doing up till now has involved the constant of integration and we refer to this type of integration as finding the *indefinite integral*, which must contain an arbitary constant *c*.

The terminology for indefinite integration is that which we have used so far, for example:

$$\int -t^2 + 3t \, dt$$

the *indefinite integral*.

When carrying out definite integration, we place limits on the integration sign, or summing sign, for example:

$$\int_0^3 -t^2 + 3t \, dt$$

the *definite integral*.

To evaluate a definite integral, we first integrate the function, then we find the numerical value of the integral at its *upper* and *lower* limits and subtract the value of the lower limit from that of the upper limit to obtain the required result.

So following this procedure for the definite integral shown above, which we used to find the distance s (area under a graph) from the velocity time graph, we get:

$$s = \int_0^3 -t^2 + 3t \, dt = \left[\frac{-t^3}{3} + \frac{3t^2}{2} + c \right]_0^3$$

$$= \left(\frac{-27}{3} + \frac{27}{2} + c \right) - \left(\frac{0}{3} + \frac{0}{2} + c \right)$$

$$s = (-9 + 13.5 + c) - (0 + c) = 4.5 + c - c = 4.5$$

Thus $s = 4.5$, we have found the *area under the graph using definite integration*!

Note that when we *subtract* the upper limit value from the lower limit value the *constant of integration is eliminated*. This will always be the case when evaluating definite integrals, therefore it need not be shown from now on.

> **Key point**
>
> When finding definite integrals the constant of integration is eliminated.

Example 4.86

1. Evaluate $\displaystyle\int_{-1}^1 \frac{x^5 - 4x^3 + x}{x} \, dx$.

2. Determine by integration the area enclosed by the curve $y = 2x^2 + 2$, the x-axis and the ordinates $x = -2$ and $x = 2$.

1. Before we integrate, it is essential to simplify the function as much as possible. So in this case on division by x, we get:

$$\int_{-1}^1 x^4 - 4x^2 + 6 \, dx = \left[\frac{x^5}{5} - \frac{4x^3}{3} + 6x \right]_{-1}^1$$

$$= \left(\frac{1}{5} - \frac{4}{3} + 6 \right) - \left(\frac{-1}{5} - \frac{-4}{3} - 6 \right) = 9\frac{11}{15}$$

Note that in this case it is easier to manipulate the upper and lower values as fractions!

Test your knowledge 4.14

1. Find the following indefinite integrals, using the basic rules:

   (a) $\displaystyle\int 4x^2 + 2x^{-3}\ dx$

   (b) $\displaystyle\int \frac{3x^{\frac{1}{2}}}{6} - \sqrt{x} + x^{\frac{3}{2}}\ dx$

   (c) $\displaystyle\int -3\sin 2x\ dx$

   (d) $\displaystyle\int \frac{x\cos 3x}{0.5\,x}\ dx$

   (e) $\displaystyle\int -0.25 e^{3\theta}\ d\theta$

   (f) $\displaystyle\int -3\log_e x\ dx.$

2. Using your results from question 1, evaluate the following definite integrals:

   (a) $\displaystyle\int_0^2 4x^2 + 2x^{-3}\ dx$

   (b) $\displaystyle\int_0^1 \frac{3x^{\frac{1}{2}}}{6} - \sqrt{x} + x^{\frac{3}{2}}\ dx$

   (c) $\displaystyle\int_0^{\frac{\pi}{2}} -3\sin 2\theta\ d\theta$

   (d) $\displaystyle\int_1^2 -0.25 e^{3\theta}\ d\theta$

   Note for questions (c) and (d) θ is in radians.

3. The acceleration of a vehicle is given by the relationship $a = 3t + 4$. Find the formulae for the *velocity* and *distance* of the vehicle, given that $s = 0$ and $v = 8$ when $t = 0$. Also find the distance travelled after a time of 25 seconds has elapsed.

4. Find the area under the curve $y = x + x^2$ between $x = 1$ and $x = 3$.

5. Sketch the graphs of the line $y = 2x$ and the curve $y = x^2$ on the same axes and determine by integration the area between the line and the curve.

2. In order to get a picture of the area we are required to find, it is best to draw a sketch of the situation first. The area with the appropriate limits is shown in Figure 4.60.

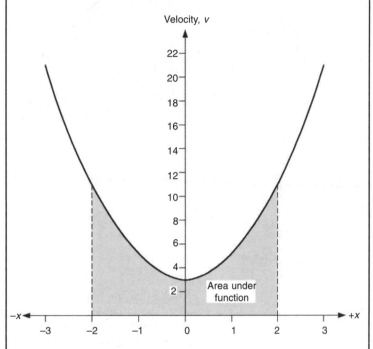

Figure 4.60 *Total area using integration, when area is above and below the x-axis*

We are required to find the shaded area of the graph between the limits $x = \pm 2$. Then:

$$\int_{-2}^{2} 2x^2 + 2\ dx = \left[\frac{2x^3}{3} + 2x\right]_{-2}^{2}$$

$$= \left(\frac{2(2)^3}{3} + (2)(2)\right) - \left(\frac{2(-2)^3}{3} + (2)(-2)\right)$$

$$= \left(\frac{16}{3+4}\right) - \left(\frac{-16}{3} - 4\right)$$

$$= 18\frac{2}{3}\ \text{square units}$$

A final word of caution when finding areas under curves using integration. If the area you are trying to evaluate is part above and part below the *x*-axis, it is necessary to split the limits of integration for the areas concerned.

So for the shaded area shown in Figure 4.61, we find the definite integral with the limits $(2, -2)$ and *subtract* from it the definite integral with limits $(4, 2)$, i.e.

$$\text{shaded area in Figure 4.61} = \int_{-2}^{2} y\ dx - \int_{2}^{4} y\ dx$$

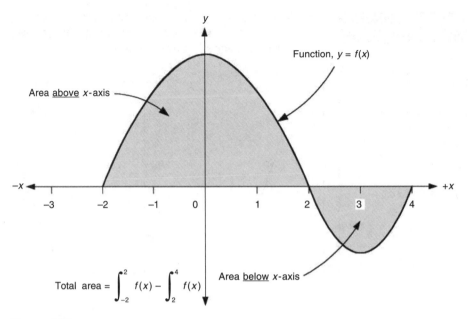

$$\text{Total area} = \int_{-2}^{2} f(x) - \int_{2}^{4} f(x)$$

Figure 4.61

Notice that the higher value always sits at the top of the integral sign. The minus sign is always necessary before the integral of any area that sits *below the x-axis*.

On this important point we finish our study of the integral calculus and indeed your study of mathematics for the unit!

Answers to test your knowledge exercises

TYK 3.1

1.

Base quantity	SI unit name	SI unit symbol
Mass	*kilogram*	kg
Length	metre	M
Time	second	*s*
Electric current	ampere	A
Temperature	Kelvin	**K**
Amount of substance	*mole*	mol
Luminous intensity	candela	Cd

2. radian
3. It decreases
4. Newton (N)
5. 9.81 m s^{-2}

TYK 3.2

1. kg m^{-3}
2. 1942 kg
3. It decreases
4. It is a ratio
5. Approximately = 18.5 kN

TYK 3.3

1. Magnitude, direction and point of application
2. (a) magnitude only; (b) magnitude in a given direction
3. $F = ma$ and for weight $a = g$ so, $F = mg$
4. 'a strut is a member in *compression* and a tie a member in *tension*'
5. Pressure = Force/Area, N m^{-2}, Pascal (Pa)
6. Degree of hotness of a body

TYK 3.4

1. Primarily, differences in structure
2. Up to 1 atomic diameter
3. Motion of molecules or atoms

TYK 3.5

1. Forces that act in the same plane e.g., plane of this paper
4. Weight (W), Normal reaction (R) and Force (P)

TYK 3.6

1. Moment = *Force × perpendicular distance from axis of reference*
2. Distance = 0 and therefore $F \times 0 = 0$
3. Use simple trigonometric ratios
4. (a) Fulcrum = point or axis about which rotation takes place
    (b) Monent arm = perpendicular distance to the line of action of the force to the fulcrum
    (c) Difference between total clockwise moment and total anti-clockwise moment
5. Upward forces = downward forces, ACWM = CWM
6. (a) A couple occurs when two equal and opposite forces acting in opposite directions have their line of action parallel
    (b) One of the equal forces multiplied by the perpendicular distance between them

TYK 3.7

1. In all cases stress = force/area
    Tensile – set up by forces tending to pull the material apart
    Compressive – set up by forces tending to crush the material
    Shear – results from forces tending to cut through the material
3. Spring stiffness $k = \dfrac{\text{force}}{\text{deflection}}$, ($N\ m^{-1}$)
5. (a) 240 kN m^{-2} = 240×10^3 N m^{-2}; (b) 0.288 GPa = 228×10^6 N m^{-2}; (c) 600×10^6 N m^{-2}; (d) 3.3×10^6 N m^{-2}; (e) 1×10^{10} N m^{-2}
6. (a) strut, resists compressive loads; (b) tie, resists tensile loads

TYK 3.8

1. By measuring the yield stress (σ_y) or proof stress
2. To cater for a 'factor of ignorance' in the design
4. (a) Upper limit of validity of Hooke's law; (b) ultimate tensile strength or greatest load; (c) start of plastic phase; (d) material is permanently deformed

TYK 3.9

1. 166.7 m s^{-1} and 504 km/hr
2. Opposition to out of balance force (N)
3. *Force = ma*

TYK 3.10

1. . . . *acceleration*
2. . . . *distance*
3. . . . *area under graph* by *time interval*
4. . . . *zero* and the distance travelled is equal to *vt*
5. . . . $\frac{1}{2}vt$
6. . . . *uniformly retarded motion*
7. . . . $s = ut + \frac{1}{2}at^2$
8. . . . *variable*
10. speed has *magnitude* only, velocity has *magnitude and direction*
11. Weight force is dependent on acceleration due to gravity, that varies with distance from the Earth. Mass is the amount of matter in a body which remains unchanged

TYK 3.11

1. (a) rad s^{-1} (b) rad s^{-2}
2. $\omega = 142.9$ rad s^{-1}
3. (a) 26.18 rad s^{-1}, (b) 21.8 rad s^{-1}, (c) 1100 rad s^{-1}
4. (a) Torque = force × radius = Fr (N m)
   (b) Point mass multiplied by radius squared, i.e. $I = mr^2$
6. (a) In rotary motion the acceleration acting towards the centre of rotation $\alpha = \omega^2 r$
   (b) Centripetal acceleration acting on a mass producing force, $F_r = m\omega^2 r$
7. Centripetal force holds the car in the turn

TYK 3.12

1. WD = force required to overcome resistance (newton, N) × distance moved against the resistance (metres, m)
2. $WD = mgh$ (N m) or joules
3. Energy can neither be created nor destroyed but only changed from one form into another
4. (a) kinetic energy into electrical energy
   (b) chemical energy into kinetic energy
   (c) chemical energy into electrical energy
   (d) electrical energy into sound energy
5. k = spring constant (N m^{-1})
6. KE (linear) = $\frac{1}{2}mv^2$, KE (rotational) $\frac{1}{2}I\omega^2$
7. Machine B, more powerful. $A = 1500$ watts, $B = 1548$ watts

TYK 3.13

1. Depends on the nature of the surfaces in contact and whether they are subject to static or dynamic friction
2. False – not true for *all* low and high speeds
3. (a) Angle of friction ϕ is angle between resultant (R) and normal force (N)
   (b) Coefficient of friction $\mu = F/N$ and related to ϕ by $\mu = \tan\phi$
4. See Figure 3.29

TYK 3.14

1. (a) 253 K; (b) 47°C
2. Thermocouple thermometer – robust
3. The amount material will expand in metres, per degree Celsius Linear expansion coefficient = α and for surface expansion = 2α and for volumetric expansion = 3α
4. Heat energy is dependent on temperature difference for flow. Internal energy U is dependent on the molecular action of the substance, which in turn is temperature dependent
5. Convection involves the mass transfer of a fluid, in addition to adjacent molecular excitation as a result of conduction
6. $c_p > c_v$ because work may be done in addition to temperature rise, when heat is transferred in a constant pressure process
7. $Q = mc\Delta T$ and for latent heat $Q = mL$ (L = specific latent heat)
8. $R = c_p - c_v$ so $R + c_v = c_p = 940$ J kg^{-1} K^{-1}
9. Increase in temperature, reduction in pressure above liquid and passing gas over surface of liquid

TYK 3.15

1. Surplus
2. An external force
3. 30 coulombs

4. 12 V
5. 288 MJ
6. 16 Ω
7. Increase in resistance
8. Resistivity is an inherent property of a material measured in ohm-metres
9. Primary cells use up the chemicals from which they are made. Secondary cells can be repeatedly charged from an external source and discharged through use
10. Polarization. Hydrogen bubble formation on positive plate causing drastic reduction in the emf of the cell

TYK 3.16

1. Conventional current flows positive to negative. Real current flow is from negative to positive
2. Constant temperature conditions
3. See Figure 3.49 and accompanying text
4. 22 Ω, 1.09 A
5. $I_T = 12.4$ A, $R_4 = 6$ A, $R_6 = 4$ A, $R_{10} = 2.4$ A
6. $\frac{1}{3} \Omega$, $\frac{1}{2} \Omega$, $\frac{2}{3} \Omega$, $\frac{1}{2} \Omega$ respectively, where p.d. across load 4 = 3 V
7. 5.76 kW

TYK 3.17

1. Iron, nickel, cobalt, etc.
2. North
3. See Figure 3.65
4. Flux Φ (weber, Wb), Flux density in Wb m^{-2} (tesla)
5. Single force × distance between forces
6. Rotating switch reversing current in each armature conductor
7. Current flowing in the coil winding
8. Movement, a closed conductor and a magnetic field
9. [See Lenz's law, page 285]
10. Magnitude of the induced emf is dependent on the relative velocity with which the conductor cuts the lines of magnetic flux
11. [See mutual inductance, page 286]
12. 1 henry = dI/dt of one ampere per second to induce 1 volt

TYK 3.18

1. Coal, oil gas, nuclear etc.
3. A traffic lights system: electrical energy input, sequenced red, amber and green lights (output). Plus heat and some sound energy as outputs

   An engineering lathe: electrical energy, switch settings (inputs). Turned work, swarf, and oil waste (outputs)

   A motor vehicle hydraulic brake system: pedal movement (input). Brake shoe/disc movement, friction, heat, noise, carbon deposits (outputs)
4.

System	Main functional components (the 'doer')	Performance checking component (the 'monitor')	Controlling component (the 'decision maker')
Central heating	Boiler and radiators	*thermostat*	Gas valve
Car	Engine and transmission	Driver's eyes and ears	*human brain*
Computer	Microprocessor and memory	*VDU*	Application program

TYK 3.19

1. (a) 16, (b) 4096
2. 7 bits
3. Sequence may include: gear lever to neutral, ignition key, key in ignition switch, set chock (if not automatic), foot on/off accelerator, turn ignition switch, power starter motor, turn over engine, engine ignites. Failure to turn or ignite engine then – check settings, battery state, coil, leads, plugs, distributor, etc.

TYK 3.20

1. Reduced need for human intervention, remote control of system outputs, improved response times, continuous monitoring of output under control, etc.
3. Error detector/comparator – Monitors system output performance against set value and automatically corrects deviations by producing an error between the desired and actual output value, which drives the system
5. Feedback transducer – Monitors output and feeds back some proportion of the output value as a signal to the error detector for comparison and control
6. Required transfer function given by $G/(1 - G)$
7. The *TF* is a generalized version of the relationship between the system input and output, which is algebraic in nature. The gain G is often a numerical value, for one or more components within the system, or with respect to the system as a whole

TYK 3.21

With respect to Figure 3.98:
  (a) the amplitude of the first overshoot = 7
  (b) the decrement; 0.75
  (c) the natural frequency; 0.8 Hz
  (d) the 10 per cent settling time = 5 s

TYK 4.1

1. Natural numbers and positive integers
2. Rational numbers
3. $\dfrac{30}{6}, \dfrac{78}{6}, \dfrac{96}{6}$
4. $-\dfrac{16}{4}, -\dfrac{28}{4}, -\dfrac{48}{4}$
5. 4 = positive integer
6. rational, real
7. $0.33\dot{3}, 0.14285\dot{7}, 1.99999$
8. (a) 9, (b) 66, (c) 39
9. $-31 = 31$
10. 11
11. 14
12. (a) -5, (b) -18, (c) 7
13. (a) 96, (b) 90
14. 80

TYK 4.2

1. (a) (2,8), (4,4), (2,2,2,2); (b) (n, n); (c) (wx, yz), (wxy, z), (wxz, y), (wxz, x)
2. ab^2c
3. (a) 32, (b) $\dfrac{8}{27}$, (c) b^2

4. (a) 70, (b) $\frac{10}{9}$
5. (a) 0.23, (b) 5080
6. (a) 3.1862×10^2, (b) 4.702×10^{-5}, (c) 5.1292×10^{10}, (d) $-(4.1045 \times 10^{-4})$
7. (a) 2.71, (b) 0.000127, (c) 5.44×10^4
8. (a) -5.0×10^4, (b) 8.2×10^{-5}

TYK 4.3

1. (a) $\frac{1}{10}$, (b) $\frac{25}{3}$, (c) $\frac{9}{10}$
2. (a) $a^5b^{-1}c^3d$, (b) $4(6x^3y^2 - xy^2)$
3. (a) $\frac{11}{9}$, (b) $3\frac{3}{10}$, (c) 2
4. (a) $\frac{3}{4}a^{-6}b^2$, (b) d
5. (a) $6a^2 + 4a - 2$, (b) $4 - x^4$, (c) $3a^3b + a^2b^2 - 2ab^3$
6. (a) $(x+3)(x-1)$, (b) $(a+3)(a-6)$, (c) $(2p+3)(2p+4)$
7. (a) $3x(x+7)(x+2)$, (b) $3xy(3xy+2)(3xy-1)$
8. (a) -0.078, (b) 0.279

TYK 4.4

1. $F = 2 \times 10^4$
2. $r = \frac{\sqrt{V}}{\pi h}$
3. $\frac{R_1 R_4}{R_3} = R_2 = \frac{(3)(6)}{8} = 2.25$
4. 0.017
5. $a = \frac{s}{n} - \frac{1}{2}(n-1)d$
6. $x = \frac{bc + ac + b^2}{b + c}$
7. $c = 4.8343 \times 10^{-6}$
8. $\frac{1}{2(x^2 - 1)}$

TYK 4.5

1. 33.17 litres
2. 20
3. $c = £23.44$
4. 35
5. $h = \frac{Vk}{r^2}$ where k = constant of proportionality

TYK 4.6

1. 157 cm^3
2. 47.1 cm^2
3. 10 mm
4. 12.73 cm
5. 4330 cm^3 or 4.33×10^{-3} m^3

TYK 4.7

1. $v = 45$ when $i = 3$
2. (a) 1, (b) 4, (c) $-\frac{1}{3}$

3. (a) 2.5,1; (b) 2,3; (c) $\frac{3}{5}, -\frac{1}{5}$; (d) 2,3

4. 13

5. (a) 7,3; (b) $-\frac{3}{5}, -\frac{4}{5}$

6. (a) $x = 3$ or $x = -1$; (b) $x = -8$ or $x = -2$; (c) 2.62,0.38; (d) ±1.58

7. $x = 2.12$ or $x = -0.79$

8. $u = -1$ or $u = +2$

TYK 4.8

1. 14
2. $\mu = 0.4$
3. $I = 0.0017R^2$

TYK 4.9

1. 10.82 cm
2. 6.93 cm
3. 48.1 m
4. (a) 27.05 m, (b) 58 m
5. $\angle A = 45.3$, $\angle B = 37$, $\angle 97.7$, $a = 37.2$, $b = 31.6$, $c = 52$
6. $\angle A = 94.78$, $\angle B = 56.14$, $\angle C = 29.08$, Area = 29.9 cm^2

TYK 4.10

1. (a) $\frac{1}{2}, \sqrt{2}, 2.0$; (b) 1,0,–1; (c) $\sqrt{3}, \frac{1}{2}, 0$

2. The graph of sin θ is produced because sin $\theta = \sqrt{1 - \cos^2 \theta}$, from the relationship $\sin^2 \theta + \cos^2 \theta = 1$

3. (a) 2.65,2.46,0.24; (b) 26º; (c) 13

TYK 4.11

1. Business and administration = 29.23%, Humanities and social science = 42.28%, Physical and life sciences = 15.74%, Technology = 12.75%

2.

x	35	36	37	38	39	40	41	42	43	44	45
f	1	2	4	5	7	5	4	7	2	2	1

3. Percentage height of column relates to average for class interval

Class interval	62	67	72	77	82	87
Percentage	6.67	18.33	30	26.67	11.67	6.66

TYK 4.12

1. $\bar{x} = 127$
2. mean = 20, median = 8.5, mode = 9
3. $\bar{x} = 38.6$ cm, mean deviation = 1.44 cm
4. $\bar{x} = 169.075$ mm, mean deviation = 0.152 mm
5. $\bar{x} = 8.5$, $\sigma = 34.73$
6. $\bar{x} = 3.42$, $\sigma = 0.116$

TYK 4.13

1. $\frac{dy}{dx} = nax^{n-1}$

2. $f(3) = 51$, $f(-2) = 76$

3. (a) $\dfrac{dy}{dx} = 12x - 3$, (b) $\dfrac{ds}{dt} = 6t + 6t^{-2} - \dfrac{t^{-4}}{4}$

4. Gradient = –1.307 to 3 decimal places
5. $x = 4$, $y = 9$
6. At $x = -2$, rate of change = –56
7. 40.7 to 3 significant figures

8. (a) $\dfrac{1}{x}$, (b) $\dfrac{3}{x}$, (c) $\dfrac{1}{x}$ then it can be seen that $\dfrac{dy}{dx}$ of $\ln ax = \dfrac{dy}{dx}$

    of $\ln x$

9. –0.423
10. 866.67 C s^{-1}, 684.2 C s^{-1}

TYK 4.14

1. (a) $\dfrac{4x^3}{3} - x^{-2} + c$, (b) $\dfrac{x^{\frac{3}{2}}}{3} - \dfrac{2x^{\frac{3}{2}}}{3} + \dfrac{2x^{\frac{5}{2}}}{5} + c$,

    (c) $+\dfrac{3}{2}\cos 2x + c$, (d) $+\dfrac{2}{3}\sin 3x + c$, (e) $-0.75e^{3\theta} + c$,

    (f) $\dfrac{-3}{x} + c$

2. (a) $10\dfrac{5}{12}$, (b) $-\dfrac{1}{15}$, (c) –3, (d) –287.5

3. $v = \dfrac{3t^2}{2} + 4t + 8$, $s = \dfrac{t^2}{2} + 2t^2 + 8t$ and $s = 1762.5$

4. $12\dfrac{2}{3}$ sq. units

5. $\dfrac{4}{3}$ sq. units

Answers to problems

Problems 3.1
1. (a) 1.1772 MN, (b) 624 kN
2. (a) 0.009 m^3, (b) 6000 kg m^{-3}, (c) 6
3. Weight of fuel in all tanks = 833.85 kN
4. (a) 336.6 N, (b) 336.6 N

Problems 3.2
1. 9.33 kN ∠59°
2. R_A = 8.7 kN, R_B = 7.3 kN
3. R_A = R_B = 8.75 kN
4. Diameter = 13 mm
5. (a) 132.6 MN m^{-2}, (b) 166 mm

Problems 3.3
1. (a) 458.3 N, (b) 458.3 N
2. 100 m
3. acceleration = 5 m s^{-2}, distance travelled s = 40 m
4. 36.5 rad s^{-1}
5. Cessna = 3000 kg, Boeing 747 = 160 000 kg
6. ω = 50 rad s^{-1}, v = 13.5 m s^{-1}
7. (a) α = 0.33 rad s^{-2}, (b) 1800 radian or 286.48 revs
8. (a) 1.57 rad s^{-2}, (b) 204.4 N m

Problems 3.4
1. KE = 123.46 kJ
2. 0.36 J
3. 400 W
4. 168 N
5. μ = 0.19
6. 15.6 m s^{-1}
7. (a) 67.5 kJ, (b) 3 m s^{-1}

Problems 3.5
1. 931 kPa
2. 2×10^{-5} K^{-1}
3. (a) $Q = mc\Delta T$, (b) 900 J kg^{-1} K^{-1}
4. (a) R = 276.6 J kg^{-1} K^{-1}, (b) c_v = 723.4 J kg^{-1} K^{-1}
5. 1.004 MJ

Problems 3.6

1. (a) kilowatt-hour (kWh), (bi) 4.167 A, (bii) 240 V
2. 52.9 m
3. (bi) 6 A; (bii) 102 V; (biii) 3.6 A, 1.8 A, 0.6 A respectively; (biv) 612 W
4. 3.4 A
5. (a) 1.5 V, (b) 2.5×10^{-3} Ω, (c) 0.8 A
6. (a) down, (b) north to south (top to bottom), (c) out of page, (d) anticlockwise
7. To determine the appropriate direction for each example you need to apply Lenz's law
8. 7.5 N
9. 0.0625 H
10. $-$ 30 V

Problems 3.7

1. Open-loop systems are dependent primarily on human intervention for their control. They have no automatic feedback from their output, therefore any deviation from the system set point cannot be automatically corrected
   Closed-loop systems have a proportion of the output feedback, which is compared with the desired output. Deviations from the set-point are automatically compensated for by the error signal which always drives the system towards the desired output setting
3. A rotary switch is a form of transducer
   An actuator is an actuation device
   A gearbox is a mechanical signal conditioning device
   A voltage regulator is a controlling device
   A motor is an actuation device
   A pressure relief valve is a controlling device
   A speed governor is a controlling device
   An amplifier is a signal conditioning device
   A visual display unit (VDU) is an output device
6. (a) the final steady-state value = 20 rev/sec
   (b) the delay time = 1.9 seconds
   (c) the 15 per cent settling time = 18 seconds
   (d) the natural frequency = 1/12 or 0.083/Hz
7. Overall voltage gain G; 0.1
8. 99.5 per cent reduction

Index